Retention of Land for Agriculture

Policy, Practice and Potential in New England

Frank Schnidman
Michael Smiley
Eric G. Woodbury

Lincoln Institute of Land Policy
Cambridge, Massachusetts

Library of Congress Cataloging in Publication Data

Schnidman, Frank.
Retention of land for agriculture: policy, practice, and potential in New England / Frank Schnidman, Michael Smiley, Eric G. Woodbury.
p. cm.
Includes index.
ISBN 1-55844-109-3
1. Land use, Rural—Government policy—New England.
2. Agricultural conservation—Government policy—New England.
I. Smiley, Michael. II. Woodbury, Eric G. III. Title.
HD266.A11S36 1990
333.76'16'0974—dc19 87-5580
CIP

ISBN 1-55844-109-3

Library of Congress Catalog Card Number: 87-5580

Printed in the U.S. on acid-free text stock.

Publication of a manuscript by the Lincoln Institute signifies that it is thought to be worthy of public consideration, but does not imply endorsement of conclusions, recommendations, or organizations that supported the research.

Lincoln Institute of Land Policy
Cambridge, Massachusetts

Lincoln Institute of Land Policy

The Lincoln Institute of Land Policy is an educational institution where students, policymakers and administrators can explore the complex linkages between public policies, including taxation, and land policy, and the impact of these policies on major issues of our society.

The Institute seeks to understand land as a resource, the choices for its use and improvement, the regulatory and tax policies that will result in better uses of land, and effective techniques by which land policies can be implemented. The major goal of the Institute is to integrate theory, practice, and understanding of land policy and those forces influencing that policy, especially taxation, which have a significant impact upon the lives and livelihood of all people.

The Institute is a school providing advanced education in land economics, including property taxation, and offering challenging opportunities for learning, research, and publication.

The Lincoln Institute of Land Policy is an equal opportunity institution in employment and admissions.

Contents

List of Tables and Figures

Tables

Figures

Foreword

This book could not have been written a generation ago. In these pages, Frank Schnidman, Michael Smiley, and Eric G. Woodbury describe the state of an art that has progressed from cave drawings to computers in a few short years.

This is not to say that public concern for the future of farmland in New England is brand-new. There is some recorded evidence of this concern as early as the Revolutionary War. The birth of the nation had left its citizens exhausted, its towns impoverished, and its agriculture in shambles. The Colonial Militia generally fought well but farmed poorly. By 1790, many of New England's best soils were depleted from years of failed stewardship.

A year later, this crisis prompted Governor John Hancock and the Massachusetts Great and General Court to establish the Massachusetts Society for Promoting Agriculture. The Society moved at once to encourage field rotation, the use of manures, and the planting of cover crops. However, by 1825, the improvements in New England farm productivity were offset by losses in comparative advantage. The Erie Canal had opened, giving hungry East Coast markets access to the richer soils of the West. For the next hundred years, if a farmer wore out his land he could always go west and find more.

It was not until the dust bowl days of the 1930s that concern for soil stewardship was given the national attention it deserved. Along with this recognition has come the gradual realization that America has been uniquely blessed with good soils and that to a very great extent its strength as a country is closely connected to the health of its agriculture.

In New England the focus of public attention on agriculture has become even more narrowly defined. In the decades following the end of World War II, soil erosion could no longer be measured in inches per acre of topsoil lost, but in regular and repeated increments of five to fifty acre chunks of land permanently converted to a "higher" use—usually residential or commercial.

As the authors describe so well, the abandonment or conversion of farmland in New England has galvanized a whole generation of farmers and city folk into action. The various courses of this action are defined with great precision in the pages that follow. Whether the issue is concern for loss of countryside and "New England character" or alarm at the impact on a basic industry and the region's food production capability, a significant following has been generated. Numerous state and municipal governments have passed bills and started programs, scores of private land trusts have emerged, and even federal policies have begun to take note that public interest in farmland is not exclusively a New England phenomenon—witness the 1985 Farm Bill. This last is particularly significant in that the most frequently heard public debate of the 1980s is not how to keep land in agricultural production, but what to do with crop surpluses. There seems to be a lurking suspicion that national crop surpluses, like international oil surpluses, are temporary. Few will dispute that, for the long haul, the preservation of America's best food-producing land is critically important.

Amidst all this public concern resides a certain level of nervousness with farmers who, quite rightly, worry about what plans others may have for their land. In New England, farmers and landowners have reacted in a variety of ways: from apathy to hostility to constructive involvement. The last of these has resulted in initiatives which have quite successfully accommodated the public welfare without shortchanging the landowner.

The common denominator for landowners and civil servants alike must be a conviction that we need not build on our best farmland. "An enlightened society must," as Robert Lemire preaches, "save what should be saved as it builds what should be built." There is still room for both—even in New England.

Frederic Winthrop, Jr.
Director, The Trustees of Reservations
Chairman, Board of Directors, American Farmland Trust
Former Commissioner of Food and Agriculture,
Commonwealth of Massachusetts
Former President, National Association of State
Departments of Agriculture

1

Evolving Governmental Role in Farmland Protection

Agricultural land protection is one of the leading land use issues for the 1980s. Federal agencies, state agencies, and local governments around the nation are worried about the conversion of farmland. A myriad of studies have been undertaken, new laws have been passed, and new legislative proposals are being considered to address the conversion issue. Additionally, public interest groups are becoming involved, and professional journals and the popular press are giving the conversion issue much attention.

Until the mid-1970s, the conversion of farmland to nonfarming uses was not a major national issue. The government paid large sums of money to purchase commodities and keep farmland out of production. Following the grain purchase of 1973 to 1974 by the Soviet Union and some crop failures in the world, commodity prices increased sharply and surpluses disappeared. These events led to a growing awareness that good farmland is limited and should be protected. Although surpluses have again surfaced, the concern about the conversion of farmland has continued.

Despite this increasing awareness, the concern for farmland protection is not universally shared. Opinions vary on how much farmland is being converted to other uses and on the impact this change could have on the nation and the world in the future. More importantly, there is also a lack of consensus on what role, if any, the government should play. That perhaps explains the caution with which the federal government began to involve itself with farmland protection.

At the state and local levels, there also has been an increasing involvement in farmland protection. Many states have adopted agricultural land retention programs of one form or another. The heightened concern nationwide parallels the concerns of New England. Although many of the issues facing New England farmers are similar to those faced by farmers throughout the United States, others are unique to the geographic, economic, and social patterns that have evolved in New England.

This chapter discusses the background of the problems and impacts which have contributed to the nationwide and statewide desire to protect farmland. In addition, it examines the evolving federal role, providing a perspective of New England in the context of agriculture as a whole. The activities of other states and their local governments are also reviewed so that they can be compared to the New England states.

Major Federal Activity of the 1970s and 1980s

The early 1970s was a time of heightened environmental awareness. Congress considered many legislative proposals during that period, including bills to protect air quality, water quality, coastal resources, endangered species, and historic resources; it passed legislation in each of these areas. One major proposal which did not pass was national land use planning legislation. The debate over land use and growth-management issues heightened the awareness generally of the impact of federal actions on land use, an awareness which was to be sharpened by the conflicts over the evolving Environmental Impact Statement requirements of the National Environmental Policy Act of 1969.[1]

U.S. Department of Agriculture (USDA)

As late as 1974, USDA studies did not consider farmland conversion a problem.[2] Increasing crop yields and large crop surpluses which resulted in expensive acreage set-aside programs and commodity price supports made the conversion issue unimportant. Rather, the government was concerned with ways to take land out of production.

Seminar on the Retention of Prime Lands. The USDA Committee on Land Use sponsored a seminar in July 1975 to gather information on the need for a national policy dealing with prime production lands. As stated by Secretary Earl L. Butz in the foreword to the seminar's background papers: The Department of Agriculture is concerned with land use alternatives and priorities, especially those that involve expenditures of Federal funds. Federal projects that take prime land from production should be initiated only when this action is clearly in the public interest. Long-term implications of various land use options on the production of food and fiber must be understood to ensure that the public is aware of the trade-offs involved.[3]

In the weeks following the seminar, the USDA Committee on Land Use examined the "work group" reports presented during the seminar and put together a series of recommendations which were an attempt at consensus among the group reports. Among those policy recommendations were:

1. USDA should take a major, defined, and well-promoted role in the national questions of utilization, enhancement and retention of agricultural lands as an advocate of retaining the maximum possible base for the production of food, fiber, and timber products and minimizing actions that will diminish the nation's capacity to produce these essential commodities.
2. USDA should promote policies to build into the federal decision-making processes clear recognition and concern for the impact of housing, transportation, defense and other federal programs on the agricultural lands inventory.
3. USDA policy should avoid and encourage others to avoid the diversion of highly productive farm and forest lands to nonproductive uses wherever feasible alternatives exist.
4. USDA should take the initiative to have prime lands for the production of food, fiber, and forest products designated as an essential natural resource to be considered in environmental impact statements.

In addition, the USDA Committee on Land Use decided to take the following actions designed to increase intergovernmental and interagency coordination:

1. USDA should take leadership in establishing a Land Resources Council, either similar to or developed as an extension of the Water Resources Council, to coordinate land resource–related programs of the federal government. An additional function could be to develop coordinated programs for providing information and financial and technical assistance to state and local governments.
2. USDA should encourage the President to initiate a national dialogue of future growth policy in cooperation with states and other units of government, with emphasis on policies essential to meeting the natural resource needs of the people in an environmentally stable manner.[4]

Secretary's Memoranda. In June 1976, Secretary Butz issued a prime-lands protection policy officially committing the USDA to incorporate agricultural land retention goals into its own decisions and to serve as the federal advocate on the issue. This departmental regulation was revised in

1978, 1982, and again in 1983. The current departmental regulation, "Land Use Policy," Number 9500-3 (March 22, 1983) commits the USDA agencies to encourage and support state and local governments and individual landowners to retain important agricultural lands, while discouraging unwarranted conversion of important land to other uses by any federal agency.

Council on Environmental Quality (CEQ)

The Executive Office of the President's Council on Environmental Quality (CEQ) oversees the efforts of other Federal agencies in preparing environmental impact statements required by the NEPA. In August 1976, CEQ issued a memorandum to all federal agencies which suggested that future impact statements assess the impacts of proposed projects on prime agricultural land.[5] Compliance was voluntary, and agencies did not have to modify their decisions to take into account the findings of their impact statements. On August 11, 1980, CEQ issued memoranda, which remain in effect today, superseding the August 1976 memorandum.[6] Now federal agencies must determine the effects of a proposed action on prime or unique agricultural lands and should have by November 1, 1980 provided CEQ with:

> identification and brief summary of existing or proposed agency policies, regulations and other directives specifically intended to preserve or mitigate the effects of agency actions on prime or unique agricultural lands, including criteria or methodology used in assessing these impacts.

In 1979, CEQ signed a memorandum of agreement with USDA to clarify relationships in undertaking the jointly sponsored National Agricultural Lands Study (NALS). This study is discussed later in this chapter.

Environmental Protection Agency (EPA)

In September 1978, the U.S. Environmental Protection Agency (EPA) adopted an internal policy:

> ...to protect, through the administration and implementation of its program, environmentally significant agricultural land from irreversible conversion to uses which could result in its loss as an environmental or essential food producing resource.[7]

Among other things, the policy mandates consideration of agricultural land concerns in developing agency regulations, standards, and guidelines, and in planning specific projects. Furthermore, EPA permit activities necessitated that an environmental impact statement:

> shall ensure that the proposed activity will not cause the conversion of environmentally significant agricultural land.

Like USDA, EPA offered to provide technical assistance to states and local governments who wish to incorporate agricultural land considerations into air quality, water quality, and other environmental planning efforts.

Congressional Action

In 1977 nine farmland protection bills were introduced in the Ninety-Fifth Congress.[8] These bills were similar in language, focusing upon inclusion of farmland conversion in the EIS process, inventorying and evaluating agricultural lands, intergovernmental cooperation, and coordination of federal activities that affect farmland.

These bills proposed to establish a commission, which would examine the farmland protection issue and make policy recomendations, and a technical assistance program. None of this legislation came to a vote in Congressional committee. Several bills were introduced in the Ninety-Sixth Congress, and Rep. James Jeffords (R-Vt.) became the primary proponent.

On February 7, 1980, The Agricultural Land Protection Act (H.R.2551) was defeated in the House of Representatives by a vote of 177 to 210, with forty-six members not voting. As amended on the floor during debate, the Jeffords bill proposed three actions:

1. A long-term study of the farmland protection issue
2. A financial and technical assistance program to help states and localities demonstrate and test methods of reducing the conversion of farmland to nonagricultural uses
3. A policy constraining federal programs and projects from adversely affecting state or local efforts to preserve farmland

Among the chief grievances that a number of federal agencies, states, and landowner or commodity group representatives expressed against the bill were fears that funding assistance for state and local demonstration programs might open the door to federal land use planning, and that the third or "consistency" policy might disrupt federal programs.

Congress did pass farmland protection legislation in 1981 as part of the omnibus Food and Agriculture Act. This legislation is discussed below.

The National Agricultural Lands Study (NALS)

In 1977 President James Earl Carter signed Executive Orders 11988 and 11990, providing for greater federal involvement in floodplain management and the protection of wetlands. The idea of an executive order on farmland protection had been considered and rejected, as was a later proposal for a Presidential commission to assess the agricultural land conversion issue and to develop future policy options. Instead, USDA and CEQ were authorized to lead a major interagency study on farmland protection, a study with the participation of ten other federal agencies. The National Agricultural Lands Study (NALS) examined the availability of the nation's agricultural lands, the extent and causes of the lands' conversion to other uses, and the various ways it might be retained for agriculture. The study began in June 1979, and the final report was submitted to the President in January 1981. During the research period, the NALS Committee held seventeen public workshops, generating substantial media coverage. These activities called considerable public attention to the issue, as did the distribution of numerous NALS reports and handbooks; concern for conversion impacts is now being expressed at professional meetings and conferences around the nation because of the conclusions of the NALS.

These conclusions presented a crisis-in-the-making scenario and called for immediate congressional and federal agency attention. The comprehensive and detailed farmland study stated that future demands for agricultural land will be increasing while supplies will be decreasing. The study claimed that 3 million acres of agricultural land are being converted each year to urban, transportation, or water uses, that millions of tons of topsoil are lost each year to erosion, and that some specialty crops are being threatened with extinction by urbanization. Moreover, according to the NALS, the demand for agricultural land will become greater because of concomitant increases in both agricultural exports (10 percent average per year) and crop production for the manufacture of gasohol. Furthermore, as this demand for land grows, the NALS predicted that increased crop yields would begin to level off. Thus, the NALS predicted that the volume of agricultural production in the year 2000 will need to be 75 percent greater than that in 1980 to meet domestic and export demands. This would translate to an additional 80 to 140 million acres of cropland brought into cultivation by the year 2000.

The NALS Final Report generated a good deal of controversy and criticism, especially concerning the statistical analysis. The Urban Land Institute (ULI) and other organizations and individuals refuted the NALS findings and argued that the land loss was substantially overstated. The ULI report questioned the NALS prediction of future agricultural land demands:[9] it claimed that unbiased analysis of the statistics shows only 900,000 acres are being converted to urban and transportation uses each year (which was later confirmed by two USDA economists),[10] and that much of that land is not prime land or even cropland. The ULI report also stated that the statistics show at least 60 million acres of cropland and 127 million acres of potential cropland are available for cultivation, which illustrates that the present supplies of agricultural land are adequate even assuming the NALS conversion rates.[11]

The Farmland Protection Policy Act (FPPA)

The NALS was finished just as the Ninety-Seventh Congress began, and supporters of federal farmland protection legislation were encouraged by the endorsement of Agricultural Secretary John Block. Strategy sessions on how to get federal legislation past Congress had begun the day of the H.R.2551 defeat, and with the NALS in hand and the support of Secretary Block, House and Senate legislation was introduced. This time, however, the strategy was to incorporate farmland protection as a subtitle of the respective omnibus farm bills, with the hope that most debate would be focused on the commodity programs and that the subtitle dealing with farmland protection would go through relatively unnoticed.

Secretary Block endorsed the policy language in the House bill, even though the Office of Management and Budget was not supportive. However, it was not the House language that was considered by the conference committee which had been formed to resolve differences between House and Senate bills.

> ...[I]t was the Senate farmland language that was finally adopted in conference. The conferees had no choice as we had earlier substituted the stronger Senate language for the House language when the farm bill was being amended on the House floor.[12]

The Farmland Protection Policy Act (FPPA) was enacted when President Ronald Reagan signed the Food and Agriculture Act of 1981.[13] The FPPA requires USDA to "develop criteria for identifying the effects of Federal programs on the conversion of land to nonagricultural uses," §1541 (a) and §1541 (b), and requires federal entities to use the criteria:

> ...to identify and take into account the adverse effects of Federal programs on the preservation of farmland; consider alternative actions, as appropriate, that could lessen such adverse effects; and assure that such Federal programs to the extent practicable are compatible with State, units of local government, and private programs and policies to protect farmland.

It also requires (in §1542) that federal entities review their own rules and regulations and consider changes if they prevent conformance with the purposes of the FPPA. A report one year after enactment was to be filed by the Secretary of Agriculture with Congress outlining federal programs which affect farmland protection and reporting the results of the §1542 reviews by other federal entities.

Also of importance was a specific limitation on any federal action that regulates the use of private or nonfederal lands or "in any way affects the property rights of owners of such land" [§1547(a)]. Finally, §1548 was a prohibition specifically stating congressional intent that no legal challenge in any way could be brought on the basis of FPPA farmland protection efforts.

Implementation Regulations under the Farmland Protection Policy Act

Although signed December 22, 1981, final regulations under the FPPA were not issued until July 5, 1984.[14] The effective date was supposed to be June 22, 1982. On August 2, 1982, USDA sent a twelve-page draft rule to all secretaries, administrators, and chairs of every cabinet or subcabinet federal agency for review and comment by September 30, 1982. Effectively, this proposal constituted guidelines for other federal agencies to use in evaluating the impacts of their actions on farmland.

Negative comments were received on these draft regulations, and the decision was made at USDA to start over again. The revised draft was reviewed by OMB and finally came out as a proposed rule in the Federal Register on July 12, 1983.[15] When the extended comment period ended, 140 comments had been received, 19 of which were from federal agencies, 22 from national organizations, 41 from state agencies, 21 from state organizations, 25 from local agencies and organizations, 4 from private companies, and 8 from private individuals. Four commentators made two submissions.

Those comments which generally supported the proposed regulations but found them inadequate argued:

1. Monitoring of agency compliance is needed.
2. Reporting of process to Congress is needed.
3. Enforcement by USDA is needed.
4. Land evaluation must be more specific.
5. Site-assessment criteria must be more stringent.
6. USDA should take a leadership role.
7. The review process should be coordinated with NEPA.

Those comments critical of the regulations argued:

1. They go well beyond congressional intent.
2. They create unnecessary red tape.
3. They unecessarily duplicate NEPA process.
4. They fail to recognize congressional intent for protection of private property rights.
5. They fail to include the congressionally mandated litigation prohibition provision.
6. Thresholds for applicability are needed.
7. Exclusions should be expanded to cover federal permits and licenses.

Some of the controversy over the scope of the proposed regulations can be illustrated by comparng the positions taken by the Natural Resources Defense Council (NRDC)[16] and the Pacific Legal Foundation (PLF).[17]

The NRDC felt that the proposed regulations did not indicate a strong enough role for the USDA in the farmland evaluation process. It wanted complete decision-making power to be given to federal program administrators, regardless of whether they had any expertise in the FPPA. They believed the FPPA had been enacted partly in response to federal officials' lack of knowledge about agricultural land preservation concerns. The NRDC also felt that the agency should routinely submit documentation of FPPA implementation to the USDA for a forty-five-day review; the USDA would have the option to comment.

In comparison, the PLF felt that the USDA's role should not in any way be regulatory and stated that while the FPPA clearly directs each agency to take farmland preservation into account in their decision-making processes, it does not mandate the USDA to establish a compulsory regulatory program for all other federal agencies.

Another area of disagreement was in the scope of consideration of alternatives to federal programs affecting agricultural land. The NRDC took the position that the statute requires alternatives to be considered and that the USDA-developed criteria should be designed with the consideration of alternatives in mind. The NRDC proposed that the final rule require the critiera to be applied early in the decision-making process, before many aspects of the project design are fixed. These criteria would be applied to reasonable alternatives as well as to the proposed action, and less adverse alternatives should be selected when appropriate.

However, the PLF felt that each agency was responsible for evaluating a particular parcel of farmland according to the criteria and for considering alternative actions. They argued that there is no statutory mandate to require either evaluating special siting requirements or using alternative sites, and that it is only after an agency has determined the adverse effects, if any, of a proposed activity on farmland that a consideration of alternative actions to lessen such adverse effects would be appropriate.

Finally, both the NRDC and the PLF felt that the land evaluation and site assessment criteria needed revision and that many terms—such as the "federal programs" that would be subject to the FPPA—needed clarification.

Many changes were made to the proposed rule in response to the comments received, and the final regulations were issued on July 5, 1984. In effect, apart from numerous technical changes, the following three substantive changes to the proposed rule significantly limited the scope of the FPPA.

1. The rule now specifies that federal agencies may not use the Act to refuse to grant assistance to programs that consume farmland. USDA officials reasoned that this would amount to interference with the use of private land.
2. The rule also specifies that any "prime farmland" that has been designated by any state or local government for commercial, industrial, or residential use, through zoning or planning, will not be covered by the Act. This land is considered committed to urban development.
3. The rule specifies that the Act does not apply to federal government activities of issuing permits or licenses on private or nonfederal lands, or approving public utility rates.

Furthermore, the USDA has decided that the FPPA "grants no express authority to the Secretary or the Department to devise enforcement or oversight procedures over other federal agencies. Nor does it assign the Department a role of encouraging other federal agencies to protect farmland."

Other important changes to the proposed rule were that the five land evaluation criteria were reduced to one. The sixteen site assessment critera were reduced to twelve and were rewritten to clarify their meaning and to make them more specific. Moreover, the rule established a threshold Land Evaluation Site Assessment (LESA) score for considering additional alternative actions, sites, or designs.

The result is that the final rule implementing the FPPA is a narrower interpretation of the Act than the proposed rule. The rule requires federal agencies to consider the farmland conversion impacts of their activities, but requires no changes as a result of those impacts.

Farmland Protection Policy Act Amendments of 1983

In an attempt to give power to the FPPA, Senator Roger Jepsen (R-Iowa) filed S.2004, the Farmland Protection Policy Act Amendments of 1983. Prefaced in the *Congressional Record* by the NRDC comments, the amendments would specifically place USDA in the leadership role with the ability to dictate review criteria to other federal entities and to require documentation of actions.[18] The amendments would have required an annual report to Congress, and taken together with loose language about USDA authority to issue regulations to carry out the Act, would have created the possibility of substantial reporting and review requirements. Also, significantly, the amendments called for the repeal of the prohibition against litigation. If passed, this would have opened every federal action affecting farmland or potential farmland to judicial challenge. These amendments were not brought to the floor for a vote.

Farmland Protection Policy Act Amendments of 1985

Amendments to the FPPA were filed as part of the Agriculture, Food, Trade, and Conservation Act of 1985, §.616. These amendments were forecast in the May–June 1985 edition of *American Farm Land,* which stated that amendments would be "offered to strengthen the 1981 Farmland Protection Policy Act, whose purpose is to stop federal tax dollars from being spent to convert farmland to non-agricultural uses."[19] The amendments require USDA to report to Congress each year, and they create an exception to the prohibition against litigation by allowing the governors in states with state farmland-protection policies to challenge in court any federal program allegedly not in compliance with the FPPA.[20] When the President signed what is now called the Food Security Act of 1985 on December 19, 1985 these amendments became law.[21]

Related Federal Activity

The President's Commission on Housing. The President's Commission on Housing in 1982 recommended repeal of the FPPA because of the potential serious and detrimental impact it could have on the cost and availability of land for housing. Citing problems with the NALS statistical allegations, it concluded:

> There is little evidence that conversion of farmland is a crisis in the making....The imposition of farmland protection regulations may create severe problems for housing at the local level....If not carefully reviewed and

implemented, the Act's regulations may foster withholding of land needed for housing development. The Federal act could also be used to justify state and local exclusionary zoning actions under the banner of farmland preservation.[22]

USDA FPPA Report to Congress. Under §1546, the FPPA required a report to Congress within one year of enactment on the progress made in implementation, including a review of the effects of federal actions with respect to farmland protection and a discussion of the results of the reviews undertaken by other federal entities of their own rules and regulations as they affect farmland.

The report, due in December of 1982, was finally submitted in June of 1983. The report listed federal agencies with programs that can contribute to farmland conversion and described how the Department of Agriculture was making the LESA system and the National Agricultural Library more accessible to state and local governments and nonprofit organizations to assist them in their farmland protection programs. The report also listed three federal laws or administrative policies with which the FPPA may conflict and explained how the review process over the proposed regulations was working. Finally, the report stated that the effect the FPPA was having with respect to U.S. farmland could not be evaluated yet.

1982 Natural Resources Inventory. The 1982 National Resources Inventory (NRI) is the most comprehensive survey of the nation's nonfederal land resources ever conducted by the Soil Conservation Service (SCS). The SCS inventoried nearly 1 million data points (compared to 70,000 sampling sites used in the 1977 NRI) over a collection period of three years. Believing that there were several problems with the 1977 land use statistics, the SCS instituted three procedural and several technological improvements to generate more accurate statistics.

The procedural changes involved a more consistent and accurate method of defining built-up versus rural land, a uniform definition of forestland, and an attempt to classify miscellaneous land uses better and to reclassify them as pastureland, rangeland, or forestland where appropriate.

In the 1982 NRI, the urban and built-up land area estimates relied completely on the sample point data, while the corresponding 1977 NRI acreages were estimated from county maps and zoning and planning commission maps. The 1977 effort overestimated the urban and built-up land areas partly because of the limited availability of the maps and problems with map scale and measurement.[23]

As a result of the inventory improvements, the 1982 estimate of land conversion to urban and built-up uses between 1967 and 1982 is about 900,000 acres per year—much less than the 1977 estimate of 3 million acres per year. The SCS is currently working to obtain estimates of the quality of rural land that is being converted to other uses.

State and Local Programs

In addition to the federal actions and studies, a variety of programs have been initiated in the states to encourage farmland protection or to slow conversion to urban uses. This range of programs usually includes the following policy objectives:

1. Reduce real property taxes on farmers:
   Current-use assessment
   Deferred taxation
   Restrictive agreements
   Circuit-breaker tax credits
   Land-gains taxation
2. Reduce estate taxes:
   Appraisal at farm-use valuation
   Deferral of payment
3. Reduce rate of conversion of farmland:
   Comprehensive planning
   Agricultural zoning
   Agricultural redistricting
   Purchase of development rights
   Transfer of development rights
   Right to farm
   Land banking
4. Integrate incentives and controls

A description of the types of state and local programs follows.[24] Table 1.1 summarizes the programs implemented by state and local governments for preserving farmland nationwide.

In the area of taxation, the states have been very active, and local governments are using a wide range of state-delegated police power to address the farmland protection issue.

Although conversion to nonagricultural uses may be arguably small in relation to national statistics, the concentration of these conversions in particular areas can have a disproportionately large impact on the balance of land use in those regions. It is at the local level where the impacts of conversion are felt, and it is here that the grassroots support for local, state, and federal action begins.

Table 1.1
State Programs for the Protection of Agricultural Land

Programs / State	*Preferential Property Tax Assessment*	*Preferential Property Tax Assessment/Deferred Taxation*	*Preferential Property Tax Assessment/Restrictive Agreement*	*Circuit Breaker State Income Tax Credits*	*Inheritance and Estate Taxation*	*Land Gains Taxation*	*Agricultural Districting*	*Agricultural Zoning*	*Purchase of Development Rights*	*Transfer of Development Rights*	*Land Use Commission*	*Land Banking*	*Right-to-Farm*	*Integrated Programs of Incentives and Controls*
Alabama	-	S	-	-	S	-	-	-	-	-	-	-	S	-
Alaska	-	S	-	-	S	-	-	-	-	S	-	L	-	-
Arizona	S	-	-	-	S	-	-	-	-	-	-	-	-	-
Arkansas	S	-	-	-	S	-	-	-	-	-	-	-	-	-
California	-	-	S	-	S	-	-	-	L	-	-	-	-	-
Colorado	S	-	-	-	S	-	-	-	L	-	-	-	-	-
Connecticut	S	-	-	-	S	S	-	-	L	L	-	-	-	-
Delaware	S	-	-	-	S	-	-	-	-	-	-	-	S	-
Florida	S	-	-	-	S	-	-	-	-	-	-	-	S	-
Georgia	-	-	-	-	S	-	-	-	-	-	-	-	S	-
Hawaii	-	S	S	-	-	-	-	S	-	-	S	S	-	S
Idaho	S	-	-	-	S	-	-	-	L	-	-	-	-	-
Illinois	-	S	-	-	S	-	S	-	L	-	-	-	S	-
Indiana	S	-	-	-	-	-	-	-	L	-	-	-	-	-
Iowa	S	-	-	-	-	-	-	-	L	-	-	-	-	-
Kansas	-	S	-	-	S	-	-	-	L	-	-	-	-	-
Kentucky	-	S	-	-	S	-	-	-	-	-	-	-	S	-
Louisiana	S	-	-	-	-	-	-	-	-	-	-	-	S	-
Maine	-	S	-	-	-	-	-	-	L	-	-	-	-	-
Maryland	-	S	-	-	S	-	S	-	L	L	-	-	S	S
Massachusetts	-	S	-	-	-	-	-	-	S/L	L*	-	L	-	-
Michigan	-	-	S	S	S	-	-	-	L	-	-	-	S	-
Minnesota	-	S	L**	-	S	-	-	L	-	-	-	-	L**	-
Mississippi	S	-	-	-	S	-	-	-	-	-	-	-	S	-
Missouri	S	-	-	-	-	-	-	-	-	-	-	-	-	-
Montana	S	-	-	-	S	-	-	-	L	-	-	-	S	-
Nebraska	-	S	-	-	-	-	-	-	L	-	-	-	-	-
Nevada	-	S	-	-	-	-	-	-	-	-	-	-	-	-
New Hampshire	-	S	S	-	-	-	-	-	L	-	-	-	-	-
New Jersey	-	S	-	-	-	-	-	-	L	L	-	-	-	-
New Mexico	S	-	-	-	S	-	-	-	-	-	-	-	-	-
New York	-	S	-	-	S	-	S	-	L	L	-	-	S	-
North Carolina	-	S	-	-	-	-	-	-	-	-	-	-	S	-
North Dakota	S	-	-	-	S	-	-	L	-	-	-	-	S	-
Ohio	-	S	-	-	-	-	-	-	-	-	-	-	-	-
Oklahoma	S	-	-	-	-	-	-	-	-	-	-	-	S	-
Oregon	-	S	-	-	S	-	-	S	-	-	S	-	S	S
Pennsylvania	-	S	S	-	S	-	S	-	L	L	-	S	S	-

Table 1.1 (continued)

State \ *Programs*	*Preferential Property Tax Assessment*	*Preferential Property Tax Assessment/Deferred Taxation*	*Preferential Property Tax Assessment/Restrictive Agreement*	*Circuit Breaker State Income Tax Credits*	*Inheritance and Estate Taxation*	*Land Gains Taxation*	*Agricultural Districting*	*Agricultural Zoning*	*Purchase of Development Rights*	*Transfer of Development Rights*	*Land Use Commission*	*Land Banking*	*Right-to-Farm*	*Integrated Programs of Incentives and Controls*
Rhode Island	-	S	-	-	-	-	-	-	S	-	-	-	-	-
South Carolina	-	S	-	-	-	-	-	-	-	-	-	-	-	-
South Dakota	S	-	-	-	-	-	-	-	L	-	-	-	-	-
Tennessee	-	S	-	-	S	-	-	-	-	-	-	S	S	-
Texas	-	S	-	-	-	-	-	-	-	-	-	-	-	-
Utah	-	S	-	-	S	-	-	-	L	-	-	-	-	-
Vermont	-	S	-	-	S	S	-	-	-	-	-	-	S	-
Virginia	-	S	-	-	S	-	S	-	L	-	-	-	S	-
Washington	-	S	-	-	S	-	-	-	L	-	-	S	S	-
West Virginia	S	-	-	-	-	-	-	-	-	-	-	-	-	-
Wisconsin	-	-	S	S	-	-	-	S	L	-	-	-	-	S
Wyoming	S	-	-	-	-	-	-	-	L	-	-	-	-	-
Virgin Islands	-	-	-	-	S	-	-	-	-	-	S	S	-	-
Guam	-	-	-	-	-	-	-	-	-	-	-	-	-	-

Notes:

S = Statute of state program. L = Local program(s).

*One town only—no results. **Metropolitan (Twin Cities area only).

Source: Combined data from 1979 National Conference of State Legislatures/Council of Environmental Quality (NCSL/CEQ) Survey, 1981 NCSL Survey, and NALS data.

Programs to Reduce Real Property Taxes on Farmers

Current-Use Assessment. Under current-use or preferential assessment, eligible agricultural land is assessed for real property tax purposes at its agricultural or current-use value, instead of its fair-market value. The effect of current-use assessment is to reduce the farm's taxes. Seventeen states have current-use assessment.

Deferred Taxation. Twenty-eight states, in addition to making current-use value assessment available for eligible land, generally require participating landowners who convert their land to ineligible uses to pay some or all of the taxes that they have been excused from paying as a result of their participation. This sanction is designed to deter landowners from converting their land and to recoup some of the revenue lost as a result of differential assessment.

Restrictive Agreements. In six states, landowners wishing to secure differential assessment must enter into enforceable agreements to keep their land in eligible use.

Circuit-Breaker Tax Credits. Two states authorize an eligible owner of farmland to apply some or all of the property taxes on the farmland and farm structures as a tax credit against the farmer's state income tax. These programs are called circuit breakers because they relieve the farmer from additional real property taxes once they exceed a given percentage of the farmer's income.

Land Gains Taxation (Capital Gains Tax). Connecticut and Vermont impose a levy on profits from the sale or exchange of land held for less than a specified number of years. The idea is to discourage the speculation and multiple sales of agricultural or open space land for development—or other purposes which would change the land use. The tax rate imposed is usually on a sliding scale based upon the number of years the land is held and upon the percentage of profit.

Programs to Reduce Estate Taxes

Appraisal at Farm-Use Valuation. Several states are using the federal estate tax law, the Tax Reform Act of 1976, as supplemented by technical corrections contained in the Revenue Act of 1978,[25] Section 2032A, to define the taxable estate for state estate tax purposes, and in most cases, to impose a state estate tax in the amount of the permissible state death tax credit.[26] Farms qualifying for Section 2032A treatment will have their estate taxes reduced in the same way as their federal taxes are reduced.

The federal estate tax law has changed considerably. Beginning January 1, 1982, an unlimited amount can be passed on to the surviving spouse; the previous law allowed only 50 percent of the value limit on the marital deduction. Also, federally qualified joint tenancy has changed. Now 50 percent of the estate is presumed owned by the spouse and there is no challenge as to who provided major consideration in the estate. Any amount can now be passed on to the surviving spouse.

Since January 1, 1982, exemptions are permitted for any heir to the estate. The automatic exemption was $225,000 (up from $175,000), and in 1987 it became $600,000. The federal changes also make it easier for farmers to qualify for the special-use valuation for their estates. State applications of those changes will vary from state to state.

Deferral of Payment. Six states have incorporated Section 6166 of the federal Tax Reform Act of 1976 into their state death tax laws, enabling the executor of an estate to defer payment of taxes attributable to the farm property for five years and then pay taxes in equal amounts over a period of ten years, or have adopted a substantially similar provision. During the deferral period, the estate pays interest at the rate of 4 percent on the deferred taxes attributable to the first million dollars of farm property, and interest at usual rates on the rest.

Programs to Reduce Rate of Conversion of Farmland

Comprehensive Planning. This type of program involves a process leading to the adoption of a set of policies regarding land use, transportation, housing, public facilities, and economic and social issues. In most states, the plan in itself is not legally binding on governments or individuals, and in a few states, zoning and major public facility plans must be consistent with comprehensive plans.

Agricultural Zoning. Zoning laws, in general, provide a legal binding designation of the uses to which land may be put, including the type, amount, and location of a development. Agricultural zoning restricts land uses to agriculture and related uses such as a farmstead. Often a large minimum lot size (20 to 160 acres) is stipulated in an agricultural zone. Agricultural zoning is the most popular and common method used by local governments to prevent the use of agricultural land for nonagricultural purposes. Zoning laws are often combined with a community plan, an urban boundary agreement, or voluntary or mandated state programs that together protect farmland. Agricultural zoning ordinances fall into two categories—exclusive and nonexclusive.

Agricultural Districting. This program involves the designation of specific tracts of long-term agricultural uses, usually coupled with benefits and assurances that improve the conditions for farming. Generally, no legal binding controls are imposed on land use. The districts are legally recognized geographic areas whose formation is initiated by one or more farmers and approved by one or more government agencies. The districts, with their benefits and obligations, are created for fixed but renewable periods of time ranging from four to ten years.

Purchase of Development Rights. This program involves the purchase of the right to develop from owners of specific parcels, leaving the owner all other rights of ownership. The price of the rights is the diminution in the market value of the land as a result of the removal of the development rights. The remaining value of the land is the farm-use value. A similar program, Purchase and Resale or Lease with Restrictions, involves the purchase of land, imposition of restrictions on use and development, and resale at market price. This program is essentially the equivalent of purchase of development rights.

Transfer of Development Rights. In the classic mandatory TDR system, districts are identified for preservation and development. Development rights are assigned to owners of land in the preservation district in a systematic manner. They are not allowed to develop but instead may sell their development rights to owners of land in the development district, who may use these newly acquired development rights to build at higher densities than normally allowed by the zoning in the development district. TDR systems are intended to maintain designated land in open uses and compensate the owners of the preserved land for the loss of their right to develop. TDR programs can reduce or eliminate the public costs of acquiring development rights by shifting the responsibility for purchasing them from the government to private developers.

Right-to-Farm Laws. These laws forbid the enactment of local ordinances that restrict normal farming practices, unless they endanger public health or safety. They also provide farmers with some protection against private nuisance lawsuits. These laws take three basic forms: they protect against local government regulations, state regulations, and private nuisance lawsuits.[27]

Land Banking. A public body purchases extensive areas of rural land at rural use values, designating some of it—such as prime farmland—for permanent resource use, selling or leasing it with restrictions on use, and selling or leasing other areas for urban development. In effect, the public acts as a large-scale real estate developer, constructs all necessary roads and utilities, and then covers its costs by selling the land at appointed values. Since the land is all publicly owned before development, the public is able to designate the future use of all land and sell it with appropriate restrictions. Thus, it could not only prohibit development on prime farmland but also provide sufficient sites for necessary urban development, in locations that would be the least disruptive to agriculture. This system is employed only in the Commonwealth of Puerto Rico.

Integrated Programs of Incentives and Controls

In many parts of the country, the problem of agricultural protection can be addressed realistically and effectively only by considering its relation to the entire system of land use and development within a given region. The need to incorporate agricultural protection into an overall strategy for dealing with growth is especially apparent in metropolitan areas where there is often intense competition for limited land resources.

A coordinated regional approach to growth management can accomplish a variety of mutually complementary objectives, such as minimizing public investment costs and focusing farmland preservation efforts on areas where agriculture is most likely to remain economically viable over the long run.

States also have the power to control the uses to which land may be put. In most states, however, most of this power has been delegated to local government, which makes nearly all decisions concerning the planning and regulation of land use. Without involving local government in any way, a state government can declare it a state policy to protect prime agricultural land and require its own agencies to act consistently with that objective. Both voluntary and mandatory state programs combine a variety of incentives and controls.

Remarks

This chapter aims at an understanding of how and why the federal government has become increasingly involved in farmland protection and summarizes the general federal, state, and local programs. This understanding will enable the reader to consider New England agricultural land protection in the larger context of agriculture in general. Chapter 2 describes the specific concerns of New England as a region and its policy responses. Following this discussion of the key issues in New England, Chapters 3 to 8 present case studies of each of the New England states: concerns, programs, and legislation. The final chapters present an overview of the New England states' experience and future options. This work does not describe in detail what is happening in each state at the local level. Rather, it focuses on the governmental role—and that of private initiatives as well—on New England farmland protection.

Notes

1. P.L. 94-52, 42 U.S.C. 4321–4347 (1970).
2. R. Otte, *Farming in the City's Shadow*, Agr. Econ. Rpt. No. 250. *Econ. Res. Serv.*, USDA (Washington, D.C.: Government Printing Office, 1974); and USDA, *Econ. Res. Serv., Our Land and Water Resources: Current and Prospective Supplies and Uses*, Misc. Publ. No. 1,290 (Washington, D.C.: Government Printing Office, 1974).
3. USDA, *Perspectives on Prime Lands* (Washington, D.C.: USDA, 1975), Foreword.
4. Ibid., 17.
5. Executive Office of the President, Council on Environmental Quality, Memorandum for Heads of Agencies, Subject: Analysis of Impacts on Prime and Unique Farmland in Environmental Impact Statements, August 30, 1976.
6. Executive Office of the President, Council on Environmental Quality, Memorandum for Heads of Agencies, Subject: Analysis of Impacts on Prime and Unique Agricultural Lands in Implementing the National Environmental Policy Act, August 11, 1980; id., Subject: Prime and Unique Agricultural Lands and the National Environmental Policy Act (NEPA), August 11, 1980.
7. Environmental Protection Agency Policy to Protect Environmentally Significant Agricultural Lands, September 8, 1978.
8. This legislation and subsequent legislation is described in detail in General Accounting Office, *Preserving America's Farmland—A Goal The Federal Government Should Support*, CED–79–109 (Washington, D.C.: Government Printing Office, 1979); W. Fletcher, "Agricultural Land Retention: An Analysis of the Issue, A Survey of Recent State and Local Farmland Retention Programs, and a Discussion of Proposed Federal Legislation," Report No. 78–117 ENR (Washington, D.C.: Cong. Research Serv., Library of Congress, 1978); and J. Zinn, "Farmland Protection Legislation," Issue Brief No. 1B78013 ENR (Washington, D.C.: Cong. Research Serv., Library of Congress, 1981).
9. The Urban Land Institute Policy Statement, "The Agricultural Land Preservation Issue: Recommendations for Balancing Urban and Agricultural Land Needs," *Urban Land* (July 1982). See also W. Fischel, "The Urbanization of Agricultural Land: A Review of the National Agricultural Lands Study," *Land Economics* 58 (1982): 236.
10. "U.S. farmland is not disappearing as fast as feared," *Miami Herald*, Jan. 1, 1985, sec. D, 8. See also Linda K. Lee, "Land Use and Soil Loss: A 1982 Update," *Journal of Soil and Water Conservation* (July–August 1984): 226–28.
11. Gregg Easterbrook, "Making Sense of Agriculture," *The Atlantic Monthly*, July 1985, 63–78. The NALS's predictions for future agricultural demands have also been criticized, since it appears unlikely that there will be significant gasohol production in the near future. Additionally, agricultural exports have stopped increasing at the predicted rate, due partly to the strength of the dollar, but more importantly due to world food surpluses: "[i]n every part of the world except for Africa and Japan food production is increasing faster than population."
12. R. Allbee, "More on Farm Bill's Evolution," Letter to the Editor, *Journal of Soil and Water Conservation* (September–October 1982): 243. Allbee was Rep. Jeffords's staff person working on the issue.
13. The Farmland Protection Policy Act, P.L. 97–98, §1539–1549, 7 U.S.C. 4201 et seq.
14. 49 *Fed. Reg.*, No. 130, 27716, Thursday, July 5, 1984, Rules and Regulations.
15. 48 *Fed. Reg.*, 31863.
16. *Cong. Rec.*, October 6, 1983, S 14677–S 14680.
17. Ronald Zumbrum, Sam Kazman, and E. Carloe Currin, "Comments on Proposed Regulations of the United States Department of Agriculture for Implementation of the Farmland Protection Policy Act," 7 *C.F.R.* Part 658, Pacific Legal Foundation (September 30, 1983).
18. *Cong. Rec.*, October 6, 1983, S 14676–S 14680.
19. *American Farm Land* (May–June 1985).

20. See *Cong. Rec.*, October 7, 1985, H8320, for the discussion surrounding the amendment. The full text of the amendment states:

> Sec. 1986.(a) Section 1546 of the Farmland Protection Policy Act (7 U.S.C. 4207) is amended by striking out the words "Within one year after the enactment of this subtitle," and substituting therefore "On January 1, 1987, and at the beginning of each subsequent calendar year."
>
> (b) Section 1548 of the Farmland Protection Policy Act (7 U.S.C. 4209) is amended by striking the words "any State, local unit of government, or" and inserting before the period at the end of the sentence: "Provided, That the Governor of an affected state where a state policy or program exists to protect farmland may bring an action in the Federal District Court of the district where a federal program is proposed to enforce the requirements of section 1541 of this subtitle and regulations issued pursuant thereto".

21. P.L.99–198, §1255.

22. President's Commission on Housing, *The Report of the President's Commission on Housing* (Washington, D.C.: Government Printing Office): 195–96.

23. Linda K. Lee, "Land Use and Soil Loss: A 1982 Update," *Journal of Soil and Water Conservation* (July–August 1984): 226–28.

24. This material is abstracted from Susan B. Klein, *Agricultural Land Preservation: A Review of State Programs and Their Natural Resource Data Requirements* (Washington, D.C.: National Conference of State Legislatures, 1982): 8–12.

25. P.L. 95–600.

26. Section 2032A contains detailed requirements that seek to limit eligibility for current valuation to bona fide farm families. A description of these details can be found in Robert Coughlin and John Keene, *The Protection of Farmland: A Reference Guidebook for State and Local Governments,* A Report to the National Agricultural Lands Study (Amherst, Mass.: Regional Science Research Institute, August 1981): 65–66.

27. Some states provide more than one form of protection.

2

Agriculture in New England

The current national concern over the conversion of agricultural land to other uses and the permanent conversion of farmland to urbanization is nowhere more strongly felt than in New England. In these modern times it is easier to grow almost everything somewhere else. Difficult physical and weather conditions, private market policies favoring large-scale agribusiness, competition from more efficient producers in other regions, and intense urbanization pressures have transformed New England from a predominantly self-sufficient agricultural region at the turn of the century to a region which must import roughly 84 percent of all its foodstuffs.[1] This dramatic shift of agricultural production away from the New England states has brought an accompanying change in the character of the New England landscape as hundreds of thousands of once-tilled acres have gone out of production, most reverting to forest, although some have been consumed by urbanization. The change in the character of the landscape has generated widespread concern among the citizens of the region and resulted in the adoption of a range of new policies to protect agricultural land. This chapter will discuss the unique characteristics of farming in New England, the problems farmers have faced, and how farmers have coped.

Background

New England is a densely populated, geographically small region. Given its high population density, food self-sufficiency would be virtually impossible, at least not with modern food consumption habits. Nonetheless, there seems to be a great amount of land available for agricultural expansion.

Even in the best of economic times, however, trends within the New England farm industry have marked a long-term decline in agriculture—a decline that is probably more pronounced in New England than in any other region of the country. Although many sectors of the New England economy have enjoyed tremendous growth in recent decades, the agricultural industry has not shared in this growth. In fact, in many of the growing urban areas, it is clear that the high-technology industrial "renaissance" has negatively affected the agricultural land resource base. Since the 1950s, as suburban office and technology parks have blossomed around the cities of New England, farmland often has proven to be the best land for development. On the other hand, many small farms have been able to continue operating because of the off-farm income made possible by the high-technology industry.

The New England economy has not always enjoyed such vigor. The high growth and low unemployment rates of the 1960s, 1970s, and 1980s are recent improvements in the economy of the region following a long period of economic retrenchment which began at the turn of the twentieth century. Prior to the recent high-technology growth, New England experienced several decades of severe economic decline, as many of the region's traditional industries became outmoded or moved to the South and West.

This pattern of economic growth and change is why New England has been termed the "American prototype of a mature economy." Economists describe a mature economy as one in which the economic base is no longer dependent on agriculture and the transition to an economy based on manufacturing and service industries has taken place long ago. In a mature economy, wages have worked themselves up to a high level in comparison with less-developed regions, the capital stock and infrastructure tend to be aged and nearly obsolete, costs of operations are high, and new job opportunities are scarce. Consequently, a region with a mature economy has generally long passed its peak years of growth and tends to be noncompetitive with emerging industrial areas, where capital stock is newer, wages are lower, average age of the population is lower, and costs to support public infrastructure and services are less burdensome.[2]

Indeed, much has been written about the decline of the New England economy in the first half of this century to the benefit of the South and West. Prior to the twentieth century, New England was well placed to supply the manufactured goods that were needed by a growing country. As the "first finished corner of the United States,"[3] New England manufactured the ploughs, saws, and axes which were used to clear and farm the rich lands from Ohio to the Rocky Mountains. And its textile industry, which formed the bulwark of the region's industrial prosperity, exported fabric around the world. As the birthplace of the industrial revolution in America, New

England's economy was the first in the country to undergo the transition from agriculture to manufacturing.

Unlike the more recent industrial renaissance, New England agriculture flourished during the period of economic growth in the nineteenth century. The farms of New England played a major role in the economic development, supplying locally produced food and fiber to the expanding mills and industrial communities. Indeed, New England's unique landscape is largely the result of the historical significance which agriculture played in the economic and physical development of the region. When the early colonists first settled, they were able to buy land, which they could not do under the old European manor system. The settlers cleared the land themselves to set up homesteads and establish croplands, pastures, and orchards.

The first New England farmers were largely self-sufficient. They grew food for themselves and their neighbors. Unlike the large, single-crop farms of today, the earliest Yankee farms produced a wide variety of farm products. As one writer described the early farms in Vermont:

> Cattle were kept to supply butter, cheese, milk, and beef as well as leather for shoes, harnesses, and other farm purposes. Sheep were kept to supply mutton and wool. Swine, turkey, geese and other poultry were kept for family use. Maple products were used in place of imported sugar. Wheat, corn, oats and other crops were grown to supply the family needs and to feed livestock.[4]

Initially, the settlers farmed only the most accessible and fertile lands in the small valleys along the coast and in the Connecticut River Valley. With its deep, rich soil and wide expanse, the Connecticut River Valley formed the backbone of New England agriculture from the earliest times. The valley also supplied some of the most suitable land for early urban development as towns grew adjacent to the most intensive farming areas. In 1796, Thomas Pownall described the farms and towns along the Connecticut River between New Haven and Hartford as "a rich, well cultivated vale thickly settled and swarming with people....It is as though you were still traveling along one continued town for 70 or 80 miles on end."[5] Even today, the Connecticut River Valley is the dominant agricultural area in New England, despite spreading highways and urbanization along its entire length.

By the beginning of the nineteenth century, cleared farmland was becoming the dominant land use in New England. But this was not yet the high point of New England agriculture. As new immigrants arrived in the cities, there followed what one writer termed a "flowering of Yankee agriculture that now appears rather magical."[6] This flowering coincided with the industrial boom in the region which began around 1810 and continued through the 1880s. The boom in agriculture which followed this commercial growth occurred in the 1820s, 1830s, and 1840s as railroads and industrial mills spread throughout the valleys and along the riversides. More and more land was cleared for farms, not only to feed the growing population, but to supply wool for the many textile mills which emerged as one of the leading industries in the region. To supply the mills, sheep were the major agricultural commodity, especially Merino sheep, which were first imported in large numbers from the Spanish Escorial flock by William Jarvis, American consul in Lisbon. Farms and villages, which once had been self-sufficient, began for the first time to make fortunes by raising sheep to sell wool to the new textile mills, as well as by raising breeding stock and meat animals. This success was not to last, however. In 1842, President William Tyler, under pressure from mill owners, altered a tariff law to allow the importation of European wool. New England farmers could not match the price of the European wool and were often pushed into poverty when wool prices failed to meet their operating costs; some farmers, however, changed to other agricultural commodities. Many farms were abandoned during the mid-nineteenth century as New England farm families joined the migration to the Western Reserve lands of the Ohio River Valley, whose soils were more favorable to intensive agriculture.[7]

Despite this short period of retrenchment in the late 1840s, generally New England agriculture continued to flourish until the 1880s as its population and economy continued to grow. The years between 1860 and 1880 witnessed the height of the region's self-sufficient farms and small villages. Indeed, the high point of New England agriculture (in terms of the number of farms and the acreage of agricultural land in production) was not reached until late in the nineteenth century. In 1880, the USDA Census of Agriculture reported that there were over 200,000 farms in the six states operating on approximately 21.5 million acres of land.[8] With the exception of Maine, whose land mass was dominated by forests, agriculture was the major use of land in each of the New England states, covering approximately 53 percent of the region's total land area.

This period of abundant, productive agricultural land was not to last long. Following the 1880s, the accelerated abandonment of farmland began. There were several reasons for this, the most notable being the development of western railroad and canal-transportation networks which quickly put New England farmers at a competitive disadvantage with larger, more fertile midwestern farms. From 1880 to 1940, the decline of farming in New England was most pronounced among the less efficient hill farms, which were plagued by poor soil, exposure to adverse weather conditions, and access problems. The valley farms, on the other hand, with their richer soils, easier access, and protected location, remained competitive until the 1940s. The decline of farming from 1880 to 1940 was thus largely the result of the poor economic viability of the less productive New England farms.[9] According to the Census of Agriculture, by 1940 only one New England state, Vermont, remained predominantly agricultural in land use. The change over this period was most dramatic in southern New England. Connecticut's and Rhode Island's farmland changed from 79 percent and 78 percent of total land use in 1880 to 49 percent and 33 percent, respectively. By 1940, farm activity in the region as a whole had declined to approximately 135,000 farms on 13 million acres, accounting for 33 percent of New England's total land area.

During this period, the mix of commodities produced by New England farms also began to change. In 1880, the farms still maintained a degree of self-sufficiency, producing a broad range of commodities to supply the farm family and the neighboring area. The leading commodities in terms of acreage included hay, oats, corn, potatoes, wheat, rye, buckwheat, and barley.

By the 1940s, however, the region's farms were devoting substantially less acreage to many of those commodities as a gradual movement toward regional specialization emerged. Except for potatoes in Maine and Rhode Island, and hay for expanding dairy farms, the competitive production centers in the Midwest and West expanded to claim many of New England's historic markets.[10] At the same time, land devoted to the production of dairy and poultry products, apples, cranberries, and other products uniquely suited to the market and geographic conditions of New England began to increase between 1880 and 1940.

Following World War II, increasing urbanization and expanding highway networks further affected the agricultural resource base of New England. This time the competition was different than previous competition from the fertile farms of the Midwest, however. In addition to the questions of economic viability of farming the hilly soils of New England, spreading urban land uses resulted in the urbanization of thousands of acres of high-quality, highly productive valley farmland. Prime agricultural land in the valleys, already cleared, well-drained, and gently graded, also had prime development potential. Since World War II, much of the conversion of land into developed acreage has come from these lands. Many New England farm owners, no matter how productive the soil or efficient the management, could not afford to resist offers to sell their land for development. Even those farmers wishing to stay in farming often found that land taxed on its possible urban development value affected production economies to a point where farming as a livelihood was nearly impossible.

Other nationwide agricultural trends outside the control of New England farmers adversely affected the production economies of Yankee farms. This resulted in abandonment of farming even in areas without development pressure. By 1970 the acreage of agricultural land in production in New England had declined substantially, to a point where none of the six states was predominantly agricultural. Even Vermont, with the highest percentage of productive agricultural land of any New England state, experienced a decrease in farmland from 62 percent of the state's land area in 1940 to 32 percent in 1970. For the six-state region as a whole, by 1970, land in farms had declined to slightly less than 5.6 million acres on 28,600 farms, representing only 14 percent of the region's total land area.

Since 1970, total farmland acreage in New England has continued to decline, although there are indications that the rate of land conversion may have slowed somewhat. Table 2.1 summarizes agricultural land use change in the region over the past century.

Substantial agricultural acreage in New England has been converted into urban uses, particularly in the more densely populated areas of southern New England. Over the region as a whole, however, most of the historic New England farmland has been merely abandoned, to become once again forest lands, which now cover over 73 percent of the region's total land area.[11] Among the six states, the largest percentage of forested lands is found in Maine—approximately 90 percent of the state's land area. Rhode Island has the least forest land, at 60 percent. Thus, in less than a century, New England has experienced a major turnaround from predominantly agricultural to predominantly forest land uses.

Characteristics of New England Farming

It has been said that farming in New England today takes place in an atmosphere of siege.[12] Judging from the patterns of agricultural land abandonment and conversion over the past three decades, the observation may be an understatement. Faced with a variety of physical and economic problems, many of the region's farm families have decided to forego farming because there is very little incentive to continue.

To understand the pattern of growth and decline of agriculture in New England and how recent efforts to protect agricultural land have evolved, the specific problems which confront New England farmers must be understood. Many of their problems are similar to those faced by farmers throughout the world; others are unique to the physical and cultural setting of the area. But the array of problems facing the New England farmer, which range from unique soil and weather problems to nationwide market imperfections, appear to be more acute than in most other regions. Furthermore, the measures which the individual farmer may take to correct these problems are extremely limited.

Table 2.1
Agricultural Land Use Change in New England, 1880–1982

Year	*Land in Farms (Acres)*	*Number of Farms*	*Change*		*Change from Total Land in Farms (1880) (%)*	*Farm Land % of Total Land in New England*
			Acres	*%*		
1880	21,483,772	207,232	—	—	—	53
1910	19,715,000	188,802	–1,768,772	–8	–8	49
1940	13,371,000	135,190	–6,344,000	–32	–38	33
1969	5,598,640	28,640	–7,772,360	–58	–74	14
1982	5,185,000	30,520	–413,640	–7	–76	13

Source: United States Census of Agriculture.

Soils

Soils are the most basic of all agricultural resources. The quality of a region's soils makes the difference between a healthy economy and an economy where costs preclude the development of viable agriculture. In his important history, *Soil and Civilization,* first published in 1952, Edward Hyams parallels the historic growth and power of major civilizations to the quality and cultural stewardship of their soil resources:

> Clever animals though we may be, we remain, our culture remains, our civilization remains, very much the creation of the soil we live on....All members of soil communities are conditioned, within the limits of adaptation possible to them, by their soil....The influence of soil on community character is by no means confined to effects upon the physical ways of life....To take extreme cases as examples, for the sake of clarity, men living on exceptionally simple and austere terms with their soils, the Arab as presented by Doughty, or Lawrence the Gaucho as presented by Hudson, the French Canadian pioneer as presented by Hemont, all these show, so clearly that explanation would be superfluous, that the character of men is profoundly modified by the nature of their relationship with their soil.[13]

The soils of New England, in both their quality and their extent, are probably the most important constraint to modern agriculture in the region. The character and distribution of those soils, as indeed are most features of the natural New England landscape, are largely a result of glacial activity which began over one million years ago. When the glacial period ended approximately 12,000 years ago, a land was left that was described as "more exciting to the eye than susceptible to the plow."[14] Most of the soils which existed prior to glacial times had been scraped away and carried to the ocean or deposited by the glacial action in great rocky moraines and drumlins. The underlying granite, shaped and polished by the massive glaciers, formed dramatic cliffs and basins where thousands of lakes, ponds, and waterfalls later formed. As the ice sheet retreated, piles of debris and jumbled rocks of various sizes—all intermingled with sands, silts, and clays—were left behind.

Few areas of good soil remained. Those areas of soil which did exist were generally strewn with boulders, not only on the surface, but throughout their entire depth. The only good, rock-free soils existed in scattered areas between the rocky hills where large chunks of ice or glacial debris had dammed the valleys to form lakes. Over hundreds of years, incoming streams deposited layers of silt and sand, and when these lakes finally drained several thousand years ago, scattered areas of land suitable for farming remained.

Unfortunately, these areas of "good" soil vary widely in chemistry, organic content, drainage characteristics, and edaphic character. In some areas, such as the Connecticut River Valley, which bisects New England and includes parts of Vermont, New Hampshire, Massachusetts, and Connecticut, the soils are as good as those found anywhere. Such areas are the exception, however. Very little of New England's soil conforms to USDA criteria for prime farmland soils.[15] This is in great contrast to areas of the Western Reserve, the Midwest, and California, where vast acreages of land may be classified as prime. Recent mapping by the USDA Soil Conservation Service found that the six states of New England combined contain only 2,297,000 acres of prime farmland soil, 6 percent of the area of the five states. By comparison, the state of Illinois, somewhat smaller in area but with a similar population density, contains 21,400,000 acres of prime farmland soils, 60 percent of the area of the state.[16]

Much of the soil in New England also is permeated with stones and boulders, which are a result of the glacial activity. These stones work their way to the surface each spring and must be removed, usually by hand, before plowing can begin. The distinctive character which the network of stone walls gives to the New England countryside was created by succeeding generations of farmers clearing these stones from their fields and stacking them along the perimeter of their land.

Despite the limited availability of high-quality agricultural soils in New England, those good farming soils which do exist are often very productive. For example, Hartford County, at the broad base of the Connecticut River Valley of Connecticut, ranked fourth nationwide in 1969 and twelfth in 1974 in the value of agricultural products sold per acre of farmland.[17] In fact, according to the 1978 Census of Agriculture, New England as a whole led the nation in production per acre.[18]

Although the most abundant agricultural soils are found in the Connecticut River Valley, many other areas of high-quality soils do exist throughout the region, albeit in scattered tracts. While New England does not possess the wide expanses of high-quality soil found in the Midwest or West, many highly productive farms do exist in the region, largely as a result of the human energy which has been devoted to making the soils productive. Even where these pockets of high-quality soil exist, however, farming has continued to decline in recent years. Other factors beside soil quality, such as small farm size, increasing urbanization with corresponding increasing land values, climate, and national farm economic conditions, are contributing to the demise of New England agriculture.

Farm Size

Beyond the limited quantity and variable quality of agricultural soils, perhaps the greatest other problem facing New England farmers is geographic characteristics. These characteristics have had a profound effect on farm size and ownership patterns. As mentioned, the glacial history of the region left deposits of good agricultural soils in small, dispersed pockets, often less than one hundred acres in size. This, in turn, resulted in a pattern of small, often isolated fields, as early farmers settled only in areas susceptible to their plow. Prior to the invention of modern farm equipment, this pattern of small soil parcels presented few problems. In those times, farms of one hundred acres or less were common, since this was all the farming family could handle with the existing technology. The limited size of contiguous, prime soil units was thus not a significant constraint to the viability of farming. Indeed, in 1880 the average farm size in each of the New England states equalled or exceeded that found in the fertile expanses of Ohio, Indiana, or Illinois.

Today, however, modern farm equipment and farm economics are based on large-scale operations. Most New England farms, settled two hundred years ago, are now too small for the efficient use of modern farm methods. In 1978, average farm size in the United States was 497 acres. In the same year, average farm size in New England was approximately 200 acres.[19] Moreover, USDA studies in 1981 revealed that the most efficient corn-producing farms in the Midwest (a commodity also found on many New England farms) were larger than the national average, about 640 acres. The study also found that efficiencies of 90 percent could be achieved on farms as small as 300 acres.[20] The average New England *commercial* farm of 170 acres, however, is clearly smaller than the current optimum size within the agricultural industry nationwide.[21]

Many of the efficiencies which can be achieved on larger farms are attributable to the use of large high-speed equipment, which requires large, gently contoured areas to operate. Unable to expand operations because of soil and topographic limitations, the New England farmer is unable to take advantage of this equipment and compete with the large, efficient farms in the West and South.

An overview of the average farm size for the entire New England region tells only part of the story, however. A closer look at average farm size for each of the six New England states illustrates more fully why the decline in agriculture in the southern New England states of Rhode Island, Connecticut, and Massachusetts generally has been greater than in the northern New England states. Table 2.2 summarizes average farm size in the two subregions. As the table illustrates, the farms of southern new England are far below the optimum size for efficient farm operations, which has had a significant impact on their viability. On the other end of the spectrum, Vermont and Maine, with the largest remaining sectors of viable agriculture in the New England region, also have the largest average farm sizes.

Landownership and Settlement Patterns

The limited tracts of prime farmland soils and small farm size are not the only limitations for farmland expansion today. In many cases, even if adjacent good farmland does exist and an efficient farmer wishes to expand operations, landownership and settlement patterns remaining from earlier times make assembly of large tracts of land difficult and expensive. Legal problems of landownership and high land costs often make the purchase and assembly of additional farmland parcels prohibitively expensive. In 1978, for example, farmland in Rhode Island had an average value statewide of approximately $2,500 per acre. For the New England region as a whole, farmland averaged nearly $900 per acre. This compares with a national average of $640 per acre in the same year.[22] Much of the higher cost of land in New England is a result of the higher population density and the premium which competing uses place on the value of the land. Thus, young farmers wishing to start, or older farmers wishing to expand their operations, often find it impossible to acquire land at prices which agriculture will support.

Farming in an Urban Environment

The high cost of purchasing land in New England is, to a large extent, a result of high population density, which places other unique burdens on local farmers. According to the 1980 census, New England population numbered slightly over 12.3 million persons, or 5.5 percent of the total U.S. population. This populace lives on roughly 62,960 square miles of land, resulting in an overall population density of 196 persons per square mile.

Table 2.2
Average Farm Size in New England (Farms with sales of $1,000 or more)

	Acres
Northern New England	
Maine	198
New Hampshire	160
Vermont	218
Southern New England	
Massachusetts	119
Rhode Island	98
Connecticut	113

Source: Crop Reporting Board, Soil Conservation Service, U.S. Department of Agriculture, December 1981.

By comparison, the nation's most populated state, California, with a land area of 156,360 square miles, has an overall population density of 151 persons per square mile. The state of Texas, with a population approximately equal to all of New England, has a density of only 54 persons per square mile.

The impact of population density on land use is particularly apparent in the southern New England states of Connecticut, Rhode Island, and Massachusetts. These three states have a combined population of 9.8 million persons, representing 80 percent of the region's total population and resulting in a population density of 713 persons per square mile. Nationwide, only the state of New Jersey has a higher population density at 986 persons per square mile. Amidst population densities such as these, agriculture in New England may be characterized essentially as farming in an urban environment. Much of the active farmland which remains today does so in the shadow of suburbia, as expanding communication and transportation systems have caused New England cities to expand tenfold beyond their historical compact centers.

Over the last few decades, agricultural land in New England has faced serious urban challenges. Expanding highways and suburbs have placed heavy pressure for development on the farms of the region, particularly those in southern New England. Although abandonment has accounted for the greatest conversion of agricultural land in New England as a whole between 1952 and 1972, in some states, notably Massachusetts, conversion to urban uses was the major cause.[23] With the difficulties of soils, farm size, and problems relating to farm marketing, New England farmers usually find it impossible to resist offers for their land which the value for urbanization can generate.

Direct pressure for conversion to urban uses is not the only problem that farms in an urban environment must face. Higher population densities and competition for the "highest and best" land uses result in higher land values. Often land values are too high to justify purchase of land for agricultural purposes. Thus, farmers are often forced to lease land. Many nonfarmer landowners, however, speculating on the development value of their land, are reluctant to grant long-term leases to farmers, who will tie up the land for long periods. Moreover, with a short lease, a farmer is reluctant to invest in long-term soil improvement measures since economic benefits may never be realized. Thus, the quality of the land for long-term agricultural uses gradually declines. Furthermore, when the owner of the land decides to sell or develop the land for another use, the leasing farmer is left to search once again for additional land.

This pattern of separation of farmland ownership from actual farm operators is a problem across the entire nation. Estimates are that nationwide fully 40 percent of all active agricultural land in the United States is farmed by a person who does not own the land.[24] This is less severe in New England, where about 80 percent of all farmland is owned by the farmer.[25] This tenure pattern is partly so because in New England most of the remaining farmers own land which has been handed down through several generations, and therefore the cost of landownership is minimal. When a New England farmer decides to quit, usually the land is either abandoned or developed, since the economics of farming in the region precludes the purchase or rental of the farm by another farmer. Furthermore, as soon as an alternative use for the land can be found, landowners often stop renting their land to farmers. In short, only farmers who already own their land can remain in farming in New England, and a higher-than-average farmer ownership pattern results.

Taxes are another problem of farming in the urban environment of New England. The high market value of farmland near densely populated areas—usually based on the value of adjacent urban uses—means higher property taxes. Estimates are that the average New England farmer pays approximately $1,800 in taxes on farm property, compared with $1,650 nationwide. This tells only part of the story. When the size of the average farm is considered, it is apparent that New England farmers as a whole pay over 2.5 times the national average per acre in property taxes. Farmers in the southern New England states of Connecticut, Rhode Island, and Massachusetts pay over five times the national average per acre. Table 2.3 summarizes the average 1980 taxes levied on farm real estate in New England.

It is clear that even if there is no immediate demand to convert a particular parcel of farmland, high property taxes in the area often contribute to make farming infeasible by reducing the economic return.

Table 2.3
Taxes Levied on Farm Real Estate: 1980 Amount, per Acre

	Taxes per Acre
Maine	$ 6.16
New Hampshire	9.51
Vermont	7.83
Massachusetts	21.44
Rhode Island	30.78
Connecticut	19.51
New England[a]	10.73
New England—Southern Tier	21.24
United States	3.85

Notes:

a. Weighted averages based on revised estimates of land in farms for the six New England states prepared by the Crop Reporting Board, SRS, USDA, December 1981.

Source: Crop Reporting Board, SRS, USDA, Concord, N.H., December 1982.

These examples illustrate that farming in a densely populated area such as New England can have several direct economic impacts on farmers and their operations. High population density also may have several less direct impacts; one of the most important of these is nuisance.

Agricultural activities which were at one time carried out in wide open spaces often begin to pose serious problems for neighboring suburban homeowners who have moved to the country. Although there has been a tremendous population growth in rural areas of New England and the United States in the last decade, farming and people seldom have mixed successfully. Urban people who move to the country often bring with them values which are not totally compatible with the less desirable aspects of rural life. They soon learn that the countryside is not as bucolic and idyllic as it looks. Normal farm practices such as spreading manure, spraying pesticide, and operating noisy farm equipment early in the morning or late at night are usually not compatible with the desires of most newcomers for fresh air and tranquility. The result is a rise in complaints and nuisance lawsuits against farmers. When a farm is creating a nuisance that threatens neighboring property owners, the farmer is usually forced to modify or cease the offensive agricultural practices, regardless of whether the farm was there first. Eventually, the farm operations can become so restricted that they become insufficient and, in a highly competitive market, unprofitable. The farmer decides to "work out and forget farming," as the benefits of farming do not offset the problems. Such nuisance problems have been common in New England and have contributed to the conversion of many farms in New England.

Climate

New England's climate is influenced by three major weather forces—northern arctic air masses from Canada, westerly influences that sweep across the Great Lakes region into New England, and southerly air flows following the Gulf Stream. The forces interact to produce one of the most varied and unpredictable climates in the United States.[26] The region generally receives ample precipitation (averaging about 40 inches of precipitation a year) but has a relatively short growing season due to its northern location. New England's climate is ideal for some agricultural products (such as cranberries and maple syrup), but it is too cool for any citrus crops and does not permit any year-round crop harvests.

The Impact of External Forces

In addition to the local problems of soils, climate, farm size and ownership, and urbanization, several external forces negatively affect the viability of agriculture in the six New England states.[27] First, the nationwide economic trends in the agricultural industry are outside the control of the small New England farmer. Most of these economic trends have to do with market imperfections.

In an unregulated market economy, the principal input groups—land, labor, and capital—are allocated among competing uses by supply and demand forces responding to price and profit signals. The process is dynamic. Resources are continuously shifted from sector to sector in response to changes in technology, resource availability, and the preferences of consumers. In the absence of market imperfections, the resulting allocation is efficient.

Land, labor, and capital have been moving from agriculture to industry for centuries in response to such natural and beneficial changes in the economy. Technological improvements in agriculture have meant that fewer resources are required to meet the growing demand for food. As productivity has increased, a smaller portion of the collective budget is spent on food, and a smaller number of farm families and acres of land are required to supply the demands of consumers. In the United States, this market efficiency has been truly astonishing, as illustrated in Table 2.4.

Most significant is the decrease of labor in agriculture, which fell from 41.9 percent of the population in 1900 to 2.8 percent in 1979. Equally interesting are the enormous productivity gains that have been achieved. As Table 2.4 illustrates, it required nearly forty times as much labor to produce a bushel of corn in 1850 as it did in 1970. Yield per acre of land in corn nearly tripled between 1940 and 1970. These gains in agricultural productivity

Table 2.4
U.S. Farm Population, Acreage and Production, 1850–1979

Year	*Land in Farms (Millions of Acres)*	*Farm Population (Millions)*	*Farm Population (% of Total)*	*Yield per Acre: Corn (Bushels)*[1]	*Man Hours per 100 Bushels (Corn)*[1]	*Yield per acre: Potatoes (Tons)*	*Man Hours per Ton: Potatoes*
1850	294	—	—	25.0	276	—	—
1900	841	29.9	41.9	25.9	147	—	—
1920	959	31.9	30.1	26.8	132	56.9	26
1940	1061	30.5	23.2	32.2	108	70.3	20
1950	1202	23.0	15.3	36.1	53	117.8	12
1960	1176	15.6	8.7	48.7	20	178.1	6
1970	1102	9.7	4.8	77.4	7	212.8	4
1979[2]	1042	6.2	2.8	100.5	3	266.1	3

[1]For 1940, 1950, 1960 and 1970, figures are averages for the preceding five years.
[2]For 1979, figures are taken from Farm Bureau News, Nov. 1980, p. 5.

Source: U.S. Department of Commerce, Bureau of the Census, *The Statistical History of the United States from Colonial Times to the Present,* New York, Basic Books, 1976, Chapter K, pp. 449–526. For 1977, *Statistical Abstract of the United States 1978,* pp. 682–727.

were achieved through the application of modern farming techniques, many of which do not lend themselves easily to the difficult farming conditions of New England.

As the demand for labor in agriculture declined, the demand for labor in industry increased and there was a population movement towards urban centers, which were the most efficient locations for industrial activity. The population movement generated demand for housing and services and again the market responded, in many cases by moving land (and therefore capital) out of agriculture and into other sectors. When the supply of labor or a parcel of land is bid away from agriculture by society, a developer, or industry, the market is making an efficient decision, as it has many times in the past, that the value of the products which the developer provides exceeds the value of the farm products.

Land Market Imperfections

Even under the best of circumstances, markets often fail to allocate resources efficiently. Several conditions are responsible for this failure. A major problem is that the market allocates land and other resources based on economic conditions and productivity gains at the national level. Public policy often further distorts the allocation process by subsidizing or penalizing certain market activities. For example, loan guarantees from the Federal Housing Authority (FHA) and the Veterans Administration (VA) have artificially produced a stronger demand for low-density housing, significantly assisted by income-tax deductions on home ownership, thereby increasing the spread of urbanization and conversion of prime farmland.

Several built-in imperfections also contribute to market failure. The most important is the failure of private decision makers to consider external values, whether those be the social value of the affected resource, such as the family farm or open space, or the cost to society of their decisions, such as decreases in the quality of life or hidden public costs of development.

Another imperfection is the short horizon on which the private market looks at resources. The tendency is to consider a resource in terms of immediate profit potential instead of as a long-term asset. Land is perhaps something of an exception. Market evaluation of land usually includes expected future return (discounted) from that resource. If a farmer expects the demand for agricultural land to increase in the future, then the value the farmer assigns that land will increase also.

A third imperfection in the land market is the tendency to look only at the pecuniary value in the bidding process. Ideally, a resource such as land should go to whoever values it most. However, in the case of farmland the competition is uneven because land development forces determine value primarily on the basis of expected return, while agricultural values include "psychic" income as well. This psychic income produces no revenues to use in bidding for the land.

The last of the built-in market imperfections is the failure of the market to distinguish between short-term conditions and long-term viability. For instance, temporarily tight credit can bankrupt many farmers who own operations which are basically sound. This is a critical aspect of farmland conversion because once farmland is taken out of production, it is very expensive to bring it back and once developed, it is essentially lost forever. Thus, a short-term economic condition can accelerate the long-term conversion of farmland to other uses. This has been a particularly vexing problem in New England.

Like farmers everywhere, most New England farmers must borrow to meet annual production expenses. Much of this financing comes through the Federal Farmers Home Administration (FmHA) and other federal and state agencies at below-market interest rates. The value of these loans, whether from a government agency or private bank, is usually established on the basis of the market value of land and other fixed assets, not on the agricultural value of these assets. Thus, it is possible for a farmer to receive a loan in excess of what the agricultural operation will support. Small New England farmers often find it increasingly difficult to meet the obligations of these loans. Estimates were that as of March 1983, 25 percent of all FmHA farm loans in the three southern New England states of Massachusetts, Connecticut, and Rhode Island were in default.[28] In such a situation, there is pressure for the lending institution, whether FmHA or a private bank, to foreclose on the farm property and sell the land at auction to the highest bidder to retrieve the lenders' investment. With loan value based on market value of the land, naturally the highest bidder will be one who seeks to use the land for nonagricultural purposes. There are many examples of such foreclosures and eventual development of farmland in New England in recent years. Chapter 9 summarizes how the New England states have addressed land market imperfections with economic incentive programs.

Commodities Market Imperfections

Trends in the marketing of agricultural products nationwide also have had a significant impact on the viability of farming in New England. Much of this can be traced to a nationwide food supply system geared to national supermarket chains which favor larger and larger sources of supply. The agricultural industry nationwide has grown to become a highly capitalized, increasingly vertically integrated, industrialized, and tightly concentrated agribusiness. In 1978, estimates were that large, vertically integrated agribusinesses, which can produce, process, and market food from seed to supermarket, accounted for 25 percent of total national farm output, and the figure was rising.[29] The production of many commodities has become tightly concentrated within a few large national corporations who are often primarily engaged in an unrelated industry outside of agriculture.[30]

This national production and marketing system stands in sharp contrast to conditions prevalent in New England, where the 1978 Census of Agriculture found that 87 percent of all farmland acreage is farmed by an individual or a family. By comparison, nationwide 77 percent of all farmland acreage was farmed by individuals or family operations.[31] As described previously, the unique characteristics of the New England region make large-scale corporate farm operations impossible except in a few areas.

Market concentration and vertical integration of the agricultural industry outside of New England have had a negative impact on the New England farmer. The control of the national food economy by large agribusinesses outside New England means uncertain and fluctuating prices for the products grown in New England, difficulty in obtaining credit, and a market dominated by a few competing outside interests. New England–based supermarket chains often decline to purchase locally during the productive season, favoring long-term, year-round supply contracts with large suppliers outside the region. This is one reason why New England as a whole imports approximately 84 percent of its food at a cost 10 to 15 percent above the national average,[32] while local farmers often have difficulty finding a reliable market for their produce.

Production Factors

In addition to farm marketing problems, several external forces affect New England farm production costs. National priorities to encourage large-scale agricultural production have resulted in an emphasis on large-scale equipment designed to operate on the large, unbroken fields typical of the midwestern states. High-speed farm equipment designed to harvest large quantities of corn in minutes and operate on automatic controls is not usually suitable to the naturally constrained smaller fields of New England. Likewise, in the dairy industry, milking parlors, automatic feeding, and high-production breeding have made efficient herd sizes far larger than can usually be directly supported by the low acreages of tillable land and pasture in New England. Thus, New England dairy farmers must obtain a substantial portion of their feed requirements from outside sources in order to compete. Through imported dairy feed, efficient, hard management, and long work days, New England dairy farmers have managed to keep pace with national agricultural trends. Average herd sizes in New England are comparable with other major dairy regions of the United States, despite the higher-than-average production costs which New England dairy farmers face.

The immense cost of large, high-production farming equipment further exacerbates the problems of the small New England farmer, whose assets are often not large enough to finance the purchase of such equipment even if the parcel of land is large enough. Only the largest farm operations can afford such equipment, since production credit policies generally require large assets to finance such purchases.

According to economists at the Federal Reserve Bank in Kansas City, "farm assets and income have become increasingly concentrated, and the trend appears to be accelerating."[33] A recent study found that in 1978 "the largest 7% of farms received more than 36% of net farm income, held 30% of all farm assets and received more than 25% of all government farm payments."[34] Indeed, in 1981, the 1.7 million farmers nationwide with gross income less than $40,000 received less than one-third of all direct payments to farms. The 700,000 farmers with gross income above $40,000 received over two-thirds of all such payments.[35] Thus, it seems that federal government policies have further encouraged large-scale farm operations. This has had a negative impact on New England agriculture since over 60 percent of all commercial farms had gross sales of agricultural produce under $40,000 in 1978.[36] Government and private policies which encourage large-scale farming have thus exerted competitive pressures on New England farming, which is beyond the control of local farmers or their suppliers. As a result, New England has not only lost farmers and farmland; it has also experienced a marked decline in other critical farm-related industries such as farm implement distributors and food processors. The loss of this critical farm infrastructure places further pressures on New England farmers, who no longer have a nearby source for their supplies. This trend has resulted in increased production costs for the dwindling number of farmers and a self-perpetuating spiral of abandoned farm infrastructure.

Agriculture's Response to Specific New England Problems

Despite the many problems which confront farmers in New England, there are some reasons to be optimistic. The most important of these considerations is the strong will which the farmers of New England have shown in their efforts to overcome their problems. Farmers have responded by limiting production to special commodities suitable to the region, direct retail marketing, and achieving higher-than-average production efficiencies. However, these responses are limited in what they can accomplish.

Special Commodities

As a response to the problems discussed above, changes in the form of agriculture have emerged in recent years. Production specialization is the most important of these changes.

Specific agricultural commodities uniquely suited to the limitations of farming in New England have come to dominate the market basket of the region. For several decades, the most significant of these has been dairy products, which generate the largest total cash receipts of all farm marketings in New England. To support this dairy industry, greater acreages of farmland in the region are devoted to pasture, hay, and other feeds than to any other agricultural commodity. To a large extent, the growth of dairy production has paralleled the rise in population of the region, as nearby dairy farms have always been necessary to supply fresh dairy products to the growing cities of New England. Cranberry production also has grown through the years; this commodity is well-suited to the glacial landscape of New England with its many bogs and wetlands. Few regions of the country can compete with New England in the production of cranberries. The region is also a net exporter of potatoes, apples, and poultry products. In 1974, Aroostook County in Maine was the leading county in the nation in tonnage and dollar value of potato production. Rhode Island agriculture has responded to changing market conditions with increasing production of high-value commodities such as turf and nursery products. In fact, Rhode Island sod farms are a major supplier of pregrown sod to consumers throughout southern New England. Figure 2.1 and Table 2.5 summarize the relative importance of the leading agricultural products in the New England region in 1981 to 1983. As Figure 2.1 indicates, dairy products generated the highest proportion of total cash receipts of all farm produce in New England, with $692 million. Potatoes, which comprise 6 percent of total cash receipts, were the field crop having the largest value in 1983, totalling $98 million.

The increase in the production of specific commodities in New England parallels a decrease in production of a wide array of other commodities. Generally, the region's farms are producing a much narrower range of those commodities produced in the late nineteenth century and a much higher yield of those specialized commodities which the New England landscape and today's markets will support.

In addition to the major commodities suitable for large-scale production in New England's difficult environmental conditions, the region produces various specialty products targeted to specific urban markets. Some of the most successful New England farmers have capitalized on their proximity to urban areas by producing specifically for the needs of the large nearby population. For example, Wilson Farm of Lexington, Mass., capitalizing on

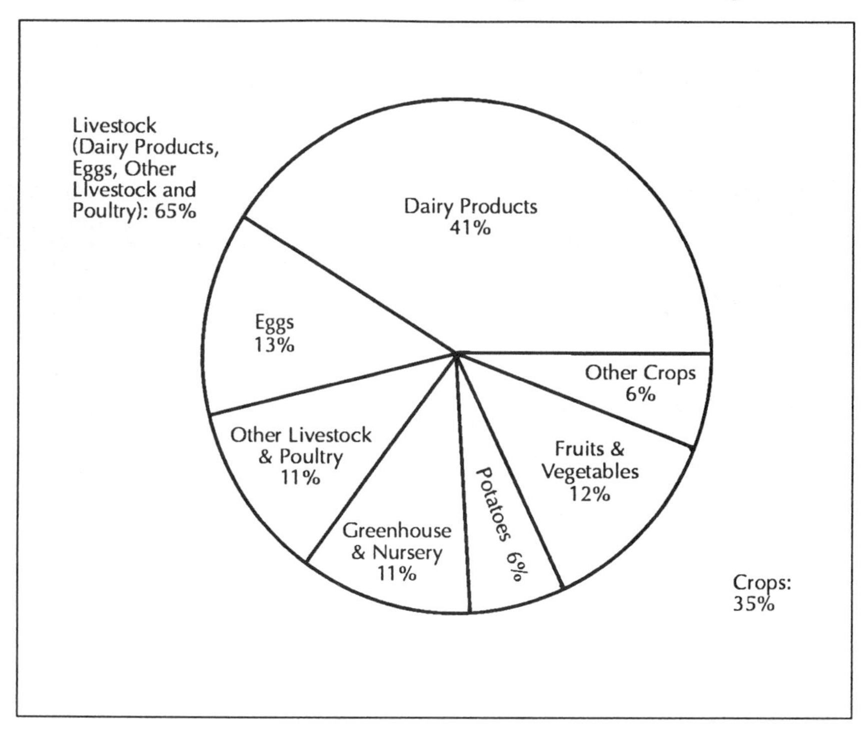

Figure 2.1. Distribution of cash receipts from New England farm marketings in 1983. *Source:* New England Crop and Livestock Reporting Service.

the cosmopolitan population of nearby Boston, is a major producer of fresh sweet basil. Basil is an essential ingredient in the cooking of many Eastern European and Middle Eastern families who live in the Boston area, and thus Wilson Farm is able to help assure the economic viability of its operation while satisfying a special need for the large urban population.

The dominant commodities produced today represent those best suited to the special problems of farming in the region. Farmers in New England also are developing special marketing responses to help assure the continued economic viability of their farming operations. For a complete discussion of programs specifically designed to correct commodity market imperfections through economic assistance programs, see Chapter 9.

Direct Retail Marketing

Although small in dollar volume, one of the most important recent trends in marketing practices has been a growing movement on the part of farmers to market their goods directly to the consumer through retail stands and

Table 2.5
Cash Receipts from Farm Marketings in New England, 1981–1983 (thousands of dollars)

Commodity	*1981*	*1982*	*1983*
Crops			
Hay	13,699	16,355	18,303
Tobacco	49,676	47,711	26,301
Potatoes	148,293	107,965	98,397
Misc. Vegetables	60,702	68,131	61,097
Apples	48,353	56,417	51,459
Berries	62,756	85,438	88,579
Misc. Fruits	1,387	3,270	3,527
Maple Products	11,859	10,289	10,190
Forest Products	16,974	17,641	19,423
Greenhouse/Nursery	165,731	178,023	189,495
Misc. Crops	25,363	26,165	26,546
Total Crops	604,793	617,405	593,317
Livestock			
Cattle/Calves	67,663	92,752	92,314
Hogs and Pigs	14,038	16,111	10,198
Sheep and Lambs	940	1,477	1,301
Dairy Products	658,231	677,123	691,751
Chickens	6,363	4,476	6,116
Eggs	229,227	213,619	215,183
Misc. Poultry	64,360	35,577	38,729
Misc. Livestock	19,350	21,315	24,159
Total Livestock	1,060,172	1,062,450	1,079,751
Total Commodities	1,664,965	1,679,855	1,673,068

Source: New England Crop and Livestock Reporting Service.

farmers' markets. More and more farmers in New England are attempting to bypass the large national distribution and marketing system and sell their products directly to the consumer, thereby capitalizing on the proximity to a large market. The high population density of the region is, of course, a natural market for the region's farm produce. The production difficulties of agriculture in New England often make it difficult for many local farmers to compete in the regional distribution and marketing system which is a part of the large-scale production and distribution network. By selling their goods directly to the consumer, however, New England farmers have eliminated the costs of the intermediary and gained more direct access to the large urban market nearby. Pick-your-own operations, retail farm stands, farmers' markets, and other methods have increased throughout New England in recent years. By 1978, the value of agricultural products sold directly to consumers for human consumption in New England was approximately $29 million, accounting for roughly 2.5 percent of the total value of agricultural products sold in the region in that year. In contrast, nationwide the direct sale of agricultural products to consumers accounted for less than 0.5 percent of the total market value of agricultural products sold. Furthermore, the 5,380 New England farms which sold directly to consumers averaged $5,400 in value of the products per farm sold compared with a nationwide average of $2,800 per farm.[37] It is also interesting to note that Massachusetts, Connecticut, and Rhode Island, with the highest population density of the six New England states, also had the highest average value per farm of products sold directly to consumers. In these three cases, the problem of farming in an urban environment has been turned partially into an asset. However, there are very few cases where such direct retail marketing efforts alone have succeeded in offsetting the many problems which confront the region's farmers.

Farm Efficiency

New England farmers have also responded to the difficulties of farming in the region by becoming some of the most efficient farmers in the country. Indeed, the efficiency of the most successful New England farmers led one writer in 1977 to observe:

> The very finest Yankee farmers—the ones who have combined the luck of owning some of the good farmland that is scattered about, with the wit, vigor, and application needed to manage complex modern operations well—these farmers are among the best agricultural practitioners in the United States.[38]

In most New England communities the dollar per acre is as high or higher than the national average. According to published findings of the New England Regional Commission, Connecticut, Massachusetts, and Rhode Island average nearly twice as much value per farm in produce as the national average. Much of this high productivity is attributable to the uniquely suited high-value specialty crops; potatoes from the Aroostook River Valley in Maine; apples from Massachusetts and Maine; maple products from Vermont; tobacco from the Connecticut River Valley; and two-thirds of the nation's cranberries from Cape Cod and the South Shore of Massachusetts.[39]

The most important agricultural product in New England is dairy. Dairy operations are an excellent example of the efficiencies which have been attained by New England farmers. An important measure of the efficiency of a dairy farm is a figure called the herd average, which represents the average annual production per cow in the dairy herd. In 1980, the national herd average was approximately 11,900 pounds of milk per cow (about 8.2

pounds make a gallon). In the same years, the herd average for New England's largest dairy state, Vermont, was 12,300 pounds.[40] And in Massachusetts, which is a small dairy state by national production standards, the dairy herd average was 4 percent to 7 percent above the national average for each of the six years from 1974 to 1978.[41] This level of production efficiency is a necessity for New England dairy farmers just to stay in business.

Most of these efficiencies have been achieved by increasing herd size and increasing technological innovations, such as larger, more efficient milking parlors and improved breeding techniques. There are limits, however, to the increased efficiencies which technological improvements may bring. As in other industries, many technological improvements in the dairy industry tend to favor agriculture outside of New England. With each advance that allows an increase in herd size, New England farmers become more and more dependent on outside feed supplies. This renders them less able to compete with dairy states like Wisconsin and New York, where larger, unbroken fields of feed grains and corn for silage are grown and harvested right on the farm. This can have a significant financial impact since feed accounts for such a major proportion of dairy production costs. In the state of Vermont, for example, where dairy products accounted for over 80 percent of the market value of agricultural products sold in 1978,[42] fully 54 percent of statewide farm production expenses were devoted to feed for livestock and poultry.[43] By comparison, in Wisconsin, the state with the largest dairy production in the nation, expenses for livestock and poultry feed accounted for only 30 percent of total statewide farm production expenses.[44] Technological improvements which allow dairy farmers to increase on-farm yield of dairy feeds such as hay and corn are an important means of reducing animal feed costs. However, many of these improvements also focus on larger units of production and acreages, further harming the competitive position of New England dairy farmers. Thus there is often a limit to the efficiencies which New England farmers may achieve. Nonetheless, to date they have done a remarkable job of maintaining competitively efficient farms.

There is an important human dimension which contributes to these growing efficiencies. That human dimension is the family farm. Recent studies indicate that farms operated by the families who own them, with the assistance of at most one additional worker, are the most efficient units of agricultural production. Although the acreage of the most efficient farm unit will vary from one region to another depending on the equipment available, most meaningful economies of scale reach a threshold at a level that can be achieved by family farming operations, regardless of size. According to former U.S. Agriculture Secretary Orville L. Freeman, quoted in the *New York Times,* "the incentive that results when the producer benefits directly from his efforts cannot be duplicated by large holdings, whether they are privately held, communal, cooperative, or state-owned."[45]

In New England, despite a national trend to the contrary, most farms remain family farms. The small units of viable farmland have made New England agriculture unattractive for investment by large corporate agribusiness. Most farm experts agree that the tradition and the long hours of family farming in New England, handed down through the generations, is the major reason for the higher-than-average efficiency of of its farms.

Despite such efficiency, the number of family farms in New England continues to decline. This trend parallels both the general decline of agriculture in the region and a national decline in family farms. Programs of the federal government have exacerbated the national trend with economic incentives and export programs which encourage large-scale corporate operations. The progressive tax structure even tends to favor large farms since larger operators who pay higher tax rates can shelter more income with each dollar of deductible costs. The result is that more and more small New England family farms are placed under increased economic pressures; the farm unit, which should be an asset because of its size and efficiency, becomes a liability. Many New England farmers are forced to abandon farming or seek nonfarm employment to augment their income.

Policy Response in New England

In order to reverse the trend of farmland conversion and slow the loss of their native agricultural industries, the New England states have led the nation in public policy efforts to protect agricultural land. These efforts grew out of the feeling of many New Englanders, farmers, and urban dwellers alike that the historic agricultural roots of the region and the unique landscape which farming creates are an important part of the social, physical, and economic fabric of the region. Indeed, the New England Yankees of lore are usually described as farmers, and they, perhaps above all others, represent one of the foundations of New England culture. The family farm and the associated community institutions have deep roots in New England, and a threat to their existence is often seen as a threat to the entire community. These sentiments are felt throughout New England, but particularly in the less urbanized states with the strongest agricultural traditions such as Vermont, parts of Maine, and New Hampshire. However, farmers in the more urbanized states of Connecticut, Massachusetts, and Rhode Island have received a tremendous amount of support from the nonfarm urban dwellers.

In recent years, there has been a growing awareness that agriculture plays an important role in the unique physical environment of New England. The physical character and settlement patterns of the region's small farms create a landscape of diversity and harmony loved by urban and rural dwellers alike. Historic farm architecture and site improvements such as stone walls, fences, ponds, and landscaping are often the most scenic architectural features of rural New England. The landscape, with farms as an integral part, has experienced a dramatic change over the last three decades, and these changes have caused New Englanders to look closely at the importance of farmland as an element of landscape character.

Agriculture is also increasingly seen as an important component of the New England economy. Despite declines, the market value of the agricultural products sold in 1983 was over $1.6 billion.[46] In 1981, the New England Regional Plan specifically identified an expanded agricultural industry as an important component of the region's future development strategy:

> This expansion will provide an opportunity to reduce the Region's trade deficit and increase the amount of import substitution. Expansion of value-added industries such as food processing could serve to foster industrial growth and higher employment.[47]

In response to the growing awareness of the value of agriculture and agricultural land, all of the New England states have initiated programs to protect agricultural land. Many such programs, however, were not initially designed to protect farmland. Rather, they were part of a characteristic general environmental consciousness dating back to the writings and activities of Henry David Thoreau, Frederick Law Olmsted, Charles Eliot, and others. Several general land management and environmental programs, which indirectly have protected thousands of acres of farmland, have grown out of this environmental consciousness through the years. More recently, several specially tailored programs to protect agricultural land also have originated 'from this environmental awareness. Like many other programs discussed in Chapter 1, programs in the New England states have been designed to reduce property taxes on agricultural land or purchase high-quality farmland for protection in perpetuity. Some states, such as Massachusetts and Connecticut, have been leaders in developing innovative initiatives to protect farmland. Other New England states have followed, after judging the success of their neighbors. Each state has responded uniquely, although many underlying similarities do exist.[48]

The following chapters discuss in greater detail the activities of each New England state in protecting agricultural land. The issues confronting the citizens and legislatures vary. The surveys therefore begin with a brief background of public concern to protect farmland, followed by a discussion of the most important programs adopted in the state and their present effectiveness. Not all of the programs are covered in each state. Generally we discuss the major efforts, which may vary from state to state. The intent has been to capture the focus of each state.

Notes

1. "Federal Food, Agriculture, and Nutrition Programs in the New England Region" (Washington, D.C.: General Accounting Office, RCED-83-36, December 2, 1982): 1.
2. Neal Peirce, *The New England States: People, Politics and Power in the Six New England States* (New York: W.W. Norton & Co., 1976): 27.
3. Chapman Paper, 4.
4. Mark B. Lapping, "The Land Base for Agriculture in New England" (Paper prepared for the National Agricultural Lands Study, October 1979): 22.
5. Peirce, *The New England States*, 19.
6. Mark Kramer, *ThreeFarms* (Boston: Little Brown and Co., in association with the Atlantic Monthly Press, 1977): 67.
7. *Ibid.*, 67–68.
8. Includes "Grand Total" of all land in farms including "improved" and "unimproved" farmland. *1880 Census of Agriculture.*
9. Mark B. Lapping, "Forests, Farms and Open Space," (Paper delivered at Lincoln Seminar 1, "Conference on Preferential Tax Treatment of Open Space," Lincoln Institute of Land Policy, Cambridge, Mass., November 23–25, 1980).
10. Lapping, "The Land Base for Agriculture," 25.
11. "Federal Food, Agriculture, and Nutrition Programs in England," 1.
12. For an interesting look at the daily life of a farm family in New England today, see Kramer, *Three Farms*, 3–107.
13. Edward Hyams, *Soil and Civilization* (New York: Harper and Row, 1976): vii, 115.
14. Peirce, *The New England States*, 19.
15. The USDA Soil Conservation Service defines prime farmland as:

    Land best suited for producing food, feed, forage and fiber crops, and is also available for these uses. It may be idle now or used for crops, pasture, hay, or forest. It is not in urban use or under water. Prime farmland has the soil quality, growing season, and moisture supply needed to economically produce sustained high yields of crops when treated and managed, including water management, according to acceptable farming methods.

    Prime farmland soils:

    - have an adequate and dependable water supply from rainfall or irrigation
    - are warm enough and have a long enough growing season for adapted crops
    - are neither too acid nor alkaline for good plant growth
    - have acceptable salt and sodium content
    - are permeable to water and air
    - lack surface stones that interfere with cultivation by machinery
    - are nearly level or gently sloping and not excessively erodible
    - do not flood during the season of use
    - are not saturated with water for long periods of time
16. National Agricultural Lands Study, "Agricultural Land Data Sheet" (NALS, June 1980).
17. Edward Micka and Vance Dearborn, *Where New England Counties Rank Agriculturally among the Top 100 Counties in the United States* (University of Maine Cooperative Extension Service, 1978); *1978 Census of Agriculture* (U.S. Department of Commerce, issued July 1981).

Figures derived from total market value of agricultural products sold for all farms (by division). A summary:

Division	*Market Value per Acre ($)*
New England	239
Middle Atlantic	225
East North Central	190
West North Central	109
South Atlantic	193
East South Central	122
West South Central	73
Mountain	34
Pacific	182

19. Farms with sales of $2,500 or more. *1978 Census of Agriculture.*
20. "Smaller is Better and Private Best for Farming," *New York Times*, March 21, 1982.
21. Commercial farm is defined here as a farm with sales of $1,000 or more.
22. Includes value of land and buildings; farms with sales of $2,500 or more. *1978 Census of Agriculture.*
23. See Chapter 3.
24. "Smaller is Better," *New York Times.*
25. *1978 Census of Agriculture.*
26. Maine Agricultural Statistics 1983–84 (Maine Department of Agriculture, Food and Rural Resources): 8.
27. This section is abstracted from Karl E. Case, "The Economics of Use-Value Taxation of Agricultural Land" (Paper delivered at Lincoln Seminar 1).
28. Interview with Chris Soliwroda, FmHA, Amherst, June 20, 1983.
29. "Seed to Supermarket," *Ag Biz Tiller* (11&12) (November 1978): 1–2.
30. For example, in 1978 it was reported that the meat industry was dominated by five large corporations nationwide: Armour, Swift, Wilson, Iowa Beef Processors, and American Beef Packers. Two of these companies are subsidiaries of large nonfood conglomerates. Ibid.
31. *1978 Census of Agriculture.* "Farms and Land in Farms by Type of Organization"—State Summary Data. Figures derived as follows:

Land in Farms by Type of Organization

	Individual or Family (Acres)	*Family-Held Corp. (Acres)*	*% of All Land in Farms*
New England	4,167,034	331,363	87
United States	686,575,506	104,083,123	77

32. "Federal Food, Agriculture, and Nutrition Programs," 1; see also Lapping, "The Land Base for Agriculture," 20.
33. "Smaller is Better," *New York Times.*
34. Ibid.
35. *Kiplinger Agricultural Letter* (Washington, D.C., October 1, 1982).
36. Includes farms with annual sales of $2,500 or more; "Farms by Market Value of Agricultural Products Sold," *1978 Census of Agriculture.*
37. *1978 Census of Agriculture.*
38. Kramer, *Three Farms*, 20.
39. "New England Regional Plan" (198): 30.
40. "Milk Production" (Washington, D.C.: USDA Statistical Reporting Service, Crop Reporting Board, February 12, 1982): 3, 10–11.

41. Ibid.; "Massaachusetts Agricultural Statistics" (Massachusetts Department of Food and Agriculture, 1979): 20-21.

Milk Production per Cow: Massachusetts and the U.S.

	U.S. Average per Cow	*Massachusetts Average per Cow*	*Massachusetts*
Year	*(lbs.)*	*(lbs.)*	*(%)*
1974	10,293	10,981	+7
1975	10,360	11,130	+7
1976	10,894	11,074	+2
1977	11,206	11,706	+5
1978	11,243	11917	+6

42. Farms with sales of $2,500 or more. *1978 Census of Agriculture.*

43. "Selected Farm Production Expenses for Farms with Sales of $2,500 or More," Vermont, *1978 Census of Agriculture.*

44. Ibid., Wisconsin.

45. "Smaller is Better," *New York Times.*

46. "Cash Receipts from Farm Marketings 1981," New England Crop and Livestock Reporting Service, USDA Statistical Reporting Services.

47. "New England Regional Plan: An Economic Development Strategy," Final Report (The New England Regional Commission, 1981): 18.

48. As we become more sophisticated in our understanding of the interrelationships among agricultural resource issues, a need has developed for a written resource which addresses saving the farmer as well as the land. An excellent publication which fills that need is George Carfagno, ed., *Farmland Preservation Directory—Northeastern United States* (New York: National Resources Defense Council, January 1986).

3

Massachusetts Programs for the Protection of Agricultural Land

Among the New England states, Massachusetts has been the leader in efforts to protect agricultural land. The rising costs of food and fuel and the abandonment or conversion of large acreages of farmland in the last thirty years has generated a widespread concern within Massachusetts about the changing character of the urban-rural landscape and limited food self-reliance. All over the state, cities and towns which once boasted a large and thriving agricultural sector have experienced a near-total decline in their agricultural industry, with a resulting loss of community character, as farmland is either abandoned or urbanized. Citizens of Massachusetts farming areas are not the only ones who have been concerned. Concern at the state level has resulted in major efforts to increase awareness of the value of greater food self-reliance and the protection of the nonrenewable resource base of agricultural land. Many of these efforts are recent, while others are part of the long history of environmental consciousness among Massachusetts citizens.

The earliest land protection programs were largely aimed at coordinating development of certain desired urban land uses while protecting specified quantities of "vacant" or open space land for planning purposes. Agricultural land uses were often classified as vacant on Massachusetts town maps, and seldom were specific state policies prepared to promote and encourage agriculture statewide.

In recent years, however, the perception of the importance of agricultural lands in the state has changed. Over the last three decades, and particularly since 1970, several new policies aimed specifically at the preservation and revitalization of Massachusetts' agriculture have been initiated. Some of these were part of larger, generalized growth policies for the Commonwealth. Others are specific statutory remedies and executive guidelines to promote agriculture. To date, these programs have had mixed results. Generally, it is too early to judge the success or failure of most of the various programs, although limited success seems apparent in several areas.

Public Concern to Protect Agricultural Land

In Massachusetts, public concern to protect agricultural land is the result of two primary considerations among the citizens and legislature of the state. The primary concern was in response to the visible conversion of thousands of acres of farmland to other uses since the 1950s. This "loss" of farmland acreage, particularly to urbanization, in turn generated apprehension among citizens over the loss of open space and changing community character, a concern which may be the major reason why farmland protection programs have received such broad support in the state. The second statewide concern has been more recent and less strongly felt. This is the concern for the direct impact which farmland conversion will have on local food production, farm families, employment, and the general economy.

Decline in Farmland Acreage

Like New England as a whole, Massachusetts appears to have had an alarming decline in the acreage devoted to agriculture in the past fifty years. In 1880, the U.S. Census of Agriculture reported over 3.3 million acres of land in farms in the Commonwealth on 38,406 farms. With a land area of 5,008,640 acres, fully 67 percent of the state's total land area was devoted to farming. By 1940, this figure had declined substantially, to 39 percent of the state's land area, with 1,938,000 acres of farmland on 31,900 farms. This represented the conversion of over 1,400,000 acres in fifty years, or 28,300 acres per year. But the most marked decline had not yet begun. By 1982, the Census of Agriculture reported total land in farms had been reduced to only 613,000 acres on 5,400 farms, a 68 percent reduction in only forty-two years, or an average annual conversion rate of approximately 33,100 acres.

These census figures tell only part of the story. The census figures for total land in farms contain sizeable acreages of land which are included in farmland parcels but are devoted to forest or other nonagricultural uses. To understand actual changes in Massachusetts agricultural land uses in greater detail, William MacConnell and John Foster of the University of Massachusetts at Amherst compiled data on the change of agricultural land over the twenty-year period from 1951 to 1971 in four primary agricultural land uses: tilled land, nursery, orchard, and pasture land. Their findings were astonishing.

In analyzing statewide aerial photographs for 1951, they found only 674,000 acres of land statewide were devoted to the above-mentioned uses. When comparing statewide aerial photographs for 1972, MacConnell and Foster found that total acreage of productive agricultural land had dropped to 432,000 acres, or a net decline of 37 percent in twenty years. The decline in productive farmland from 1951 to 1971 thus averaged about 12,000 acres per year.[1] Table 3.1 summarizes these findings. As the table indicates, more than one-third of Massachusetts' productive farmland in 1951 went out of production in twenty years. If roughly comparable data from the 1982 Census of Agriculture are used, there has been a further decline in active farmland acreage to less than 335,000 acres since 1971.[2] If these rates of decline were to continue, estimates are that land-based agriculture in Massachusetts would disappear within thirty years, except for a few acres used for nursery and cranberry production.

The decline of active agricultural acreage is only part of the picture. To understand recent changes in agricultural land use in Massachusetts, information regarding the converted uses of former agricultural lands is needed. Based on the research of MacConnell and Foster, Table 3.2 summarizes the average annual conversion of agricultural acreage to various uses in the twenty-year period ending in 1971. It shows that an annual average of 5,300 acres of farmland was converted to new urban uses between 1951 and 1971. Urbanization thus consumed 40 percent of all the land removed from agricultural production in the state. The second largest new use of 1951 farmland during the period was "abandoned and unused" land, accounting for an average annual total of 4,200 acres per year. Review of these statistics reveals that urbanization has been the dominant cause of the decline in active farmland acreage in Massachusetts in recent years. Furthermore, prime agricultural land is well-drained, not too steep, and already cleared;[3] therefore, such tracts of land often have been the first to be developed. The conversion of agricultural land to urbanization is more serious than losses to abandonment or reforestation, since the best farmland is lost from agriculture forever. Once stripped of its topsoil and developed, it is nearly impossible to return farmland to its original productive capability. Abandoned farmland, on the other hand, can be brought back into production if it becomes economically feasible to do so, although in New England this is unlikely.

Today, most of the land remaining in agriculture in the Commonwealth has excellent productive capacity. The process of abandoning land with low or moderate productivity is nearly completed, and losses are slowing down.

Table 3.1
Statewide Estimated Net Change in Acreage Used for Each Agricultural Purpose in Massachusetts: 1951–52 to 1971–72[a]

Agricultural Use	*Total Acreage (1971)*	*20 Year Change*		*Annual Change*
		Acres	*%[b]*	
Tilled	245,100	–120,700	-33	–6,100
Nursery	5,100	+3,300	+183	+200
Orchard	13,300	–11,800	-47	–600
Pasture	168,800	–112,800	-40	–5,600
Total	432,300 [c]	–241,700	-36	–12,100 [d]

a. Acreage figures rounded to nearest 100.
b. Percent change from 1951 acreage.
c. Addition of cranberry acreage would increase this total
d. Net annual change. A total of 13,300 acres of active agricultural land was converted to other uses each year between 1951 and 1971 (see Table 3.2). However, 1,200 acres per year of new agricultural land was created from other nonagricultural uses, resulting in a new annual change of –12,000 acres.

Source: John H. Foster and William MacConnell, *Agricultural Land Use Change in Massachusetts* (Amherst: University of Massachusetts, January 1977).

Despite this abating pace, the result is that, compared to 1940, when approximately 40 percent of Massachusetts' total land area was in farmland, by 1970, only 14 percent was actively farmed. By 1980, the figure was estimated at about 12 percent.[4] Although farmland conversion is not continuing at the same rate as previously, farmers on the best remaining lands still must contend with the disadvantages of small, irregular fields and scattered areas of excellent soil. In a period when the increasing efficiency and economies of scale of the large western farms demand farm expansion and mechanization, many Massachusetts citizens fear that poor economic conditions will continue to plague farmers and lead to continued gradual urbanization or abandonment of farmland unless effective steps are taken to improve the competitive position of Massachusetts farmers. State agriculture officials now feel that Massachusetts must retain most of the remaining acreage of agriculturally productive farmland to keep the existing agricultural economy viable. As a result, in the last decade, several new farmland protection programs have been established.

Concern over Loss of Open Space and Community Character

As the conversion of farmland to other uses (particularly urbanization) became apparent in the late 1960s, the citizens and legislature of Massachusetts became increasingly concerned over the resulting loss of open space and dramatic changes in community character. Initially, public reaction to

Table 3.2
Average Annual Conversion of Agricultural Land, 1951–1971 (Acres change per year)

	Land Use in 1971				
Land Use in 1951	*Abandoned*	*Urban*	*Forest*	*Other*	*Total Annual Conversion*
Intensive Agriculture	1,700	3,000	900	800	6,400
Pasture	2,500	2,300	1,800	300	6,900
Total	4,200	5,300	2,700	1,100	13,300

Source: John Foster and William MacConnell, *Agricultural Land Use Change in Massachusetts 1951–1971* (Amherst: University of Massachusetts, January 1977).

the conversion of active agricultural land was focused not so much on the loss of land for food-production purposes. Rather, much of the concern was over the loss of open space and a perceived negative change in rural community character. Particularly in the more populated areas of eastern Massachusetts surrounding Boston, where spreading urbanization was a major reason for the conversion of agricultural land, issues of uncontrolled community growth and loss of rural character were raised in towns which had always been farm communities. Since farmers represent a small minority of the largely urban population of Massachusetts, the average citizen did not feel the conversion of farmland directly; it was perceived as a general deterioriation in the visual quality of the landscape. As Tim Storrow, director of land use for the Massachusetts Department of Food and Agriculture, observed, "I think more people have been alarmed at the tremendous change that has occurred in the general landscape of Massachusetts over the last twenty years."[5] Considering the important role that farming has played in the landscape, it is easy to see how the conversion of agricultural land to urban uses or to abandonment was perceived by urban and rural dwellers alike as a negative change in the character of the landscape.

This widespread concern for the general character of the environment was nothing new. Perhaps more than any other New England state, Massachusetts has had a long history of environmental consciousness and concern for the relationship between humans and nature. Beginning in the nineteenth century, the writings of George Perkins Marsh, Henry David Thoreau, and others were a significant beginning to the conservationist spirit, generated by a growing recognition of the effects of rapid technological change. The strength of this spirit was unique to Massachusetts, already one of the most densely populated states in the country by the mid-nineteenth century. Unlike most areas of the growing United States, the Bay

State's available land supply was fully allocated long ago. Therefore, an attitude of stewardship toward the land became more important at an earlier date than elsewhere in the country.

Later, the work of individuals such as Frederick Law Olmsted and Charles Eliot furthered the tradition. At various times, external forces also gave breathing time from the effects of rapid growth. Periodic postwar slumps of the local economy and the flow of industry and development to the southern states gave all of New England a modest reprieve from the pressures of growth on environmentally sensitive lands. The early conservationist background gave rise to some of the earliest conservation programs in the nation, some of which have become useful vehicles of more recent farmland protection efforts. The spirit of the early conservation tradition partially explains why the initial legislative concern to protect farmland was targeted for open space and environmental protection and not for the more direct purposes of agriculture—food production. Today, Massachusetts has become a leader in farmland protection efforts, primarily because of these early landscape and open space concerns.

Concern over Loss of Food Production Potential

Since the mid-1970s, the declining acreage of farmland and rising food prices have generated new concerns in the legislature over the loss of the local food-producing ability of the state. In 1979, a study undertaken at the University of Massachusetts at Amherst found that only 12 percent of the state's total food consumption is actually produced in Massachusetts. Even this figure does not take into account that 73 percent of the current production of animal products, including dairy and meat, depends on feed grains, forages, and supplements imported from other regions (notably Canada, New York state, and the Midwest). If these imported animal feeds are considered, Massachusetts agriculture supports only 7 percent of the annual statewide consumption of the four major commodity food groups.[6]

It is interesting to note that total cash receipts from farm marketings in Massachusetts in 1983 amounted to nearly $366 million. Figure 3.1 illustrates the relative importance of major commodity groups to total cash receipts in 1983. It shows that crops account for 63 percent of Massachusetts total cash receipts. Of all the agricultural land required to produce these crops, over 85 percent was devoted to the production of hay and corn for silage, used primarily in dairy production. Table 3.3 summarizes Massachusetts' major crops for 1983.

The decline of food production potential has caused increasing concern in recent years as awareness of Massachusetts' dwindling food resources and limited self-sufficiency has grown. Beginning in 1973, in response to

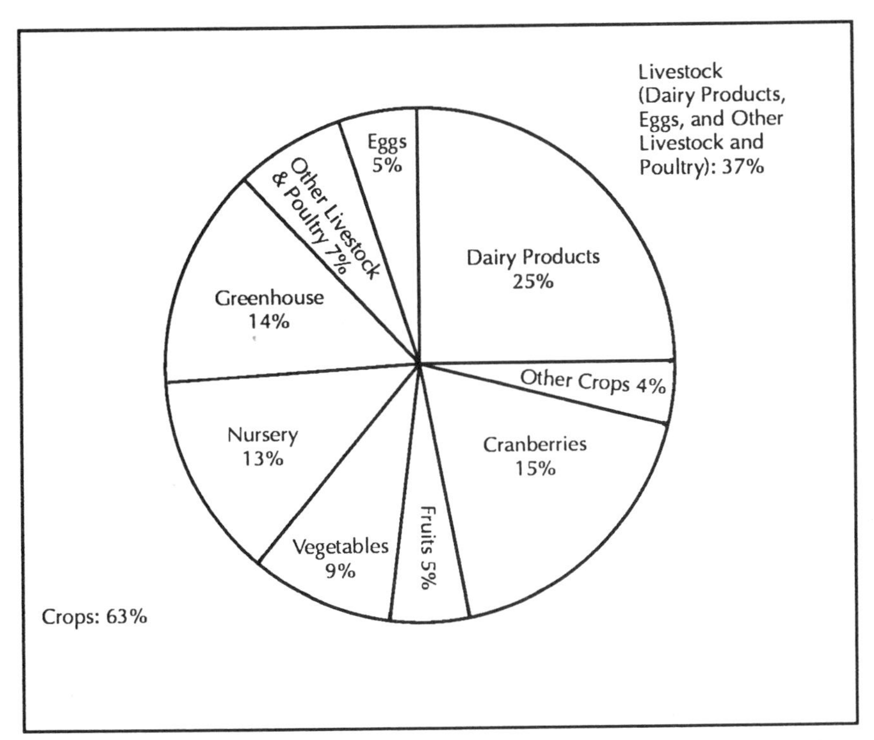

Figure 3.1. Distribution of cash receipts from Massachusetts farm marketings in 1983. *Source:* New England Crop and Livestock Reporting Service.

transportation disruptions caused by a trucking industry strike and Midwest blizzards, then-Governor Francis Sargent appointed an Emergency Commission on Food, a nonpartisan citizens' group representing agriculture, labor, education, government, business, and consumer interests. Chaired by Dr. Ray Goldberg of the Harvard School of Business Administration, the Commission published the "Report of the Governor's Commission on Food" in 1975, following extensive studies. The Report highlighted the Commonwealth's tenuous food supply situation and recommended that the state take action to develop a long-term food policy and related programs. One year later, in 1976, the Massachusetts Department of Food and Agriculture and the Executive Office of Environmental Affairs published "Policy for Food and Agriculture in Massachusetts," which was endorsed by then-Governor Michael Dukakis. The Food Policy identified the fundamental causes of Massachusetts food problems and recommended remedial programs for the state to follow in an effort to address

Table 3.3
Massachusetts Crops: Acreage, Yield, Production, and Value (1983)

Crops	*Acres Harvested*	*Yield per acre*	*Total Production (1,000)*	*Value of Production*** *(1,000 Dollars)*
Corn for Silage	39,000	17.0	663 tons	19,956
Hay, all	123,000	2.54	313 tons	28,483
Potatoes	3,400	190	646 Cwt.	4,005
Tobacco	425	1806	768 pounds	3,475
Maple Syrup	—	—	26 gallons	551
Apples, com'l*	—	—	2,310 42-lb crates	16,403
Peaches	—	—	35 48-lb crates	782
Cranberries	11,200	130.4	1,400 barrels	74,022
Sweet Corn	8,700	92	800 Cwt.	10,640
Tomatoes	570	165	94 Cwt.	2,820
State Total	186,295	—	—	161,137

Notes:
*Production is the quantity sold or utilzed.
**Value relates to marketing season or crop year.
Source: New England Crop and Livestock Reporting Service, USDA, Concord, N.H.

declining food production. The many programs recommended included research and development, marketing and consumer assistance, and agricultural land protection. One of the most important of these recommendations was a proposal to initiate a program for the state purchase of development rights on farmland. This program is discussed fully later in this chapter.

When it was published, the "Policy for Food and Agriculture" was widely acclaimed and became a model for other New England states. It focused, for the first time, on the statewide concern over farmland conversion not only as an issue of declining open space and changing community character, but also as an issue of food production and a balanced economy. Many new programs to protect agricultural land have been adopted in recent years as an outgrowth of this policy and the production capability. However, although declining food production potential has become a concern, most observers in Massachusetts agree that concern over loss of open space and community character are the primary reasons that farmland protection programs have received such broad-based support in the state.

The following sections discuss the various programs to protect agricultural land in Massachusetts and their effectiveness, beginning with the earliest major conservation programs which have sometimes protected farmland indirectly.

Conservation Programs and Early Efforts to Protect Agricultural Land

The earliest conservation programs developed in Massachusetts were not directed specifically at farmland protection, but rather at a wide range of issues concerning the conservation of the Massachusetts environment. Open space, wildlife, rivers, lakes and ponds, forests, town and rural character, recreation, and public health were the major foci. Farmland protection was occasionally an indirect benefit of those efforts. Agriculture was one of the few productive land uses that was encouraged and desired because of its compatibility with early conservation programs. In several cases, these early programs have become important models for more recent farmland protection efforts.

Trustees of Public Reservations Act (1891)

One of the earliest conservation programs was the Trustees of Public Reservations Act of 1891.[7] This special statute allowed the establishment of a private, charitable organization to protect the unique and scenic areas of the Massachusetts landscape. The effort began in 1890, when Charles Eliot, working through a committee of the Appalachian Mountain Club, called a meeting to discuss landscape deterioration in Massachusetts and to propose measures to counter undesirable trends which had become apparent to several concerned citizens. It was decided that a new organization should be formed with several goals, including acquiring sites of outstanding natural beauty or historic interest, stimulating public understanding and education, and developing systematic and formalized procedures for dealing with landscape change in future years. Eliot's committee proved very effective, and in 1891 the Massachusetts legislature adopted an Act establishing the Trustees of Public Reservations.

The Trustees was created "for the purpose of acquiring, holding, arranging, maintaining, and opening to the public, under suitable regulations, beautiful and historic places and tracts of land within the Commonwealth."[8] At its first meeting, the Trustees decided to proceed in four directions:

1. Secure and publish facts regarding open spaces in Massachusetts.
2. Collect and publish laws relating to the acquisition and maintenance of those areas.
3. Work with existing park commissions in towns to encourage cooperative action in meeting regional open space needs.
4. Initiate a legislative inquiry into the question of open space preservation.

Substantial progress was made to meet these objectives. The work of the organization over the next seventy years laid the groundwork for many important land protection programs, including the Conservation Commission legislation of 1957 which, in turn, has become an important vehicle of farmland protection in the state. The Trustees of Public Reservations has been instrumental in assuring permanent preservation of more than 22,000 acres of open space throughout the Commonwealth.[10] Many acres of high-quality farmland have been protected because of their open space or historic value.

In 1925, the Trustees and other groups interested in open space needs sponsored a conference in Boston. The result was the appointment of a new Governor's Committee on the Needs and Uses of Open Space. In 1929, this blue-ribbon committee submitted a report entitled "Open Space Plan for the Commonwealth of Massachusetts." Based on the thinking of Olmsted, Eliot, and MacKaye, the plan proposed statewide patterns of urban expansion in harmony with the topography of Massachusetts, the needs of future metropolitan populations, and the availability of open space lands. In a public statement responding to the final Report of the Committee, Governor Allen concluded:

> We in Massachusetts have become a City and Town people to a degree which perhaps we hardly realize. Ninety-four percent of our population are urban dwellers. And 3,500,000 to 4,150,000 of the people live within 40 miles of the Boston State House. So it is unnecessary to emphasize the importance of providing facilities for our citizens to win health and recreation for soul, mind and body in the beautiful countryside which Nature has given to Massachusetts. The incomparable beauty of Nature for which there is no synthetic substitute is not a fad of the few, but a natural appetite in every normal human being. Beauty is as necessary as bread if man is to develop to his highest possibilities.[11]

The plan designated large areas of the Massachusetts landscape, much of it in agriculture, that should be acquired and held for future open space needs. Although farmland was not specifically identified as a type of open space to be protected (primarily because so much of it existed at that time), the proposed Open Space Plan was one of the first statewide responses to the growing concern for the scenic environment of Massachusetts. Thus, the stage was set for future open space conservation programs which included farmland protection for its open space values even though programs for farmland protection were not established at this time.

In the mid-1970s, the Trustees of Reservations became more directly involved in farmland protection efforts with the establishment of a new subsidiary affiliate, The Massachusetts Farm and Conservation Lands Trust (MFCLT). Among its responsibilities is to protect agriculturally productive land and the rural landscape throughout the Commonwealth.[12] Originally established in 1973 as the Land Conservation Trust and renamed in 1980, the MFCLT is a private charitable trust established to "acquire and permanently protect important tracts of agricultural land within the Commonwealth."[13] Using both a special revolving fund and bank lines of credit, the MFCLT can buy farmland quickly. It then holds the property in its own name, ultimately placing ownership of an agricultural protection restriction or easement with the Commonwealth, a municipality, or a local land trust when the availability of funds makes this possible. The farm itself can then be resold to a qualified farm buyer at a lower price which will help assure its operation as an economically viable farm unit.

In addition to its farmland protection program, the MFCLT maintains a $800,000 Land Conservation Revolving Fund, which is used to help the Trustees of Reservations acquire land for open space protection purposes. Loans from the revolving fund are available to other Massachusetts conservation organizations.[14] By October 1981, the MFCLT had already become an effective instrument of farmland protection efforts in Massachusetts, a direct result of general environmental protection and conservation efforts begun long ago by the Trustees of Reservations and other concerned citizens. The results of trust activities will be discussed more fully later in this chapter.

As the conservation spirit grew and a tradition of concern for the Massachusetts environment emerged, other programs and policies evolved. Like the Trustees of Reservations, these early efforts were not aimed at farmland protection, but rather stemmed from general concern over environmental quality and the deterioration of open space in the Commonwealth. Also, like the Trustees, many of these other general conservation programs were instrumental in later farmland protection efforts.

The Massachusetts Forest and Park Association (Early 1900s)

Established at the beginning of this century, the Massachusetts Forest and Park Association (MFPA) was a quasi-public organization created to promote better management of private forest lands. The organization supported the establishment of a state agency to carry on this work and, in 1913, sponsored legislation to create Town Forests in Massachusetts. The legislation was designed originally to grow timber on a commercial basis as a source of revenue for towns as well as a means to manage and protect valuable open space.[15] Later, the increasing concerns of the Massachusetts citizens to protect forest and open space lands resulted in the passage of the Forest Tax Law in 1971,[16] a differential tax-assessment law designed to encourage management of forest land by taxing such land at its value as a productively managed forest rather than at its full market value for development or other uses. The MFPA was instrumental in the passage of this law. Although it was not developed specifically to protect farmland, it can protect forest land in areas of high-quality agricultural soils or on farm parcels with significant forest acreage. In 1973, a similar differential assessment law was adopted for agricultural land. This program is discussed in a later section. Here again, the general conservation programs of the MFPA helped provide the conservation background for the establishment of important recent farmland protection programs.

Conservation Commission Act (1957)

Perhaps the most significant program to emerge from general public concern to protect the Massachusetts environment was the Conservation Commission Act of 1957.[17] This act was the first of its kind in the United States, and several New England states soon followed the Massachusetts example. The Conservation Commission program has been a significant vehicle at the local level for farmland protection efforts throughout the Commonwealth, and several features of the program distinguish it from previous citizen conservation programs. The most important feature of the program is that the idea and the legislation originated from efforts at the grassroots level.

The legislation evolved from the efforts of a few concerned individuals to save a marsh which was threatened by a "drain and fill" operation in the town of Ipswich. These citizens began to search for legal authority to prevent the project. At that time, a state statute provided for the creation of Industrial Development Commissions which gave communities power to acquire land for purposes other than park, recreation, or watershed protection. Using this precedent, Representative John Dolan, an Ipswich resident and member of two pertinent legislative committees—the Committee on Conservation and the Committee on Towns—set to work with groups from Ipswich to draft comprehensive enabling legislation that would allow any town or municipality in the Commonwealth to take action to prevent resource deterioration. A bill was filed at the beginning of the 1957 legislative session, and in March of that year the Conservation Commission Act was adopted.

By the end of the first year, only two other communities had taken advantage of the law. After three years, only nineteen communities had established commissions. Surveys indicated that most towns were unaware of the new law, and concerted statewide educational campaigns were initiated. By 1969, 275 of the state's 351 towns and municipalities had Conservation Commissions;[18] by July 1981, the total was 345.[19]

Generally, the protection and preservation of natural resources is the specified area of concern for local Conservation Commissions. This is a broad mandate, and several local Conservation Commissions have been actively involved in farmland protection. According to the *Environmental Handbook of Massachusetts Conservation Commissions:*

> The Conservation Commission Act does not require that land be kept entirely in its natural condition. Although the Act makes indirect references to "conservation and passive recreation," the basic standards under which the commission holds the land are summed up in the phrase "maintain, improve, protect, limit the future use of, or otherwise conserve and properly utilize open spaces in land and water areas." Many commissions feel that properly conducted agriculture is a proper and beneficial use of public open space because it preserves the wildlife habitat, the historical New England landscape, and the dwindling stock of local farmlands.[50]

Indeed, Conservation Commissions in some towns have had measurable results in protecting farmland. Statistics on protected farmland are not available, but the Town of Lincoln, just west of Boston, provides a striking example. The Lincoln Conservation Commission, in concern with a local nonprofit land holding and development trust, the Rural Land Foundation (RLF), has been instrumental in guiding growth away from the town's farmland. As Robert Lemire, the commission's chairman for several years, wrote in 1976,

> I knew from my experience as chairman of Lincoln's Conservation Commission over the past 10 years that guiding growth away from a town's farmland made eminent sense as part of a program for preserving physical character while struggling to keep taxes within reach of rural residents, maintaining integrity of natural systems and providing regional open space.[21]

Over the past twenty years, the Lincoln Conservation Commission has placed a high priority on the protection of active cropland, taking a leading role in coordinating the town development process. The Commission has mapped and identified all farmlands, encouraged landowners to sell agricultural preservation restrictions and other easements, and in some cases, actually purchased farmland. On certain occasions, through negotiation, the Commission has assisted private developers to assemble adjacent parcels of land to make cluster-type development possible, placing all development on the portions least suitable for agriculture, thereby saving the active cropland. The Commission takes an active role, working with the developer to obtain the necessary special permits and planning board approvals to achieve these positive results at little or no cost. As a result, sizeable tracts of land have been retained, in perpetuity, for agricultural or open space use, to the benefit of local farmers and town citizens alike.[22]

Although the Conservation Commission Act has provided a means to help protect agricultural land in towns like Lincoln, generally the gains have been modest. Mostly, this is because the primary focus of the Act was to protect other natural resources. Furthermore, until recently, only a few Conservation Commissions in Massachusetts had identified productive cropland as a land use priority on their town plans. Like cities and towns across the country, most Massachusetts towns have tended to view the future of farmland within their borders as housing, industry, recreation, parks, or other active uses. Moreover, as discussed in Chapter 2, until recently land that was not designated an actual "use" often was identified on city and town maps as "vacant," implying that agriculture was not a land use to be considered. As concern over the loss of farmland has grown, many towns in other areas of New England are becoming more actively involved in farmland protection, learning from the examples set by towns such as Lincoln. Experience indicates that such farsighted local planning is probably the most effective, most equitable, and least costly method of protecting viable agricultural land, and the Conservation Commission Act has been instrumental in encouraging towns to undertake such local planning. However, the applicability of this approach is limited since relatively sophisticated local ordinances and planning techniques are required. In many rural communities of New England, as elsewhere, local government is not equipped to use these techniques.

The Conservation Commission Act of 1957 was not in itself adequate to ensure that conservation efforts would be undertaken. Since its adoption, several important amendments and related programs have been established, which have improved the mechanisms of the Conservation Commission Act and set the stage for state programs which specifically target farmland protection and use Conservation Commissions as the local liaison agencies. Some of these acts and programs include:

Self-Help Conservation Program (1960).[23] This program gave state financial assistance through the Commissioner of Natural Resources to assist local communities with Conservation Commissions in acquiring land and planning for outdoor recreation facilities. The funds may be used for the direct purchase of land in fee simple or easements, or to aid planning efforts.

Conservation Easement Act (1961).[24] This Act modified the 1957 Conservation Commission Act, enabling commissions to acquire less than fee simple interests in land and water resources. As a result, Conservation Commissions could accept or purchase development rights, easements, covenants, "or other contractual arrangements limiting the future use and development of particular resources." The legal instruments of this early program have subsequently become an important part of more recent farmland protection programs. The 1977 Agricultural Preservation Restriction Act (APR),[25] which provides for the state purchase of development rights on high-quality farmland, was a direct amendment of this Act. The APR program is today the most direct farmland protection program in the Commonwealth and will be discussed more fully later in this chapter.

Conservation Easement Act (1969).[26] This legislative action amended the previous Conservation Easement Act and protected "conservation and preservation restrictions held or approved by appropriate public authority, providing for public restriction tract indexes at registries of deeds and clarifying certain statutory provisions relating to restrictions."[27]

All of these programs have granted important enabling powers to local conservation commissions and have established legal mechanisms for the conveyance of specific agricultural protection restrictions and other easements which are an important part of recent agricultural protection programs. Without the previous existence of both the Conservation Commission Act and the conservation easement and restriction program, the current form of agricultural preservation restriction programs may not have evolved.

Recent Agricultural Land Preservation Programs

Beginning in the early 1970s, as the substantial decrease of land in agriculture became more apparent to citizens and officials of the Commonwealth, programs were targeted specifically at the conservation and protection of agricultural land. Unlike the earlier programs which were primarily directed at conserving open space and other natural or historic resources, these new programs were designed to curb the conversion or abandonment of active agricultural lands.

As the public began to perceive a problem of farmland conversion, initial reactions were focused not so much on the loss of land for food production as on the loss of open space and community character. William King, Farm Real Estate Consultant to the Massachusetts Department of Food and Agriculture, described the situation:

> We had the pressure of population, of course, and the first Earth Day was held here in 1970. With the environmental consciousness that we all have and the statistical picture that was openly displayed by census figures, it's no wonder that the citizens and legislators got up and said, "Let's save some of this open space before it is all lost."[28]

In 1972 a Becker Opinion Poll, commissioned by Massachusetts Citizens to Save Open Space, indicated that "the general public was well disposed toward farmers and farming, toward preserving agricultural land, and favored the retention of rural life as part of the New England culture."[29] The poll provided a strong indication to the legislature, state agencies, and other concerned groups that specific action to protect agricultural land would be supported. This marked the beginning of recent efforts to protect farmland in Massachusetts. Since that time, there has been much farmland protection activity in Massachusetts, and the state has led the region in the development and continued support of new programs to encourage agriculture and protect its land resource.

Among the first of these actions, in the same year as the Becker Opinion Poll, the Massachusetts legislature amended the state eminent domain law to require a public hearing before agricultural property is condemned.[30] The landowner is entitled to present evidence to the municipal board that there is alternative land, without occupied buildings and not used for agriculture or farming, which is available for the intended public use. This provision is aimed primarily at eminent domain taking proceedings by local and regional agencies, since it does not apply to takings on behalf of the Commonwealth of Massachusetts, by the Department of Public Utilities, or for highway purposes. The Supreme Judicial Court of the Commonwealth ruled, however, that after the hearing the appropriate body may condemn the land even if an alternative site is available when that body acts in its "best judgment in light of all the relevant circumstances."[31] The effect that this law has had in protecting Massachusetts farmland is unknown.

As public concern grew in the early 1970s, several institutional forces, including the USDA Extension Service in conjunction with the University of Massachusetts at Amherst, and the Massachusetts Farm Bureau Federation, representing farmers and the farm industry, also became actively involved in farmland protection efforts. Additionally, the Massachusetts Department of Food and Agriculture became more than a regulatory agency by taking an active role in the promotion of farmland protection and production activities. To this end, the Department of Food and Agriculture provided additional legislative guidance in 1974 with the adoption of the Massachusetts Farming and Gardening Act.[32] This legislation established the Division of Agricultural Land Use within the Department of Food and Agriculture "with the responsibilities of achieving better farming and gardening uses of publicly owned agricultural land."[33] Since that time, the Department of Food and Agriculture has become increasingly involved as a farm advocate and has established a wide range of programs to stimulate agricultural production, marketing, consumer awareness, and, of course, farmland protection.

Among the recent marketing programs, a new "Massachusetts Grown and Fresher" logo was developed for statewide marketing purposes and to aid consumer identification of locally grown products. The Department of Food and Agriculture also has assisted in the establishment of over forty-five farmers' markets throughout the state to increase direct retail sales opportunities for farmers. Brochures are published by the Department identifying farms that offer "pick-your-own" fruit and vegetables. An annual farm tour is held for state legislators and the public to increase awareness about farming activities in the state and what may be done to encourage them. All of these programs have helped to disseminate information to consumers, farmers, and government officials.[34]

In addition to these marketing and promotional efforts, the Division of Agricultural Land Use has become the most active farmland protection agency in the state, coordinating most of the state's recent farmland protection activities and serving as an advocate for the adoption of further legislative programs. Farm industry officials believe this total package of marketing, promotion, and land use programs is needed to maintain a viable agricultural economy in Massachusetts in the years ahead.

The Farmland Assessment Act (1973)

As a result of growing citizens' awareness, increasing institutional efforts, and the results of the Becker Opinion Poll, a "Farmland Assessment Referendum" seeking an amendment to the State Constitution to allow use-value tax assessments of farmlands was placed before the voters on the November 1972 ballot. The measure was ratified by a three-to-one margin and carried every precinct in the Commonwealth, urban as well as rural, thereby establishing Constitutional Amendment 99. The following year, 1973, the legislature enacted the Farmland Assessment Act[35] over minor dissent. This Act was the first specific farmland protection tax incentive legislation to be adopted in Massachusetts.

The Farmland Assessment Act, often referred to as Chapter 61A, sets up a special taxation scheme for land actively devoted to agricultural or horticultural use. Like differential taxation programs in other states, the purpose of this program is to decrease the property tax burden on a farmer's operations which is caused by the increasing market value of the land. Land which meets certain acreage and economic productivity requirements is assessed at its *use* value, rather than its full market value or speculative value for development. Thus, Chapter 61A encourages farming by providing an economic incentive, or more accurately, by removing the negative effect of high property taxes. In 1973, prior to enactment, estimates were that as much as 41 percent of farm income in Massachusetts was spent on property taxes, compared with a national average of 8.1 percent.[36] The law was designed to remove this disincentive to farming in the Commonwealth.

The Act begins by defining agricultural and horticultural lands. These definitions today serve as the legal reference for defining agricultural lands for other agricultural protection programs in the state as well. To meet the eligibility requirements for special tax assessment, the law defines the following conditions which must be met:

1. The parcel must be actively devoted to agricultural or horticultural uses or a combination of both during the two previous tax years.
2. The farm parcel must contain a minimum of five contiguous acres.
3. There must be minimum annual farm sales (gross revenue) of $500 plus $5 for each additional tillable acre above the minimum five acre parcel size.

The local board of tax assessors annually determines whether the land qualifies for special assessment, in accordance with state-mandated guidelines. The landowner must file an application to the board of assessors by October 1 of the year preceding the applicable tax year. Assessments are then made on the basis of value schedules published each January by the State Farmland Valuation Advisory Commission (FVAC), which was created by the Farmland Assessment Act. The FVAC works in conjunction with economists at the University of Massachusetts at Amherst to establish annual farmland use valuation schedules. These value schedules are a unique feature of the Massachusetts law.

Farmland value tax assessments are made according to land use classes, that is, according to the agricultural commodity that is produced on the parcel of land, and the value of the land in producing that commodity (tobacco, sweet corn or other vegetables, fruit, hay, cranberries, etc.). A yearly value schedule assigns values to each one of these land use classes, based on USDA indices of value, rental surveys, and other factors. In contrast, several other states base farm value assessments on a soil productivity rating scheme. If the soil from one area to another is similar and the climate acceptable, regardless of the type of crop produced, assessed value will be similar. In Massachusetts, it is felt that a soil productivity basis would imply that only certain crops should be grown on certain lands and thus would discourage the free choice of a farmer in the commodity use of his land. Although there may be a connection between soil productivity and actual commodity use, Massachusetts officials feel that this connection is not absolute.

Once the property is accepted under farmland assessment and the current-use valuation has been determined by the local board of assessors, a lien is recorded in the county registry of deeds. This constitutes a lien for taxes which is released when the taxes are paid. If the landowner decides to sell or convert the land to an urban use within a specified time period, this document requires that penalty taxes be paid to the municipality.

The penalty provisions of the Farmland Assessment Act are an integral part of the program's impact. They tend to discourage participation, particularly by landowners in marginal farming areas where development pressures offer significant financial rewards, or when older farmers without heirs seek to retire from farming and sell their land for nonfarm uses. In these cases, farmland owners have been reluctant to subject their land to the penalties imposed by participation in the program. However, once participation has begun, the penalties are designed to encourage farmland protection. The penalties include:

Conveyance tax. A landowner is liable for a conveyance tax—up to 10 percent of the selling price—if the use of the land is changed within ten years of acquisition. This tax graduates from 10 percent in the first year to 1 percent in the tenth year. A person who owns the land for ten years or more is not penalized under this provision. The conveyance tax provision was designed to discourage land speculation by essentially nonfarm development interests.

Rollback tax. Although the conveyance tax becomes inoperative in ten years, there are additional tax penalties to discourage conversion of farmland which has received tax benefits. The rollback tax amounts to the tax savings accrued during the preceding five years. If the rollback tax in the case of the sale is greater than the conveyance tax, then the rollback tax applies.

In addition to the tax penalties, as an added farmland protection measure, Chapter 61A provides that a farmland owner must notify the local municipality of his intentions to sell his farmland. The municipality then has sixty days to meet any purchase offers. Because of the slow workings of the state and local bureaucracy, sixty days is considered too short by many officials who would like to see the municipality's right of refusal period extended. The right of first refusal portion of the Massachusetts law is unique among the New England states. In fact, few states in the nation have this provision in their use-value assessment laws.

The Farmland Assessment Act has been used primarily in suburban towns where pressures from spreading urbanization are greatest and the tax impact of high land values on farm real estate are most severe. In 1980, estimates were that it was operative in 30 to 45 percent of the communities in the Commonwealth,[37] and there has been very little experience with the sixty-day option period for municipalities. Since fewer than half of the communities in the state have taken advantage of the Act, it is unclear how much it has helped to save farmland.

Inequities and the unpredictability of the tax-assessment program have had a negative impact on participation in Chapter 61A. For example, in 1979, the Massachusetts Supreme Court mandated that all cities and towns reassess all property within their jurisdictions to 100 percent of market value. This revaluation process was due to be completed initially by mid-

1982 and updated periodically in following years. When towns and cities completed their revaluation process, it was expected that increased activity with Chapter 61A would be seen since statewide values under Chapter 61A are based on 100 percent valuation. Unfortunately, the applicability of these statewide farmland values at any point in time is extremely varied. With over 350 taxing jurisdictions in the Commonwealth, the Department of Revenue has a nearly impossible task to oversee all jurisdictions and assure that they are all assessed at 100 percent of market value at the same time. Thus, in a town which has not yet been revalued at 100 percent of market value, a farmland owner who participates in Chapter 61A will be reassessed to the current 100 percent farmland value, while the remainder of the community may be assessed at something less than that figure. The alternative for the farmland owner, of course, is not to participate in the program. However, even if the assessed value of the town is low, the tax rate in Massachusetts is often very high, and a farmer who does not participate in the program may have a less than 100 percent valuation but a high tax rate. Thus, the farm owner is caught in the middle with no effective method of tax relief, and Chapter 61A has not provided the intended incentive to protect farmland. In a neighboring town which has conducted a revaluation, the same type of land may have significantly lowered taxes because the overall assessed valuation of all properties has gone up, the tax rate consequently has been lowered, and qualifying farmland then benefits from the lower use values permitted by Chapter 61A. Thus, not only is there potential inequity from town to town, but there is also a higher relative tax burden on farmland in Massachusetts than in other agricultural states. These inequities are among the reasons for the lack of participation in the program and its subsequent ineffectiveness in protecting farmland.

To make the Act more applicable to farmland throughout the state, various alternatives have been proposed. One method would be a provision in the Farmland Assessment Act to permit an equalization rate calculation similar to that used in New York State in farm value assessments. The equalization rate, determined each year by the Massachusetts Department of Revenue for each city or town, when multiplied by the farm value assessment in towns of less than 100 percent valuation, would bring farmland assessments in the town in line with other levels of assessment in the town.

An alternative proposal, which would put more of a burden on the state as a whole, would be the use of a "circuit breaker" or threshold on farm property taxes. Any taxes paid above a certain threshold would be applied as a credit on a farmland owner's state income tax. Calculations could be made on a per-acre basis, a high-quality farmland acreage basis, or as a percentage of a farmer's adjusted gross income. These and other proposals are being discussed to improve the applicability and results of the Farmland Assessment Act as it relates to assessed valuations in the Commonwealth. To date, however, no legislative action has been taken.

Despite its limitations in many areas of the state, the Farmland Assessment Act has been used widely in the suburban towns surrounding Boston. As population in these towns has grown and complaints from landowners about inequitable assessments have increased, local assessors have been more active in keeping assessments up to date.

In addition to the problem of applicability in towns at less than 100 percent valuation, there are other problems and misconceptions among farmland owners and the general public, which may be a reason why the Farmland Assessment Act is not broadly used. These problems include a loss of municipal revenue, shift of tax burden to other landowners, and increasing pressure on the land supply available for development.

Loss of municipal revenue is one of the most common complaints by opponents of the Farmland Assessment Act. Although there may be some loss of local revenue, most observers in Massachusetts feel that these losses have been minimal. Supporters of the program point out that farmlands demand fewer public services for police, fire, sewer, water, and the like on a per-acre basis than most residential and commercial land uses, and hence increased public outlays in farm areas are not required. Indeed, some officials believe that if the land is kept open, the burden of public services is lowered for everyone, thus resulting in a savings of precious municipal revenues. In any case, so far the problem of lost municipal revenues has not been so significant as to cause the state to reimburse local municipalities for lost revenues as Vermont has done.

Shift of the tax burden to other landowners is also a commonly heard complaint against Chapter 61A. As farmers are sheltered from the burdens of increased property taxes, critics of the program point out that these burdens must fall inequitably on other property owners. Although the burden of supporting local services may be shifted from farmland owners to others, supporters point out that this shift is not unfair. Indeed, it seems clear that Massachusetts farmland owners who are enrolled in the program are paying their fair share since (a) the value of their property is based on the 100 percent value as determined each January by land use classes and (b) as previously discussed, a farmland owner's "fair share" is much less than landowners of other types of land uses because their need for local services is much lower.

Furthermore, the political climate in Massachusetts before the passage of the Farmland Assessment Act was itself a justification of why it would be fair for Massachusetts farmland owners to accept a lower property tax burden. In the 1972 Constitutional referendum, the citizens of the Commonwealth mandated that farmland should be assessed at a lower rate, primarily based on its value in agricultural use for open space and community character. This referendum, by implication, thus recognized the farmland owner as a sort of unpaid landscape gardener and mandated that the society should repay this public benefit financially. Other land uses must therefore carry any additional property tax burdens which are created as farmers are sheltered from the burdens of increased property taxes.

Finally, pressure on the available land supply is sometimes viewed as a problem created by the Farmland Assessment Act because as farms come under the program, the supply of available land for development is decreased. This results in a value increase of nonparticipating land because of greater development pressure on adjacent land. Nonparticipating landowners, therefore, may reap an unfair economic advantage.

In Massachusetts, this is not widely considered to be a problem. In fact, in 1976 the short-lived Massachusetts Office of State Planning, when conducting local-growth policy surveys, found that most outlying suburban and rural towns with large acreages of farmland "available" for development generally desired to control and restrict growth, while urban centers containing large areas of abandoned and underutilized land sought to increase growth. Thus, if the Farmland Assessment Act does cause a restriction in available land for development in rural areas, this would assist the controlled growth drives of suburban and rural communities and encourage growth on available land in the cities. In any event, considering the limited amount of farmland remaining in the state, the effect of Chapter 61A on the available land supply is very limited. Until significant acreages of land are included, the effect will continue to be minimal. If genuine needs for developable land are found, land always can be removed from farmland assessment and developed.

In theory, the Farmland Assessment Act should be quite effective. In practice, however, critics in Massachusetts say there is little evidence of a reduction in farmland buyouts since the inception of the program. Like use-value assessment programs in most states, the Act works best when the farmer desires to remain in agriculture and speculative pressure on the land is just beginning. If the farmer is not committed to remain in agriculture or if development pressure is too strong, then the Farmland Assessment Act is not strong enough. Given recent economic trends in agriculture, a single tax program cannot be expected to improve the Massachusetts farmer's confidence in the viability of agriculture. The farmers' risks in committing themselves to agriculture, particularly in urbanizing areas, are not compensated by the rewards provided by the Act. In areas with several marginal farms, farmland owners are not willing to relinquish their option to sell if market conditions render farming uneconomical. Thus, because Chapter

61A does not fully address major economic problems facing the agricultural industry in Massachusetts, it can be effective only as part of a larger package of programs.

Even supporters of the program concede that it is not functioning as well as it should and has been abused. Farms are often subdivided into large residential parcels containing five acres or more of cropland. New landowners apply for and are granted tax relief under the Act even though the larger, more economically viable parcel of farmland has been subdivided and destroyed. Thus, the Farmland Assessment Act is often considered more of an "open space" program than a farmland protection program. As one Department of Food and Agriculture official commented,

> [The Farmland Assessment Act] is primarily an open space bill and for anybody who has five acres and thinks it's great to save it, 61A keeps land open. In some cases, though, it's a waste of land and resources. Some five-acre plots shouldn't be kept open whereby others should. If it's five acres surrounded by a 200 acre farm, it should be saved.[38]

Despite the problems and limited use of the Farmland Assessment Act, the program is generally considered to have made a positive contribution to farmland protection in Massachusetts. In recent years, it has become an important part of a larger package of current and pending agricultural land protection programs, which include state and local land use planning programs, innovative zoning techniques, direct protection techniques such as purchase of development rights of agricultural land, and agricultural incentive programs—all of which will be discussed later in this chapter.

Resource Management Policy Council (1972) and Growth Policy Development Act (1975)

During the early 1970s a comprehensive National Land Use Policy was under active consideration by Congress. Although Massachusetts always had been regarded as experienced and generally progressive in the area of land planning, land planning activities had been delegated almost exclusively to the 351 cities and towns and twelve regional planning commissions of the state. Thus, when interest in comprehensive national and statewide land use planning began in the early 1970s, Massachusetts lacked any central planning agency or facility.

In 1972, then-Governor Francis Sargent established the Resource Management Policy Council to begin to address the need for state-level planning. The Council was chaired by the Secretary of Communities and Development and included cabinet officers from Administration and Finance, Educational Affairs, Environmental Affairs, Human Services, Manpower and Economic Affairs, and Transportation and Construction. The

Council reported directly to the Governor. In the first year, several special task forces were appointed, including a Task Force on Rural Development.

The Council had only begun to study the implementation of a coordinated planning process for Massachusetts when, in January 1975, Michael Dukakis succeeded Sargent as governor. Although Dukakis dismantled the Resource Management Policy Council, legislative and administrative concern for coordinated planning had become quite strong. Indeed, Dukakis later wrote that he "entered office with strongly held ideas about planning and economic growth."[39] He immediately established an "informal task force of cabinet secretaries and respected professionals outside state government to explore what the duties and functions of a new Office of State Planning would be."[40] Soon afterward, he set up the new agency. Although this agency was subsequently disbanded by Governor Edward King when he was elected four years later, during its short life, the Office of State Planning was instrumental in preliminary efforts to develop a statewide rural policy and in the passage of the Growth Policy Development Act of 1975.[41] Many of the agricultural land protection programs in existence today, and some still under consideration, came about because of the state planning activities begun in 1972 with the Resource Management Policy Council and stimulated by the Growth Policy Development Act.

Prior to adoption of the Growth Policy Development Act, legislation creating a "Special Commission Relative to the Effects of Present Growth Patterns on the Quality of Life in the Commonwealth" was organized in 1973. The Special Commission, known as the Wetmore-McKinnon Commission after its two cochairmen, submitted its findings in July 1975. Most important was the recommendation for a program of bottom-up planning whereby local Growth Policy Committees would be formed in cities and towns to determine local attitudes towards growth. From these local statements, a statewide growth policy was to be adopted. The Massachusetts Growth Policy Development Act of 1975 implemented the recommendations of the Special Commission by authorizing the establishment of the local Growth Policy Committees.

The response from local communities was impressive. Over 330 of Massachusetts' 351 cities and towns, representing 98 percent of the state's population, responded to the "Local Growth Policy Questionnaire," which had been prepared by the newly created Office of State Planning. By 1976, the state's thirteen regional planning agencies had compiled these questionnaires into a set of coordinated "Growth Policy Statements" for each region. The Office of State Planning summarized these statements in September 1977, in a report entitled "City and Town Centers: A Program for Growth." The report compiled and coordinated the objectives of the local and regional growth policy statements and found broad consensus at the local level for a statewide perspective on growth.

The local Growth Policy Statements identified a broad array of concerns among the local municipalities of the state, ranging from a desire for local home rule to the need for increased economic development and job opportunities. On the subject of community character, however, there was broad consensus. With few exceptions, the statements expressed considerable alarm over the conversion of farmland and the diminishing share of the state's economic activity that farming constituted. As "City and Town Centers: A Program of Growth" reported:

> This alarm is based not on the belief that Massachusetts could be or should be a major food producing state or largely self-sufficient for its food supply. Rather, it is based on the view that limited resources such as prime farmland ought to be cultivated or preserved for a time when they may be needed. Increasing farming would help the Massachusetts economy. Every job saved or created now constitutes economic progress. And finally, Local Growth Policy Statements tied the protection and cultivation of farmland to the preservation of rural character of the state and emphasized the importance of this as a complement and contrast to the urban character of its cities and suburbs.[42]

The statewide Growth Policy Report made two specific "Action Recommendations" to assist in agricultural preservation efforts:

1. Establish an acquisition of development rights program.
2. Authorize a transfer of development rights (TDR) program.[43]

The TDR program already had been authorized by the legislature with the New Zoning Act of 1975. The program to acquire farmland development rights was subsequently adopted by the legislature in 1977 and is today the most direct state intervention in farmland protection.

The Zoning Act (1975)

Although it was not specifically amended to protect agricultural lands, the New Zoning Enabling Act[44] of 1975 (now called The Zoning Act) granted new and innovative powers to towns and cities to control growth. New powers which allow towns to plan for the protection of agricultural lands and other local resources are among the benefits of the Act.

Although the Act was adopted in 1975, before the completion of the statewide Growth Policy Report, the need for new and innovative zoning legislation had become clear earlier in the decade from the debate over controleld growth which the citizens had expressed in the local growth policy statements. The need for new zoning enabling legislation had been expressed as early as the 1973 legislative session, when bills were introduced seeking to change nearly every section of Chapter 40A, the existing Zoning Enabling Act.

Initially, public meetings were held throughout the state to help formulate objectives for a new zoning law. Based on the comments and suggestions received from these meetings, several objectives were defined:

1. Standardize the procedural aspects of zoning.
2. Encourage innovation and flexibility in community zoning.
3. Broaden the purpose of zoning to reflect environmental, aesthetic, and historic conditions.

In accordance with these objectives, new legislation was drafted during the next three legislative sessions.[45] On December 22, 1975, Governor Dukakis signed into law Chapter 808 of the Acts of 1975, which became effective on January 1, 1976.[46]

The New Zoning Act granted new and innovative powers to local municipalities which can aid local efforts to protect agricultural land. Indeed, since 1975 several Massachusetts communities have adopted special zoning measures under the Act to protect farmland and rural character. The most important of these zoning innovations include:

Cluster Zoning. This zoning provision is becoming increasingly popular among towns wishing to protect agricultural land while still allowing new development. Under cluster zoning provisions, many acres of high-quality farmland have been permanently protected in some Massachusetts towns by encouraging higher-density clustered development in areas not suitable for agriculture and placing permanent development restrictions on lands most suitable for agriculture. An advantage of cluster zoning is that it allows farmland owners to realize the development value of their land while still protecting the prime agricultural soils of the farm parcel. The town of Lincoln has protected many acres of farmland primarily by using cluster zoning for new residential developments.

Incentive Zoning. This allows density bonuses or other incentives in exchange for the protection of certain amenities or desired land uses, such as farmland, wetlands, or low-cost housing. Incentive zoning can be used in conjunction with cluster zoning to encourage cluster developments which protect farmland.

Performance Zoning. This category allows land use determinations to be made based on the performance characteristics of the land and performance attributes of the development. Prime agricultural land could have performance characteristics that limit its use to agriculture only. Performance zoning has had limited use in Massachusetts because of the highly complex nature of a typical performance zoning ordinance.

Critical Resource District. A town or city may determine farmland to be a critical resource. This allows a town the right of first refusal if farmland is to be sold.

Transfer of Development Rights (TDR). This zoning provision allows towns and cities to establish a program whereby development rights on farmlands may be sold to development interests for use in other specified areas of the town more suited to development. By 1982 only one Massachusetts town, Sunderland, in the Connecticut River Valley, had established a TDR program. The town adopted it in 1974, but due to difficulties in administration and problems with the marketability of development rights, the program had not yet been exercised by 1985. Also, much of the growth pressure of the early 1970s which caused the town to adopt the ordinance has been reduced in recent years. Local officials are optimistic, however, that the TDR program will be useful in guiding future growth away from farmland if growth pressures increase in coming years.[47]

In addition to these zoning innovations, the Zoning Act also mandated exemptions from restrictive zoning regulations for certain agricultural uses. In general, a Massachusetts municipality no longer can prohibit an agricultural use except in clearly defined circumstances specified in the Act. All local zoning bylaws that prohibit agriculture in a city or town must be changed. In rural districts where agriculture is a specified use, agricultural uses cannot be prohibited or subject to unreasonable regulations. In residential or other districts, an agricultural use of more than five acres cannot be prohibited or subject to unreasonable regulations. Communities may, however, prohibit or regulate agricultural uses on parcels of less than five acres in such areas.

The Zoning Act also made allowance for existing agricultural buildings which are currently zoned as nonconforming uses. Normally, reconstruction or expansion of a nonconforming use is not allowed. The Act exempts structures used for agricultural purposes from this restriction. The reconstruction or expansion of such structures can be required to fit local dimensional regulations, however, as long as the requirements do not restrict the use of the building such that agricultural uses are no longer possible.[48]

It will take many years for the full impact of these zoning innovations to be realized in Massachusetts. Nevertheless, it is clear from experience to date that in communities where citizen concern is translated into action, such local planning and zoning mechanisms have been the best method of protecting agricultural land. Such programs have provided equity for farmland owners while still protecting the best land for agriculture and open space at minimum cost to the public. Unfortunately, most towns in New England and throughout the country are not yet prepared to deal with the sophisticated legal and land use planning functions which such programs often require.

Agricultural Preservation Restriction Act (1977)

The Agricultural Preservation Restriction (APR) Act[49] is the most important and most direct statewide farmland protection program to date. The Act grew out of the recommendations of both the Policy for Food and Agriculture of 1976 and the Massachusetts Growth Policy Report of 1977. The APR program was adopted by the legislature in December 1977 by amending Mass. G.L. Ch. 184, the Conservation Restriction Act, and is similar to programs in other New England states known as purchase of development rights programs (PDR). Under the APR Program, the Commonwealth, local government, or both jointly may purchase the development rights for agricultural land and thereby restrict the future development of the land in perpetuity. The Massachusetts program was modeled after a similar program in Suffolk County, New York, which was funded by a bond issue in the mid-1970s.

When originally adopted, the Agricultural Preservation Restriction Act had overwhelming support in the legislature. The measure passed—unanimously in the Senate, and with only five dissenting votes in the House—another indication of the level of nonfarm support for farm measures in Massachusetts since the early 1970s.

Awareness of the decline in farmland acreage and concern for the changing character of the Massachusetts landscape were clearly stated by Massachusetts citizens in the local Growth Policy statements and the Office of Planning's 1977 summary Growth Policy Report. This mandate for action from local governments and the recommendation that the legislature establish an acquisition of development rights program was a powerful mandate to the legislature to authorize the program.

When adopted in 1977, the Agricultural Preservation Restriction Act was funded initially by a $5 million bond issue to purchase farmland development rights. The owner of good farmland is paid the difference between the "market" value and the "agricultural" value of his land. In return, the landowner places an "Agricultural Preservation Restriction" on the land which requires that no development except for agricultural purposes occur on the property. The owner retains all other rights of ownership, including the right to privacy and the right to sell or lease the property. The restriction applies to all subsequent owners of the property.[50]

The Massachusetts system is totally voluntary and works by a case-by-case review of applications received by the Massachusetts Department of Food and Agriculture. Program staff, municipalities, and the Agricultural Land Preservation Committee (ALPC)[51] all take part in the final decision for public purchase of development rights. Decisions to purchase are made on the basis of five criteria:

1. The suitability of land for agricultural use as determined by soil classification and agricultural viability criteria
2. The contribution which the farmland parcel makes to the environmental quality of the area
3. The cost of the development rights to the public body
4. The degree to which the acquisition would serve to protect the agricultural potential in the Commonwealth
5. The degree of threat of losing the farmland parcel from agriculture

Soil quality is the most important consideration of the suitability of the land (criterion 1). For a farmland parcel to be accepted, soil types must be of high-quality agricultural potential as determined by the USDA Soil Conservation Service. The amount of such good, versatile cropland in a farmland parcel is investigated carefully.

Viability for farming is also an important consideration under criterion 1. The Committee evaluates the continued farm viabilty and likelihood that the land will be used for farming. If it is an operating unit, the farm should be large enough to continue to exist economically as a farm, or there should be other agricultural land in the area which could be rented or eventually purchased to make it viable.[52]

The degree of contribution which the farmland parcels make to a good environment in the area through water retention, wildlife habitat, recreation, and open space amenities also is considered. Nearly all farms contribute to the environment in these ways, however, and experience has indicated that unless a parcel's contribution is unique, these benefits do not assist in differentiating among applications.

Two criteria evaluated by the ALPC relate to the dollar amount that the Commonwealth finally pays for the preservation restrictions. First, the ALPC is required under the provisions of the Act to pay no more for a preservation restriction than the value established by independent appraisal. If a landowner is willing to sell development rights for less than fair-market value, however, the Committee may view the application more favorably. In some cases where land values are extremely high, a bargain sale is necessary for project approval.

Second, in certain instances the Committee is allowed to favor projects where the local community is willing to share in the cost of the acquisition and become a coholder of the restriction with the Commonwealth. The ALPC seeks such contributions wherever possible since it allows state monies to go further and permits the burden of enforcement and surveillance of the restriction to be shared with the municipality. Among the many qualified applications to the program, contributions are most often requested from municipalities in suburban areas where land values are highest. In such cases, where the benefit to the town may be greater than it is to the state, the tax base is usually adequate to permit funding assistnace without serious penalty to the local tax rate.[53] Once a municipality has become a coholder, it also has the opportunity to vote on any petition by an individual for land use changes or to repurchase the development rights. To date, local contributions have been received for nearly 50 percent of the program participants, and this amount has been sufficient to cover more than 90 percent of the administrative costs of the program.

The final criterion for farmland purchase is the degree of jeopardy of losing the farmland from agriculture. Often this is the deciding criterion among farms which have already met the other criteria. The degree of jeopardy from development underlies a primary purpose of the program: to curtail the immediate conversion of farmland in Massachusetts.[54]

The goals of using these criteria may be summarized as follows:

1. Save, as far as funding permits, those top-quality farms which would otherwise be irreversibly lost from farming.
2. Reduce the pressures for conversion and slow down the rate of attrition of farmland by all possible means and programs.
3. Provide continuing annual funding in an amount which will preserve each year those farms which would otherwise be lost.[55]

By close adherence to these principles, the Committee and others concerned with agricultural land protection believe that funding for a successful program can be reasonably accommodated in a tight state budget.

The application and evaluation procedure begins with an application by the farmland owner to the Commissioner of the Massachusetts Department of Food and Agriculture. The landowner must also file a copy of the application with the local Conservation Commission of the town or city where the farm parcel is located. This filing allows the municipality to react to the proposed purchase and assist in evaluating the application. Following the initial application by the landowner to the Commissioner, a field inspection of the property is conducted by the state program staff. The application is analyzed and generally ranked according to the selection criteria. Top-ranked applications are reviewed and nominated for appraisal before the Agricultural Land Preservation Committee (ALPC).[56]

A secondary consideration is the geographic distribution of the farmland parcels in question. Generally, it is felt that a reasonable geographic distribution across the state should be attained to encourage the retention of agriculture wherever it is still feasible and to provide benefits for citizens in all parts of the state. William King, farm real estate consultant to the Massachusetts Department of Food and Agriculture, discussed this objective:

> At first, we tried to pick up one farmland parcel in every county where agriculture was represented so that every county could say that they had at least one purchase. Now we're more conscious of trying to fill in those areas where we had initial parcels and purchase contiguous parcels or those in close proximity. We are continuing to look for new areas where blocks of farmland can be assembled. For example, we have a block of about 800 acres under restriction in the northern part of Worcester County in the Leominster/Lunenberg area. There were four property owners who were all 60 years of age or older and thinking of retirement. It was good farmland, in a viable area, and three of the properties were within the city of Leominster. They all had development pressure and met the other criteria for purchase. All four transactions have been completed and we now have four other applications from the same area. We hope to continue assembling farmland parcels in this way throughout the state.[57]

To date, ninety-eight applications have been received from Hampshire County. Twenty-nine of these are from participants in the program (paid and recorded). Twenty-four are under appraisal and five are awaiting committee action. Moreover, forty-two applications have been received from Hampden County; eight of these have become participants, five are under purchase and sale, and eleven are under appraisal. In all, there have been forty-two applications from Franklin County; ten projects have been closed, eleven are under appraisal, and one is under purchase and sale.

Program interest has increased substantially in the Pioneer Valley, as have development pressure and land values. Concerned town officials in the Connecticut River Valley have been very helpful in increasing the awareness and feasibility of the APR program to prevent the conversion of farmland to development. Of twenty new applications received in January 1986, eight were located in Franklin and Hampshire Counties. Applications from the Berkshire Counties are also increasing because of the interest and support of the Berkshire Natural Resources Council and the increase in development pressure.[58]

The appraisal procedure consists of three appraisals:

1. A fair-market value appraisal based on its highest and best use, usually for residential subdivision.
2. An "after" value appraisal: the value of the restriction once the APR has been placed on it. This appraisal involves the use of economic profiles of the highest and best agricultural use for the land. It also makes use of comparable sales of restricted land.
3. A review appraisal which confirms or changes the findings of the other two appraisals. The value of the APR is the fair-market value minus the after value.

Because of the lack of recent farmer-to-farmer sales and farm appraisal experience in Massachusetts, it was necessary to establish economic profiles by which local appraisers could evaluate the agricultural values of the selected farms fairly and insure uniformity and consistency of appraisal practices throughout the state. To assist in establishing these guidelines, a private appraisal consultant was contracted to prepare a handbook of appraisal procedures.[59]

Following appraisal, the Commonwealth's offer to purchase is discussed with the applicant. If there is agreement between the two parties, the Committee votes on final approval, a purchase-and-sale agreement is signed, and processing for payment begins with other involved agencies. A title search is conducted, and if no defects are found, purchase is made and an Agricultural Preservation Restriction is recorded in the local registry of deeds.[60] In instances where municipalities contribute to the purchase and become coholders of deed restrictions, instruments describing the dollar amount of the contribution and the coholder rights and responsibilities are recorded in addition to the restriction.

Since the initial passage of the Act in 1977, more than 14,390 acres have been protected in Massachusetts. By January 1986, there were nearly 16,000 additional acres under appraisal. Beginning in 1978, the Massachusetts legislature appropriated $5 million a year for each of the first four years, $20 million in 1982, and another $5 million in 1983 for a total of $45 million to fund the program.[61]

When the first $5 million was appropriated, ninety-three applications were received from landowners in twelve counties and eighty municipalities, representing 10,565 acres of farmland, from which nineteen farms (1,695 acres) were selected. The total asking price for this land was $5.4 million, but the development rights were actually purchased for about 75 percent of the asking price.[62]

By January 1986, over $25 million of the $45 million in available funds had been expended. The program had received 560 applications statewide (representing approximately 52,000 acres in 158 cities and towns), from which restrictions on 153 properties had been purchased or were under agreement. Development rights on an additional 195 properties are under appraisal or subject to negotiation which would, when completed, totally deplete the available funding for the program.[63]

Due to pressures on the state budget following the passage of Proposition 2 1/2 in 1980, most informed observers were doubtful that additional funding for the APR program would be forthcoming.[64] Despite the economic difficulties, on July 13, 1982 Governor Edward King signed legislation authorizing the additional $20 million for the program. This appropriation made the Massachusetts APR program the largest and most active PDR program in the country. Supporters of farmland protection point to the large legislative appropriations as evidence of the strong general support for agricultural land protection in Massachusetts. In a state where over 83 percent of the population is urban[65] and only 0.1 percent are farmers, the nonfarm urban population has, in recent years, demonstrated a remarkable willingness to take the often expensive measures necessary to protect what limited farmland remains.

Cities and towns are encouraged to participate in the APR Program, and by January 1986, local contributions amounted to $996,000. Proposition 2 1/2 has had some negative effect on local contributions, but overall local support remains strong, and many towns are annually appropriating modest amounts for the program.[66]

Despite such strong nonfarm support for the program, future funding remains an open question. Because it was established as a trial program, predictable annual funding never has been assured. This has caused some difficulties both in future planning for the program and in current selection of properties for acquisition. Experience in the first round of applications demonstrated the need for funding appropriations well in advance of selection to make the decisions easier. The most difficult decisions were between those under immediate and obvious jeopardy and those somewhat better farms whose jeopardy was not as immediate. Generally, the Committee reported that the best farms in immediate jeopardy were selected first in the hope that additional funding would soon be available for the purchase of the remaining eligible farms.

Equally difficult for the APLC were the many questions from applicants, towns, interested parties, and legislators, who were curious to know when and if certain applications were being accepted. Farmland owner applicants could not be advised how long it would take to get a decision and consequently were unsure how long to delay pending sales. Towns likewise did not know whether or not they should appropriate funds to become coholders of restrictions.[67]

These problems were considerably relieved when the legislature provided the additional appropriations of $5 million in 1979, 1980, 1981, and 1984 and $20 million in 1982. The ALPC was then able to schedule many of the qualified applicants in accordance with their jeopardy. A few applicants were able to wait, allowing a few crisis cases to be moved up for immediate action.

According to the ALPC 1981 "Report of Progress," the assurance of predictable annual funding would permit the program to operate on an annual rather than an ad hoc basis. The report summarized the importance of this annual funding:

> Only so many farmers die or retire in any given year and if these could be accommodated by the program, through annual funding, those applications not in immediate jeopardy could be postponed, farmers could wait until retirement to apply, and the costs could be fitted into a tight state budget.
>
> Best of all would be the psychological effect on the farmers as it would give them assurances of the program's being there when needed and inspire them to have faith in the future of farming in the Commonwealth, thus slowing down the rate of conversion. Farmers are now plagued by what has been termed the "impermanence syndrome" which implies that they recognize that when they retire or if for any reason the farm must be sold, it cannot remain a farm. Consequently, toward the end of their career they let the farm run down, rather than investing money in buildings, equipment, and fertility of the land which they would not get back if the farm were developed. With confidence in the future of farming, they can invest adequately in their operation, farm more profitably, and be assured of a market for the farm as a going concern when they retire. Concomitantly, their children, agricultural school graduates, and other aspiring farmers will find farms to be bought at prices which farming will justify. Our state and county agricultural schools are enjoying record enrollments, and many young people wish to farm but they cannot find land to purchase at a price which farming will support.[68]

Despite the successes of the Massachusetts APR Program, there are problems. The most pressing one is cost. As of December 1982, development rights to Massachusetts farmland were averaging approximately $1,600 per acre and represented between 35 percent and 90 percent of full market value of the land, depending on the farmland's location and other factors.[69] Even in the best of times, the cost of operating the program can be a heavy burden to the fiscal resources of the state. If agricultural protection restrictions are required to protect all of the roughly 500,000 acres of commercially viable farmland remaining in the state, a total of $800 million will be required.[70] If funded at a rate of $20 million per year, forty years will be required, which is not such a long time if predictable annual funding is available. The funding, of course, will save the land, but it does not guarantee that farmers will farm it, particularly if farm economics deteriorate further to drive the remaining farmers out of business.

Obviously, not all of the farmland remaining in Massachusetts will require the purchase of development rights to assure its protection for the future. Much farmland outside the influence of cities may never feel pressures for development. Other farmland has too low a land value to make subdivision economically feasible, and some farmland lies in floodplains. Many of the properties not funded by the APR program fall into these categories.

Furthermore, other programs, such as creative use of local planning and zoning, can be used to protect many additional acres of Massachusetts farmland. Such alternative programs should be encouraged because of the immense costs of a PDR program and the difficulties of relying upon PDR as the primary method of farmland protection. The experience in Massachusetts also illustrates that for a PDR program to be effective, predictable long-term funding, in increments which the state can afford, is useful and may help mitigate the problems of cost which a short-term, one-time program will face.

In an effort to find and assure a source of permanent annual financing in Massachusetts, several alternative methods of funding have been suggested, including a one-quarter of one percent real estate transfer tax and a Graduated Capital Gains Conveyance Tax. The latter measure seeks to reduce speculation in farmland by taxing a percentage of the gross future proceeds from the sale of land owned less than six years. Although it is not intended as a revenue-raising program, it is believed that monies will eventually be raised which could be allocated to the APR program.

Continued funding of the APR program for the near future has been included in a legislative proposal seeking an additional bond issue of approximately $20 million per year for five years. This proposal is unlikely to succeed in the current period of economic austerity. Thus, the high cost of the APR program will remain the primary obstacle to the success and long-term reliability of the program.

Moreover, there are other problems with the APR program. In some cases, because of the lengthy review and approval procedures, the program has been unable to respond as quickly as needed. When a farmland parcel is in immediate jeopardy of development, delay has made retention of the land for agriculture difficult. The Massachusetts Farm and Conservation Lands Trust (MFCLT) has been instrumental in mitigating this problem. One of its primary purposes as been to act as an interim landholding organization for the APR program. As a private, nonprofit organization, the MFCLT can act quickly to purchase important parcels of farmland threatened by development or abandonment. Drawing upon The Trustees of Reservations' Land Conservation Revolving Fund, the Trust can sell the preservation restriction for the agricultural portion of the parcel to the state and the restricted agricultural portion to a farmer. Thus, the MFCLT is able to move much more quickly than the state program and help assure the protection of viable agricultural land that is in immediate jeopardy of development.[71] Over 1,400 acres on eighteen farms throughout the Commonwealth had been purchased in this manner by 1985.

The APR program is, to date, the most active and direct intervention undertaken by the Commonwealth to protect agricultural lands. To a large extent, it is the centerpiece of the array of farmland protection efforts in the state and has received the greatest amount of attention in the state in the last few years. Since its passage, however, several other legislative and executive activities in Massachusetts have helped slow the conversion of valuable agricultural land. It is widely felt that no single program will be adequate to assure farmland protection. Of all the New England states, Massachusetts has gone the farthest in establishing the wide range of programs that are needed, although most of the programs initiated since 1977 do not intervene as directly as the APR program. Several of these are discussed below.

Governor's Executive Order 193 (1981)

An important action affecting substantial farmland acreage throughout the state was Executive Order 193, signed by then-Governor Edward King in March 1981. It directs "all relevant state agencies to seek to mitigate against the conversion of state-owned agricultural land" and align their policies to prevent the loss of state-owned farmland. Public projects should not be constructed on state-owned agricultural land if alternative land areas exist.

Prior to this order, there was no effective mandate requiring state agencies to assure the protection of agricultural land resources under their control. Although programs such as the Farmland Assessment Act and the APR Program sought to stimulate the private sector to protect farmland, the public sector had not taken steps to protect its own lands. This directive requires all state agencies to adopt the following policies:

1. State funds and federal grants administered by the state shall not be used to encourage the conversion of agricultural land to other uses when feasible alternatives are available.
2. State agency actions shall encourage the protection of state-owned agricultural land by mitigating against the conversion of state-owned land to nonagricultural uses and by promoting soil and water conservation practices.
3. The Secretary of Environmental Affairs shall identify state-owned land suitable for agricultural use based on soil types, current and historic use for agriculture, and absence of nonfarm development.
4. Surplus state-owned land, identified as suitable for agriculture, shall remain available for agricultural use when compatible with agency objectives.[72]

The acreage of Massachusetts land subject to Executive Order 193 is significant. In 1981, the Commonwealth owned over 9,000 acres of agricultural land, much of it under the control of the Department of Corrections and the Department of Mental Health. Both agencies once used farming activity as work therapy for inmates and patients. The remainder of the state-owned farmland is part of state parks, fish and wildlife preserves, and similar areas. Although a small portion of this land is still used by state agencies, most is rented to farmers and some lies idle. Taken together, publicly owned lands represent a considerable farmland resource in the state.

King's Executive Order seeks to assure that these large farmland holdings are not unnecessarily converted to nonfarm uses. In implementing the order, the Massachusetts Department of Food and Agriculture has urged that programs be designed either to assure farm use by the controlling state agencies themselves or to lease the land to farmers on a long-term basis. Furthermore, in situations when sales of surplus land to the private sector may be necessary or desirable, the relevant agency has been encouraged to sell the land with an agricultural protection restriction in place.

Executive Order 193 is the most effective interagency coordination program developed so far to protect state-owned farmland in Massachusetts. As the 1981 Agricultural Lands Preservation Committee report of progress stated, "it would seem incongruous for the state on the one hand to invest large sums in the preservation of farmland by restriction and on the other hand to let already owned land be irreversibly lost through development."[73]

The Agricultural Incentive Areas Act

This legislation was considered to be the most important of the recent farmland protection proposals to be introduced into the Massachusetts legislature. As originally introduced in May 1981, the Agricultural Incentive Areas Act would have encouraged local governments to combine and coordinate many existing programs such as use-value assessment, agricultural preservation restrictions, and others, within specially defined agricultural incentive areas or districts.[74] Furthermore, the legislation was designed to remove several government-imposed regulations and economic disincentives on lands within the districts. The legislation was modelled after the program of agricultural districts adopted in New York State in 1971 and had the following goals:

1. To encourage towns to plan for agriculture, to locate and map their farms and productive lands, and to plan and zone accordingly, providing places for public works and necessary development on lands of low productive capability.
2. To raise farmer morale by generating town consciousness and sympathies for farmers and their problems.
3. To establish freedom from nuisance ordinances in designated areas and to give farmers confidence in the future and permanence of farming.

4. To encourage towns to establish priorities for the orderly purchase of agricultural preservation restrictions as needed, with town funding assistance wherever possible.

To accomplish these goals, the Massachusetts law would grant new powers to local communities to plan for agricultural land uses. At the same time, the law would place restrictions on the activities of various government agencies, thereby assuring farm operators a more predictable farming future. The key provisions of this program would include:

a. Limitation of local regulation. No local government shall enact any regulations or ordinances which would restrict the use of normal farming practices within a designated Agricultural Incentive Area.
b. Limitations on the power of communities to impose betterment assessments or special assessments on land within an Incentive Area.
c. Eminent domain proceedings. No land within an agricultural incentive area may be taken by eminent domain without consent of the owner, except after hearing and an EIS proving that there is no feasible alternative. This provision does not apply to federal eminent domain proceedings.
d. City, town or state first refusal option. In cases of intended sale of agricultural or horticultural land within an Incentive Area, the city, town, or state shall have, for a period of 180 days, first refusal option to meet a bona fide offer to purchase said land. The option may be exercised as (i) purchase of all or part of the fee, (ii) purchase by a private land trust or charitable organization authorized by the Commissioner of Food and Agriculture.

Perhaps most important, the Incentive Areas Act would have allowed the board of selectmen or mayor of a municipality to establish an Agricultural Incentive Areas Committee for the purpose of "investigating, delineating, and establishing an agricultural incentive area in the municipality." Accordingly, the committee would locate and map all viable agricultural land in the community, identify potential incentive areas, coordinate affected landowners, hold public informational meetings, and submit recommendations to local conservation commissions, planning boards, selectment or mayors, and regional planning authorities. Thus, much local coordination of farmland protection efforts would be placed in local control.

Farmland protection advocates in Massachusetts believe the agricultural incentive areas concept could offer a flexible tool at the local level for controlling the loss of agricultural land and operations. It would have several advantages. First, it would be a locally determined overlay type of district which operates on the basis of a cooperative, rather than inflexible, government-mandated zone. Second, it would address problems of farm operations, as well as problems of the land. Third, it could adapt to many local problems at the local level. Fourth, and perhaps most importantly, it could serve as a coordinating mechanism for other farmland protection programs. As William H. King commented:

> There is no better umbrella of programs than the Agricultural District (Incentive Areas) concept. I don't know of a more efficient way to place several aspects of the agriculture retention problem under one piece of legislation. It provides enough positive incentive that if agriculture is going to take place, it does so in a general environment where it can survive. But, it doesn't preclude development from occurring either. In many agricultural areas where you don't yet have the tremendous urban pressure to convert farmland, often you find first pressure to speculate on the land and keep it idle. In this situation, the more stringent programs like zoning or PDR or TDR are not required because they aren't relevant. But Agricultural Districts, which restrict the application of nuisance laws, eminent domain laws and betterment taxes and, at the same time, coordinate positive incentive programs such as use-value assessments, development rights purchase, etc., are a useful approach. The Agricultural District idea combines all of these programs in one package and keeps one phrase in peoples' minds.[75]

The Agricultural Incentive Areas Act, in modified form, was passed by the Massachusetts legislature in 1985.[76] This Act is substantially similar to the 1981 Senate Bill 2218. However, the phrase that municipalities should "plan for agriculture" and the limitation to the municipality's power of eminent domain over lands within the agricultural incentive areas has been deleted. Further, the state and municipality now have only sixty days to exercise the right of first refusal. The Act also clearly states that land can only be included in an incentive area with the landowner's approval, and that the landowner may withdraw the land from the program by notifying the local Agricultural Incentive Area Committee and the Agricultural Lands Preservation Commission. Two years after notification, the land is released from the incentive area classification.

The processes of instituting the Agricultural Incentive Areas Act are lengthy. The municipalities must adopt the provisions of the Act by a majority of the city council vote with the approval of the mayor. They then must appoint a Commission who will map all of the viable agricultural lands, hold town meetings, and obtain the agricultural landowners' approval and certification by the Commissioner of the Agricultural Lands Preservation Committee of the Department of Food and Agriculture. Thus, it will be some time before the results of the Agricultural Incentive Areas Act can be assessed. Some feel that the process for adopting the areas and for including specific forms is so cumbersome that the Act will be useful only in the long run, if at all, unless it is considerably strengthened.[77] The Act, however, appears to be a significant step toward increasing local contributions to agricultural land protection, and it may become an important internal part of the "web of programs" approach used in Massachusetts.

Pending Agricultural Protection Programs

Growing awareness in the last few years that the protection of agricultural land is a complex problem has resulted in a surge of new legislative activity designed to deal more comprehensively with the problem. Indeed, from 1973, when the Farmland Assessment Act was passed, to the present, new proposals to protect farmland have emerged almost every year. The result of this intensive activity is the evolution of a web of programs which individually do little to protect farmland but collectively offer great hope by approaching this multifaceted problem with a multifaceted solution.

These proposals range from amendments to Chapter 61A seeking to reduce the five-acre minimum requirement to three acres,[78] to proposals requiring state institutions to purchase Massachusetts-grown foods.[79] Many of these proposals stand little chance of approval. The major thrust of future legislative efforts to protect farmland probably will be in only two areas.

One important direction will be targeted at the large supply of state-owned agricultural land. Several proposals are currently under consideration. The most important of these will require that in the event important state-owned farmland is declared surplus and disposed of, an APR will be placed on the land prior to disposal.[80] Although Executive Order 193 recommended it, specific legislation is needed to require it. In addition, several other legislative proposals would authorize specific sales and leases of surplus state-owned farmland to towns, private institutions and farmers, with APRs in place. With Executive Order 193 and other proposals, the Commonwealth will be placing a high priority on the protection of farmland directly under its authority.

The second major thrust of future farmland protection efforts in Massachusetts probably will be to create a large role for private interests in the APR program. Legislative proposals are under consideration which seek to allow private trusts to be coholders of APRs. While the Department of Food and Agriculture is allowed to accept APRs as gifts, agriculture officials believe there are more trusts who would seek to raise funds to become coholders of APRs if allowed by law. This, of course, could reduce the public burden of the program and make the PDR programs a much more effective centerpiece of farmland protection efforts.

By June 1985, none of these proposals had been adopted; however, it is clear that a major effort in the late 1980s will be made to strengthen the role of the private sector in farmland protection in the state and provide incentive for participation in the APR program by many diversified groups, thereby increasing the "web" of programs in Massachusetts.[81]

Conclusion

Massachusetts' long history of environmental consciousness has made the state a leader in efforts to protect the limited active agricultural lands remaining in the Commonwealth. Prior to the 1970s, the protection of farmland was not an important priority, and most agricultural land protection was an indirect result of other conservation and environmental protection programs. Recent farmland protection programs in the state undoubtedly would not have been possible without the long history of environmental concern among the citizens of Massachusetts, nurtured by organizations such as the Trustees of Public Reservations, the Massachusetts Forest and Park Association, and town Conservation Commissions.

In the early 1970s, growing citizen awareness of the tremendous decline of farm acreage in the state resulted in the rapid adoption of a range of new legislative programs and executive actions aimed specifically at the protection of Massachusetts' dwindling farmland resources. Most of these recent farmland protection activities have taken place at the state level, resulting, since 1982, in the widest range of programs of all the New England states. Many of these were undertaken in a piecemeal approach as understanding of the problem which confronts Massachusetts farmers grew throughout the decade. Indeed, only by the late 1970s had a truly comprehensive approach to farmland protection been sought.

Since the late 1970s, the centerpiece of statewide farmland protection efforts has been the Agricultural Preservation Restriction Act, a program of state purchase of farmland development rights. The Massachusetts program is the largest and most successful program of its type in the nation, and its experience indicates that a PDR program can work effectively in protecting large areas of farmland within the constraints of a tight state budget. However, the costs are immense. It is noteworthy that at the state level, the legislature chose to purchase the farmland and open space to be protected, even at high cost, rather than resort to restrictive regulatory mechanisms such as statewide zoning, which Vermont and Oregon have done. These programs often place the burden of farmland protection inequitably on a few farmland owners.

Much of the recent farmland protection activity in Massachusetts has taken place at the state level. The Massachusetts experience indicates that great opportunities for success in farmland protection exist at the local level, however. Although their numbers are still small and their successes still modest when viewed on a statewide basis, a few cities and towns in the Commonwealth have taken strong initiatives to protect their farmland resources. The most successful of these local programs have combined various state conservation and farmland protection programs with sound local land use planning and zoning. These cases illustrate that large acreages of farmland can be retained without placing unfair economic burdens on farmland owners and at little or no cost to the public.

Careful local planning by Conservation Commissions, town officials, and citizens—combined with innovative leadership at the state level—can produce an effective farmland protection program. The Massachusetts experience also indicates that such efforts at the state level should include a wide range or web of programs to assure success when confronted with the many and varied problems of farmland protection.

Notes

1. John H. Foster and William MacConnell, *Agricultural Land Use Change in Massachusetts 1951–1971* (Amherst: University of Massachusetts, September 1976): 9.
2. Includes "Total Cropland" and "Woodland Pastured" and "Pastureland and Rangeland other than Cropland and Woodland Pastured." *1982 Census of Agriculture*, vol. 1, part 21 (Department of Commerce, issued September 1984): 1.
3. Of course, not all prime farmland is cleared. The USDA Soil Conservation Service estimates that approximately 150,000 acres of prime farmland in Massachusetts are covered with forest. William H. King, Mass. Dept. of Food and Agriculture, interview, February 14, 1983.
4. William H. King, comment, Lincoln Seminar 1, "Conference on Taxation and Preservation of Farms, Forests and Open Space," Lincoln Institute of Land Policy, Cambridge, Mass., November 23–25, 1980.
5. Tim Storrow, comment, October 5, 1981.
6. Anne Marie Chickering, *Toward Greater Self Reliance: An Assessment of Massachusetts' Food Production Potential* (Amherst: University of Massachusetts, Department of Plant and Soil Science, August 1979): 18.
7. Andrew J.W. Scheffey, *Conservation Commissions in Massachusetts* (Washington, D.C.: The Conservation Foundation, 1969): 21.
8. Ibid., 22.
9. Ibid.
10. Ibid.
11. "Report of the Governor's Committee on the Needs and Uses of Open Space" (Boston: 1929): 1.
12. *New Trust Seeks to Protect our Farmlands* 1(2) (Milton: The Trustees of Reservations, n.d.): 1.
13. Massachusetts Farm and Conservation Land Trust, *1980 Annual Report* (Beverly: MFCLT, 1980): 1.
14. *New Trust*, 11.
15. Scheffey, *Conservation Commissions in Massachusetts*, 26.
16. M.G.L. c. 61.
17. M.G.L. c. 40, §8c.
18. Scheffey, *Conservation Commissions in Massachusetts*, 11.
19. Christi Murphy, Massachusetts Association of Conservation Commissions, comment, July 17, 1981.

20. *Environmental Handbook for Massachusetts Conservation Commissioners* (Medford, Mass.: Tufts University, Mass. Association of Conservation Commissions, 1978): 24.
21. Robert A. Lemire, *The Economics of Saving Massachusetts Farmland* (Medford, Mass.: Tufts University, Massachusetts Association of Conservation Commissions, February 4, 1976): 1.
22. Russell Barnes, Town of Lincoln, comment, October 7, 1981.
23. M.G.L. c. 132A §11.
24. M.G.L. c. 40 §8c.
25. M.G.L. c. 132A §11A.
26. M.G.L. c. 184 §§31–33.
27. Preamble to Chapter 666 of the Acts of 1969, reprinted in Robert J. Ellis and Alexandra D. Dawson, *Massachusetts Conservation Commission Handbook* (Boston: Mass. Association of Conservation Commissions, 1973): 55.
28. William H. King, comment, Lincoln Seminar 1.
29. *Farmland Preservation in Massachusetts: Report of Progress* (Boston: Dept. of Food and Agriculture, March 1981): 2.
30. M.G.L. c. 79 §5B.
31. *Del Prete* v. *Board of Selectmen of Rockland*, 351 Mass. 344, 346 (1966).
32. M.G.L. c. 20 §13 et seq.
33. *Farmland Preservation in Massachusetts*, 2.
34. William H. King, "Massachusetts Encourages Agriculture and Preserves Farmland," *Aglands Exchange* 2(3) (Washington, D.C.: Agricultural Lands Project, January–February 1981): 3.
35. M.G.L. c. 61A.
36. Gregory C. Gustafson and L.T. Wallace, "Differential Assessments as Land Use Policy: The California Case," *Journal of the American Institute of Planners* 41 (November 1975): 386.
37. William H. King, comment, Lincoln Seminar 1.
38. William H. King, comment, Lincoln Seminar 2, "Conference on Agricultural Preservation in New England: A Review of the Experience to Date," Lincoln Institute of Land Policy, Cambridge, Mass., October 28–29, 1981.
39. Michael S. Dukakis, "State and Cities: The Massachusetts Experience" (Cambridge, Mass.: Lincoln Institute of Land Policy, 1980): 2.
40. Ibid.
41. Chapter 807 of the Acts of 1975. The Growth Policy Development Act was the first of a series of legislative actions which ultimately led to the preparation of a statewide growth policy by the Office of State Planning. The program grew out of the legislative efforts of Sen. William Saltonstall of Manchester and Rep. John Ames of Easton. Its roots extend to the environmental movement of the 1960s and the activities of the Sargent and Dukakis administrations.
42. "City and Town Centers: A Program for Growth" (Boston: Massachusetts Office of State Planning, September 1977): 74.
43. Ibid.
44. M.G.L. c. 4BA, §1 et seq.
45. With involvement from such groups as the Joint Committee on Urban Affairs, the Massachusetts League of Cities and Towns, the Massachusetts Association of Home Builders, and the Massachusetts Farm Bureau Federation. The final bill was introduced early in the 1975 legislative session, adopted by the House in ninety days, and by the Senate near the end of the 1975 legislative session. The Zoning Act passed by an overwhelming majority in both chambers.
46. See Barbara E. Gray, "Chapter 808, Massachusetts' New Zoning Act: How It Came to Be," in *A Guide to Massachusetts' New Zoning Act—Chapter 808 of the Acts of 1975* (Amherst: University of Massachusetts Cooperative Extension Service, March 1978): 3–5.
47. Max Gowen and Robert MacKenzie, "Preserving Prime Agricultural Land," in Lawrence Susskind, ed., *The Land Use Controversy in Massachusetts: Case Studies and Policy Options* (1975): 81–92.
48. Christopher Lane, "The New Zoning Act's Mandatory Requirements," in *A Guide to Massachusetts' New Zoning Act—Chapter 808 of the Acts of 1975*, 11.

49. M.G.L. c. 132A §11A.

50. An owner who sells development rights to the state and/or municipality and sells the residual interests to another farmer should be receiving the same remuneration as he or she would have received from a developer. The land thus remains in farming.

51. The Committee is the body established by the APR Act to oversee administration of the program. It is composed of four ex officio menbers: the Commissioner of Food and Agriculture, who chairs the Committee; the Secretary of Environmental Affairs; the Secretary of Communities and Development; and the Chairman of the Board of Food and Agriculture. The Committee also includes four public members and two nonvoting members—the dean of the College of Food and Natural Resources at the University of Massachusetts and the state conservationist of the USDA Soil Conservation Service.

52. Most farms in Massachusetts are small by national standards, averaging only 115 acres. *1978 Census of Agriculture*, Preliminary Report, Massachusetts (U.S. Department of Commerce, issued May 1980): 1. Many farms are able to exist only by renting nearby land to give them economies of scale. They are thus in a very tenuous position if the lessor decides to sell the rented land, requiring the lessee to find other land to rent, to purchase land at development prices, or to give up farming altogether. In evaluating the viability of a farmland parcel for continued farming, the Agricultural Land Preservation Committee investigates opportunities to purchase restrictions on both the farm unit and the rented land necessary for the economic viability of the farming operation. The farmland operator could use the proceeds to purchase the farmland he rented.

53. William H. King, comment, Lincoln Seminar 2. In the first year of operation, the program was very successful in eliciting local funds to assist in purchasing preservation restrictions. However, Proposition 2 1/2, the property-tax-cutting measure approved by the voters in November 1980, greatly impaired the ability of towns to provide funding assistance. Prior to Proposition 2 1/2, the APR program gained town and city contributions in 50 to 60 percent of the cases in the first year of operation (1978). Surprisingly, many local conservation funds were built up before Proposition 2 1/2, and some towns were still able to provide contributions in 1981 and 1982.

54. *Farmland Preservation in Massachusetts*, Appendix B, 2.

55. Ibid., 5–6.

56. With more qualified applications than available funding, the Committee has been faced with difficult decisions in the selection and approval process since each farm landowner's situation is unique. Direct comparisons between, for example, a dairy farm in Berkshire County and a vegetable farm in Essex County are impossible, and the ALPC must exercise a great deal of discretion in the selection process. In an effort to achieve the most effective and equitable use of available funds and assist in the selection of farms to be purchased, the ALPC has established additional considerations to supplement those specified in the legislation.

57. William H. King, comment, Lincoln Seminar 2.

58. Christine Sullivan, Senior Land Use Planner, Massachusetts Department of Food and Agriculture, to Lincoln Institute of Land Policy (LILP), February 12, 1986.

59. Ibid.; *Farmland Preservation in Massachusetts*, 4. The "cost of development" approach looks at the property as a developer would look at it. If the land under appraisal cannot be feasibly developed, there are no development rights. If the land can be feasibly developed, the value of the development rights is the difference between the value of the land if fully developed (based on market demand) and the value of the parcel as farmland. Farm value is based on the farm income potential of the parcel as indicated by the previously established budget values of typical crops and farms in Massachusetts. The private appraisal firm has been involved in farmland appraisals in New Jersey and Suffolk County, New York.

60. The legal mechanisms of the APR conform to those previously established by M.G.L. c. 132A §§11A–11D and M.G.L. c. 184 §§31–33, which defined the nature and mechanism of a preservation restriction and allowed Conservation Commissions and other public bodies to accept or purchase development rights, easements, etc. Under these provisions, which were amended in 1977 to add agricultural preservation restrictions, the "Commonwealth, cities and towns, and charitable corporations and trusts, may purchase the right to restrict land and

water areas for use as conservation, agricultural, or preservation sites." Conservation Law Foundation of New England, Inc. Agricultural Land Preservation Project (November 1980): 11.

61. Christine Sullivan to LILP, February 12, 1986.

62. A few applications were rejected as ineligible under the law and one was already restricted under a deed of gift. Several more were deemed to have little likelihood of approval as they lacked sufficient statewide agricultural significance when compared with the other applications. Of the nineteen selected, four were not included due to an inability to reach a satisfactory agreement or other reasons; three others which required immediate action were substituted. These farms have had APRs placed upon them.

63. Christine Sullivan to LILP, February 12, 1986.

64. William H. King, comment, Lincoln Seminar 2. Proposition 2 1/2 not only affected town budgets, it also affected the state budget. Several thousand state employees were laid off as a result of the measure, and most major programs were cut back. One of the arguments against additional funding for the APR Program was that the money was not needed and that it was not being spent fast enough. The ALPC is very conscious of how it spends the available funds, taking time to work out all of the details of selection and approval. Thus, it takes time actually to spend the money. It is not, however, for lack of worthwhile projects.

65. U.S. Bureau of the Census, press release CB81–138 (1981).

66. Christine Sullivan to LILP, February 12, 1986.

67. *Farmland Preservation in Massachusetts*, 7.

68. Ibid., 7–8.

69. William H. King, "Massachusetts Encourages Agriculture," 3.

70. The *1978 Census of Agriculture* found 535, 675 acres of land in farms were productive commercial operations (with sales of $2,500 or more). This figure includes that land used as a buffer to nonfarm uses, for protection from vandalism, for farm residence, for water recharge in cranberry bogs, and the like. *1978 Census of Agriculture*, 361.

71. When an important tract of agricultural land comes on the market and is threatened with nonagricultural development, MFCLT can quickly buy the property using both the revolving fund and bank lines of credit. The Trust then holds the property in its own name, ultimately selling the APR to the Commonwealth, the municipality, a local land trust, or a combination of these. The farm itself can then be resold to a qualified buyer at a price which will permit operation as an economically viable farm. Occasionally, the Trust will purchase a farm which includes substantial acreage not essential to its operation. In these cases, the Trust may sell this portion of the property to a nonagricultural buyer for a limited, controlled use. The Trust uses the proceeds of this sale to reduce the sale price of the APR to the Commonwealth or reduce the sale price of the farm to a farmer. The MFCLT draws upon two financial sources for farmland purchases. The Land Conservation Revolving Fund was initially established by the Trustees of Reservations and has been supplemented by contributions from several foundations and other sources. This fund allows the MFCLT to borrow funds for option payments, purchase-and-sale agreement deposits, fees for appraisals and land capability studies, and the purchase price itself. When a project is completed, MFCLT repays these funds to the revolving fund. Interest is charged at 1 percent above the current money market rate. The Trust also maintains commercial bank lines of credit for actual purchases. Money is loaned by a Massachusetts commercial bank for periods ranging from thirty days to one year. Aside from assisting with the APR program, the Trust also assists local Conservation Commissions and trusts with land conservation projects. The MFCLT can provide real estate negotiations expertise and loans from the revolving fund. "Background Information and Operating Procedures" (Beverly: The Massachusetts Farm and Conservation Land Trust, February 1981): 1–3; Wesley Ward, associate director, MFCLT, personal communication, March 21, 1986.

Since 1980, the Trust's ability to function quickly as an intermediary in the sale of preservation restrictions to the Commonwealth has made the difference between saving or losing several prime farmland parcels. By December 1985, the Trust had undertaken seventeen farmland protection projects involving a total of 1,568 acres, 676 of these acres in active cropland (typical of New England farmland, the remaining acreage of the properties was mostly wooded). The total purchase price of the seventeen properties was $3,925,500 or an average of $2,504 per acre. Sale of APRs on the farmland to the Commonwealth returned $2,572,925 to the Trust. In addition, sales of the actual restricted farmland brought $944,905, and sales for limited development brought $634,800, resulting in a total return to the MFCLT of $4,152,630, which was adequate to meet overhead, legal, and interest expenses and return the original $3,925,500 in purchase costs. Wesley Ward, personal communication, March 21, 1986.

To a large extent, the activities of the MFCLT depend on continued funding and public support of the APR program. With the future funding of the APR program in question, the Trust is seeking more private support in the form of financial contributions and land donations.

Periods of relatively high interest rates also increase the dependence of the MFCLT on the APR program. If the Trust purchases a piece of farmland property when the interest rates are even modest, at high interest rates, the carrying costs arguably become much too high unless the state APR program purchases the restriction immediately. The Trust usually prefers to have the APR program identify a project the MFCLT can close on immediately to be assured of a quick sale of the preservation restriction. If there is also a farmer who will purchase the restricted land and farm building, the Trust is able to recoup its funds in a manageable period of time while streamlining the often slow state administrative process to achieve the goal of farmland protection.

72. Executive Order 193, Preservation of State-Owned Agricultural Land by his Excellency Governor Edward J. King (March 1981).

73. *Farmland Preservation in Massachusetts*, 10.

74. Senate Bill 2218 (1981 session).

75. William H. King, comment, Lincoln Seminar 2.

76. Agricultural Incentive Areas Acts of 1985, Chapter 613.

77. Wesley T. Ward, Associate Director for Land Conservation, The Trustees of Reservations, Massachusetts, to LILP, March 21, 1986.

78. Senate Bill 1904 (1982 session).

79. Senate Bill 55 (1983 session).

80. Senate Bill 1225 (1983 session).

81. Tim Storrow, comment, March 2, 1983.

4

Vermont Programs for the Protection of Agricultural Land

Vermont's landscape is admired by all who visit the state. The rolling Green Mountains, covered by forests and farms, run through the center of the state, giving a dramatic verdant background to the valleys on the eastern and western slopes of the state. Lake Champlain, with hundreds of farm- and forest-studded islands, sits in the northwest corner of the state, buttressing New York and Canada. On Vermont's eastern border flows the Connecticut River, which separates the state from New Hampshire. Throughout the state, small villages with their white-spired churches are surrounded by a landscape of historic farms with elegant farmhouses and rustic barns. Indeed, much of the idyllic quality of Vermont derives from its predominantly rural agricultural character.

Vermont differs from the other New England states because agriculture always has been and remains the major industry. Vermont remained a rustic state even during the nation's heyday of industrialization at the end of the nineteenth century. Although Vermont's pastoral landscape has been dramatically affected by the development of some forty ski areas and numerous second-home subdivisions in recent decades, the state has preserved much of its charm. Particularly in the north, Vermont remains largely unblemished by modern development.

Like Massachusetts, in recent years Vermont has been faced with a substantial decline of farmland acreage. Also like Massachusetts, public concern to protect the remaining farmland in Vermont has grown significantly. Unlike Massachusetts, however, Vermont has not developed the broad range of programs to grapple with the problem. In fact, until the early

1980s, Vermont had relied almost solely on a single comprehensive land use statute known as Act 250 to address the problem of farmland protection.[1] Because Act 250 was not specifically targeted at farmland protection, the results in farmland protection have been poor. Whereas by 1982 there was reason for cautious optimism in Massachusetts, in Vermont the issue remained unsolved and a source of great debate among the concerned citizens and the legislature of the state. As a result, in the early 1980s, a range of programs, some similar to those of Massachusetts, were beginning to emerge. Few, however, have had time to show major results.

The Vermont experience in farmland protection and Act 250 is unique among the New England states. Indeed, few states in the entire country have enacted comprehensive statewide land use laws. The Vermont experience illustrates the potential problems and opportunities of relying largely on a single, comprehensive, statewide land use program as a vehicle of farmland protection.

Largely, reliance on Act 250 was not effective because it was politically difficult to make statewide regulations work. An important shift of focus and source of optimism, however, were created by the Agricultural Land Task Force (ALTF) report in 1983.[2] The ALTF recommended local planning and protection initiatives coupled with state guidelines and supportive incentives to encourage farmland protection by individual landowners. This chapter will discuss the Vermont experience in farmland protection and some of the recent programs which have evolved in response to the continuing decline of farmland acreage in the state. Not all the topics of the previous case study will be covered. The most significant programs and issues, and those that have made Vermont unique, will be addressed.

Public Concern to Protect Agricultural Land

Despite the generally well-preserved rural atmosphere of Vermont, recent decades have witnessed dramatic pressures for new urban development throughout the state. These development pressures, combined with declining economic conditions for Vermont farmers, have caused a dramatic and ongoing visual change in the character of the Vermont landscape. As in Massachusetts, much of the concern in Vermont began in response to this visible change in the character and quality of the landscape. One aspect of this change was the decline of active agricultural acreage. The desire of Vermonters to adopt comprehensive land use planning legislation and many other programs was not specifically based on farmland protection. The conversion and abandonment of agricultural land, however, was one important consideration in many of the early efforts to protect Vermont's unique rural character. Also prompting public and private efforts to protect farmlands was the realization that the transformation in the Green Mountain State's landscape posed environmental problems, adversely affected the native lifestyle, and brought new economic problems to landowners.

Decline in Farmland Acreage

Vermont currently has more land in farms than any other New England state, but the decline in farmland acreage has been substantial in recent years. The state has approximately 5,930,000 acres of total land area.[3] In 1880, some 82 percent of this land, or 4.8 million acres, was in farm use.[4] By 1940, this figure had dropped to 62 percent, or approximately 3.7 million acres, an average decline of approximately 20,000 acres per year. Since 1940, however, the annual rate of decline has nearly tripled. In 1978, the Census of Agriculture reported that only 30 percent of the state's land area, or 1,633,000 acres, remained in farms, an average annual decline of over 54,000 acres per year since 1940. The number of farms also dropped over the same period, from over 35,000 in 1880 to 24,000 in 1940 to less than 6,000 in 1978. Today, as in most other New England states, the rate of decline in farms and farmland acreage appears to be slowing down. Table 4.1 summarizes the declining acreages of land in farms in Vermont from 1880 to 1982.

Although Vermont today still has a larger percentage of its total land area devoted to agriculture than any other New England state, in 1977 the USDA Soil Conservation Service estimated that only 28 percent of the state's primary agricultural soils were devoted to cropland use.[5] Much of the former tilled farmland which is no longer tilled has returned to forest, which covered approximately 66 percent of the state in 1977.[6]

The transformation from farm to forest uses has occurred mainly in the interior areas of the state, where hilly terrain has made farming an uneconomic land use. Yet there is no dispute that conversion of farmlands to urban uses also has caused some of this change. In the 1960s, ski resorts and second-home developments, with their requisite support services, mushroomed. Vermont farmers found that to meet rising property tax burdens caused by the rising market value of their land, yielding to market demands for development was often their best opportunity. This has contributed to a significant increase in urban land uses and a corresponding decline in active cropland uses in recent decades. Analyses undertaken by the USDA Soil Conservation Service and the Vermont Department of Agriculture indicate that the area of urban and built-up uses increased by 53 percent, or about 100,000 acres between the years 1967 and 1977.[7] Figure 4.1 illustrates the relative percentages of Vermont's land base devoted to major land uses in 1977.

Table 4.1
Statewide Change in Acreage of Farmland and Number of Farms in Vermont, 1880–1982

Year	*Acres*	*Number of Farms*	*Change*	*% Change*	*% Change from 1880*	*Average % Change per Year*	*Average Farm Size*
1880	4,882,588	35,522	—	—	—	—	137.5
1910	4,664,000	32,709	-218,588	-4.5	-4.5	-0.15	142.6
1940	3,666,835	23,582	-997,165	-21.4	-24.9	-0.71	155.5
1950	3,527,381	19,043	-139,454	-3.8	-27.8	-0.38	185.2
1959	2,945,343	12,099	-582,038	-16.5	-39.7	-1.8	243.4
1969	1,915,520	6,874	-1,029,823	-35.0	-60.1	-3.5	278.7
1978	1,633,049	5,852	-282,471	-14.7	-66.6	-1.60	279.1
1982*	1,574,441	6,315	-58,608	-3.5	-67.8	-0.90	249.3

*The 1978 Agricultural Statistics were revised in 1982 to compare more accurately to the 1982 and previous U.S. Agricultural statistics.

Source: U.S. Bureau of the Census, Census of Agriculture.

It is noteworthy that land area devoted to urban uses in the state increased by one-and-one-half times while land devoted to agriculture and woodland declined over the ten-year period. Urban and built-up uses have increased, but data on the type of land uses they have replaced are not available. Generally, urbanization has not been the major reason for the conversion of agricultural land in Vermont, but a major transformation of the Vermont landscape has occurred.

Despite increased urban development in the 1960s and 1970s, agriculture remains an important part of the state's economy. The state produces a variety of farm goods, including dairy, maple, and apple products. Over 15,000 people are employed in farming or farm-related industry.[8] In fact, about one-fifth of the state's revenues is derived directly from agriculture,[9] mostly dairy products. Table 4.2 and Figure 4.2 detail cash receipts from farm marketings in Vermont and the importance of dairy as the state's primary agricultural commodity. As Figure 4.2 indicates, dairy production accounted for 82 percent of total farm cash receipts in 1985, making Vermont agriculture essentially dependent upon a single commodity. Vermont is one of the most intensive dairy states in the country, averaging over twenty milk cows per square mile. Only Wisconsin exceeds this figure, with approximately thirty-one milk cows per square mile.[10] Such dependence has created an increased concern with the continued viability of agriculture in Vermont, particularly in view of increasing federal efforts to reduce dairy price supports in the 1980s. Despite the important role that agriculture plays in the Vermont economy, and because agriculture is essentially a one-crop industry, the state still must import over 80 percent of its food products.[11]

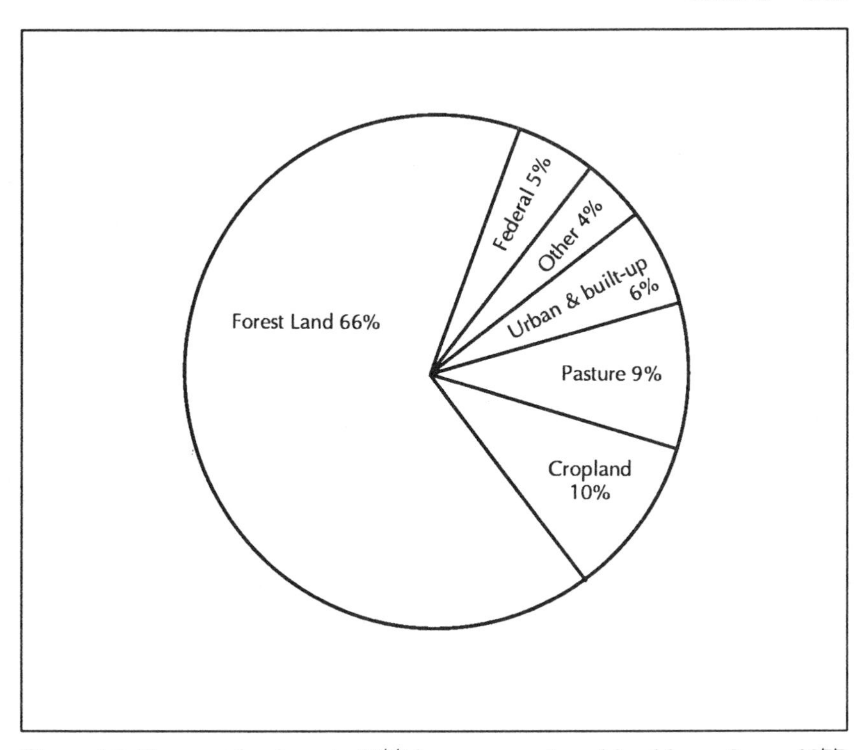

Figure 4.1. Vermont land use in 1977 by percent of total land base. *Source:* 1977 National Researces Inventory (NRI).

Just as the agricultural industry has survived in Vermont, so has the rustic lifestyle of many Vermont residents. In 1960, some 63 percent of the population lived on farms or in small town areas, compared to 24 percent in the rest of New England.[12] Although population grew by 14 percent from 1960 to 1970,[13] one of the highest growth rates in the country, by 1980 only 33.8 percent of the population was considered urban, giving Vermont the most rural population in the nation.[14] Today, many farm residents no longer depend on agriculture for their sole source of income, and part-time farming has become prevalent, especially with newcomers. Yet the independent family farm remains an important element of Vermont society. According to the 1978 Census of Agriculture, 88 percent of all Vermont farms are operated by an individual or a family. A 1982 study prepared by the Vermont State Planning Office revealed that 10 percent of the state's labor force is self-employed, the highest percentage in New England. Of these 22,640 self-employed residents, independent farmers make up the largest single group.[15]

Although the problems of farming in New England have taken their toll in Vermont in the past decades, those farms that did survive are among the

Table 4.2
Cash Receipts from Farm Marketings in Vermont, 1981–1983 (thousands of dollars)

Commodity	*1981*	*1982*	*1983*
Crops			
Hay	4,661	5,840	6,906
Potatoes	835	720	452
Misc. Vegetables	2,585	2,887	2,610
Apples	6,188	7,092	7,088
Berries	404	540	497
Misc. Fruits	14	14	14
Maple	8,806	7,755	7,859
Forest Products	5,270	5,200	5,363
Greenhouse/Nursery	2,231	2,231	3,025
Misc. Crops	1,915	2,025	2,085
Total Crops	32,909	34,304	35,899
Livestock			
Cattle/Calves	28,557	37,889	44,124
Hogs and Pigs	2,135	3,463	1,635
Sheep and Lambs	202	309	331
Dairy Products	322,983	330,445	337,040
Chickens	131	92	161
Eggs	6,188	5,968	5,472
Misc. Poultry	249	236	576
Misc. Livestock	1,666	1,707	1,799
Total Livestock	362,111	380,109	391,138
All Commodities	395,020	414,413	427,037

Source: New England Crop and Livestock Reporting Service.

most successful in New England, largely because of the dominance of high-value dairy production. In 1978, the Census of Agriculture reported that one-third of all Vermont farms had sales over $40,000, the highest percentage of successful farms of any New England state. Only 55 percent of all Vermont farmland was fully owned by farm operators, however, well below the average ownership rate of 63 percent for New England as a whole.[16]

Vermont agriculture remains an important part of the state's economy, despite increased urban development in the 1960s and 1970s. It has a larger percentage of its total land area devoted to agriculture than any other New England state and has maintained its rural atmosphere. Nonetheless, there has been a substantial decline in agriculture and farmland acreage in recent years. This decline, brought about by development and increased property tax burdens, has generated concern among Vermont's citizens over the loss of open space and rural community character.

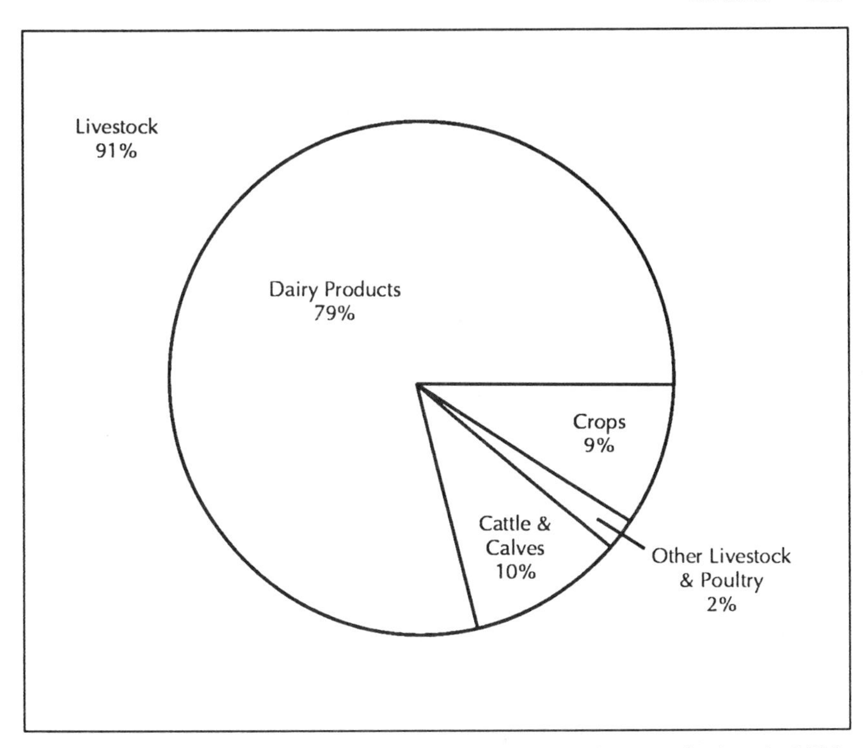

Figure 4.2. Distribution of cash receipts from Vermont farm marketings in 1983. *Source:* New England Crop and Livestock Reporting Service.

Concern over Loss of Open Space and Community Character: Vermont's Development Crisis

The conversion of agricultural land and corresponding change in landscape character became an issue of serious concern among the state's residents beginning in the early 1970s. The concern over farmland conversion was a natural outgrowth of the heightened concern over landscape change. In fact, the early efforts in the state to protect agricultural land were part of a more comprehensive effort to protect the existing character of the landscape and control the large-scale development that had occurred during the1960s.

The increased rate of development in Vermont, as in other areas of New England, was made possible largely as a result of the new interstate highways that suddenly made the state readily accessible to people living in the densely urbanized centers to the south and east. Much of Vermont was now only a three-hour drive from New York City and Boston. City and surburban residents, exposed routinely to the noise and congestion of city life, were lured to this rural environment. At first, the potential revenues to be derived from the influx of industry and tourism were actively sought and Vermont promoted itself as "the Beckoning Country." Vermont citizens quickly learned, however, that these econmic benefits had undesirable hidden costs. Second-home developments began to make their mark as outsiders began moving into the state. As late as 1960, Vermont's population was still only 390,000; there were more cows than people.[17] By the beginning of the 1970s, the population had increased by 14 percent to 445,000.[18]

Far more alarming was the specter of future change which this influx brought. In the late 1960s, a subsidiary of International Paper Company announced its intention to build vacation homes on 20,000 acres near Stratton in southern Vermont. This aroused fear that other large corporations, able to afford the spiraling land prices that development pressure was creating, also would see Vermont as a good location for large-scale development projects. The new interstate highways, rapid population growth, and pressures for large-scale development had created what one writer described as a "development crisis" in the minds of Vermont citizens.[19] By the close of the 1960s, there were three major concerns motivating the legislature to act to protect the Vermont landscape:

1. The effect that development was having or would have on the integrity of the environment
2. The effect that development was having or would have on traditional Vermont culture and ways of life
3. The effect that development was having on land values and, concomitantly, on the tax burden of the Vermont landowner[20]

Effect of Development on the Environment

Of special environmental concern was the health threat posed by sewage from the new large developments. With sewage treatment systems virtually nonexistent, untreated overflow from poorly planned septic systems sometimes ran down hillsides and into streams, sometimes from one development to another. Many second home and recreational developments were built in mountainous areas with thin soils on granite bedrock that clearly could not support large-scale multiple septic systems. In the Mt. Snow area alone, for example, it was reported that seventy-three different developers were operating in 1969. Only minimal health controls existed in the area at that time, and those were often poorly enforced.[21] Indeed, during this time, of Vermont's 237 cities and towns, only 75 had zoning regulations and only 20 had subdivision controls.

The problem of environmental degradation had become so acute by 1969 that the Vermont health department issued emergency subdivision regulations to protect the citizens from health hazards associated with uncontrolled development. Human beings were not the only species endangered: the life cycle of several natural lakes and streams in the state was upset by the increased organic content of sewage run-off,[22] by the diversion of streams to create artificial lakes, and by erosion caused by land clearance and other development-related activities.

Effect of Development on the Way of Life

A concern not as easy to define, but perhaps even more important to the average Vermonter, was a vaguely xenophobic and somewhat resentful attitude toward the newcomers, who were seen as representing different values. Vermont's pioneering effort in land use legislation derives, in part, from the strong ties that bind native Vermonters to the soil. Land use regulation found greater grassroots support in Vermont than in most other states. As Governor Thomas P. Salmon commented in 1973, "we were always aware that we had something special, worth preserving."[23]

A large part of the aversion to land development stemmed from a resentment toward those who would exploit rather than appreciate the land. The visual and environmental impact of large-scale development also brought with it the fear that the state's tourist industry might suffer. Some of the newcomers shared the concern for the landscape and, once their own niches in the state were secured, they also sought to preserve that which initially had attracted them to Vermont. Even they, however, were viewed with some trepidation by native Vermonters, who feared that foreign ideas and values would begin to be represented in their political process.

Effect of Development on Land Value and Taxes

Concern over the spiraling price of land and the accompanying increase in property taxes also soon caused Vermont citizens to want to protect their rural way of life and, in particular, agriculture—which had always been such a substantial, although diminishing, part of the economy. In many areas, the production economics of farmers were greatly affected by these changes in land values. Naturally, however, the rising land prices caused by development opportunities did not upset everyone. Farmers and others who were offered a handsome profit for their holdings were often more than willing to sell. For those holding onto their land, however, the tax burden (which was based upon a statewide assessment established at 50 percent of fair-market value) often became incredibly burdensome.[24]

As in Massachusetts, there were two primary reasons why development was responsible for drastic increases in Vermont's property taxes: it fostered the rise of land values on which the tax was based,[25] and the resulting population growth created a need for increased local public services which were supported to a large extent by property taxes. This increased burden on public services was magnified by the tendency of new citizens to demand from Vermont's local governments the type of services they had experienced in more highly urbanized and wealthy areas.

Thus, the concern to protect agricultural land grew out of a larger concern over the effects that rising land values and taxes were exerting on Vermont citizens generally. A more specific concern for the farmer was also a part of this general economic concern. Even though high property tax was not the only reason that agricultural land uses were diminishing, it was a significant economic issue. The value at which Vermont land was assessed increased by 12 percent in 1968, by 16 percent in 1969, and by 9 percent in 1970.[26] Thus, those farmers remaining on the land found an already difficult farm economic situation was exacerbated. Fortunately, the tax problems could be improved directly or indirectly through the state legislature.

By the early 1970s, legislative efforts for farmland-protection laws were primarily in response to concern over three issues: negative environmental effects of development; the changing life-style in Vermont; and economic problems, notably rising property taxes. To be sure, farmland protection per se was not the primary objective of this broad consensus for improved land use controls. At the end of the decade, however, these general concerns were supplemented with other more specific concerns to protect agricultural land.[27] The evolution of motives for agricultural land protection in Vermont produced several legislative responses over the 1970s.

Conservation Programs and Early Efforts to Protect Agricultural Land

Initially, legislation addressing the issue of agricultural land protection in Vermont was designed to protect the landscape quality of the state. In response to growing development pressures, the state adopted one of the most far-reaching development review and permit laws in the country, known as Act 250. In a few situations, Act 250 worked to protect farmland from unnecessary conversion. In general, however, it failed to address the issue adequately. Various tax incentive measures, when applied to a broad range of land uses, also helped the farmland protection efforts. When Act 250 was first passed, local planning efforts—including purchase of development rights and zoning mechanisms—and private land trusts did not have a measurable impact. These local and private programs are now gaining significance, however, because they are targeted specifically at farmland protection, unlike Act 250.

Act 250 (1970)

When it was adopted in 1970, the Land Use Development Law, or Act 250, was one of the nation's most comprehensive state land use laws.[28] By the early 1980s, modest success in controlling and directing growth had been achieved. As a primary tool of farmland protection, however, Act 250 has proved disappointing. This is primarily because it was not specifically designed to protect farmland.

Events which led to the passage of Act 250 began in response to the development crisis. In May 1969, Governor Deane Davis appointed a seventeen-member Commission on Environmental Control, composed of businessmen, industrialists, developers, conservationists, and influential legislators, chaired by State Rep. Arthur Gibb. Governor Davis charged the Gibb Commission with exploring "how we can have economic growth and help our people improve their economic situation without destroying the secret of our success, our environment."[29] Predictably, the enemy was perceived as the large out-of-state developer, and as the Commission embarked on six months of bimonthly meetings, it settled upon "land misuse" as the priority issue. The proceedings received wide coverage by the press, and by the time the Commission had developed its recommendations, public interest was high. Many of the Commission's recommendations were soon reflected in the bill which established Act 250.

Act 250 made its way with surprising ease through the Vermont legislature in 1970, passing almost unanimously. One reason for its success is believed to be the importance of environmental issues at the time, an importance that transcended party lines, despite a well-organized lobbying effort against the bill mounted by developers and realtors. Another reason was the participation of key legislators on the Gibb Commission and in the drafting of subsequent legislation. Perhaps most important, however, was the support given by the people of Vermont, a support somewhat surprising given the traditional Vermonter's sense of independence and resentment toward government interference.

The support of citizen groups is credited with playing an important role in shaping public opinion to assure the passage of Act 250. Several citizens groups, most of which were homegrown and not part of national organizations, prepared surveys to learn Vermonters' attitudes on environmental problems, publicized their findings, and encouraged citizen participation. One of the most important of these groups was the Vermont Natural Resources Council (VNRC) which today has grown to become one of the most active advocates of farmland protection and sound land use policies in the state. In recent years, this organization and others have sought, with mixed results, to improve Act 250 as a vehicle of farmland protection.

This is not to say that Act 250 received unqualified support. Indeed, Democratic House Minority Leader Thomas Salmon, who soon became governor, led an active campaign against the bill and sought to amend it with provisions that would have seriously curtailed its effectiveness. Despite such opposition, however, the bill eventually passed the House by an overwhelming margin and passed the Senate with certain amendments that even strengthened it.[30]

As originally enacted, Act 250 established seven (later increased to nine) District Environmental Commissions (DEC). Specified categories of proposed new development are required to apply to their respective DECs for project review approval.[31] Hearings before the relevant DECs are held on virtually every major development in the state, affording opponents of the development an opportunity to voice their concerns. Appeals from the DECs were to be taken first to a state-level Environmental Board established under Act 250, and then to the state Supreme Court.[32] In determining whether to grant a development permit, a DEC was to use ten criteria. Only one of these (criterion 9) bore any direct relation to the goal of protecting agricultural land.[33] Actually, as this criterion originally read, the goal of farmland protection was not readily apparent. Criterion 9 simply stated that the commission was to ensure that development would be "in conformance with a duly adopted development plan, land use plan or land capability plan." The preparation of such plans by the State Environmental Board was mandated by Act 250. Specifically, the Board was directed to prepare three statewide plans: an interim land capability and development plan, a capability and development plan, and a state land use plan. It was in these plans that the concern for agricultural land was made explicit.

Interim Land Capability and Development Plan. The Interim Plan, a joint effort on the part of the Environmental Board and the state Planning Office, consisted of a series of maps for each county and a general statement of policies. One series of maps illustrated the natural resource opportunities with the state, including those lands with either a high or a limited but still significant potential for agricultural use. The Interim Plan was to be used by the District Environmental Commissions to limit development on lands with high natural resource opportunities. However, because future plans were required to be consistent with this first plan, the Board refrained from being too specific. Following hearings in which reaction to the Interim Plan was most favorable, the plan was adopted by the State Environmental Board in February 1972.

Land Capability and Development Plan. The second plan, submitted by the Environmental Board in 1972 and approved by the legislature in April 1973, was far more detailed than the first. The plan had been mailed out to each household in the state before debate began in the legislature, and public favor again had played a large role in the legislature's decision to adopt it.

The central part of the new plan consisted of a number of amendments to the ten original criteria established under Act 250. One of these amendments, known as criterion 9-B, was designed to make development on farmland far more difficult. To build on prime agricultural land, an owner must show that development is necessary to realize a "reasonable return on the fair market value" of the land.[34] Farmland protection advocates saw criterion 9-B as a major component of statewide efforts to protect Vermont farmland. However, in the decade following adoption of the Land Capability and Development Plan, criterion 9-B has proven to be a hollow protection. Much of the debate in recent years has focused on the definition of "primary agricultural soils" and what constitutes a "reasonable return" on the fair-market value of the land. These two terms have proven very difficult to define and subsequently to support in the courts.

The Land Use Plan. The last of the three plans mandated by Act 250 proved the most controversial. It was not a regulatory plan like the first two plans (setting criteria for the District Environmental Commissions to evaluate proposed developments). Rather, the Land Use Plan was designed specifically to point out to developers, landowners, and citizens the areas of the state in which growth was desired and the areas in which it was not. In general, the plan submitted to the state legislature early in 1974 sought to set density limits on various areas and "determine in broad categories the proper use of the lands in the state,"[35] including agricultural lands, on the basis of the prevailing or future use of the land. Growth would be encouraged in those areas already relatively densely settled, while lands with high agricultural potential and other natural resource areas would be allowed only one building for every twenty-five acres.[36]

The plan proved to be very unpopular. In particular, the density limitations in rural and natural resource areas turned the public against the Land Use Plan.[37] While landowners may have decried the high tax burden, they were often quite happy with the appreciation in the value of their land which development would bring, and were unwilling to see this value destroyed by the adoption of statewide zoning. Opposition was so great at the public hearing held by the special House committee to which the plan had been referred that it failed to be reported out of committee. State Senator Arthur Gibb explained the difference in attitude to the two approaches as follows: "When you're talking about regulation, you're talking about regulating the big bad outside developer; when you talk about a plan, you're affecting all of the land in Vermont."[38]

Expressing the opposing view encouraging the adoption of the Land Use Plan, the State Planning Board made the following argument: "Many of the regulatory and non-regulatory programs that would flow from, and be strengthened by the state Land Use Plan are needed now, not some time in the future. These include such important programs as property tax reform, public investment planning, and agricultural and forest land conservation."[39] As it turned out, however, the Land Use Plan was not needed as a catalyst for these three programs, all of which were pursued after the demise of the plan. Although Act 250's Land Use Plan was never adopted, the regulatory mechanisms of the Land Capability and Development Plan have been in place for over a decade.

In general, Act 250 has not proven to be an effective means of protecting Vermont's agricultural land. Indeed, the adoption of the farmland criteria has had little deterrent effect on the continued conversion of farmland; it has resulted in few successes and many abuses and failures. Except for a handful of cases, there have been few examples of development-permit denials on agricultural grounds.[40] The ineffectiveness of Act 250 in farmland protection efforts is largely attributed to two important weaknesses in the law: the ten-acre loophole and definitional problems associated with criterion 9-B.

Ten-Acre Loophole. Prior to 1983, one major reason for the lack of effectiveness of Act 250 in protecting farmland was thought to be the jurisdictional limits under which Act 250 operated. A permit under the Act was only required for developments of ten or more lots, each of which occupies ten or fewer acres. Thus, if a landowner anticipated serious problems gaining permit approval, it was relatively easy to avoid the permit process entirely by subdividing and by developing on larger lots. This type of large-lot subdivision results in a pattern of spreading urbanization that actually can consume far greater acreages of agricultural land than more dense forms of development which require a permit under Act 250.[41] The jurisdictional limit of Act 250 has come to be known as the ten-acre loophole and has been the source of powerful controversy in Vermont. Shortly after adoption of Act 250, it became clear that this provision may be causing more harm than good as a land use policy. As early as September 1972, the *Boston Sunday Globe* reported:

> A major loophole has been found in Vermont's environmental control laws which allows developments to spring up without any form of state approval.
>
> The loophole is causing a proliferation of development in Vermont, and legislative action must be taken to correct the action, environmental officials said.[42]

Despite such warnings by the *Boston Globe* and others, more than a decade passed before the law was amended. Prior to 1983, when the legislature finally amended Act 250 to close the ten-acre loophole, a 1981 study by the Chittenden County Regional Planning Commission indicated that the loophole had been used extensively by developers to set the scale and pace of their development. Between 1973 and 1981 the development of building lots of ten acres or larger increased from 1 percent of all lots developed to 13 percent, and accounted for over 40 percent of the land used for new single-family lots in Chittenden County. In southeastern Vermont, the Windham County Regional Planning Commission reported that over 3,700 acres in a six-town area were subdivided in similar fashion between 1971 and 1982.[43]

When initially passed in 1970, the legislative intent of the ten-acre jurisdictional limit was to free individual homeowners from the burden of obtaining a state permit if they desired to sell a portion of their land. Indeed, much recent farmland development has been individual family residences. Some argue that these single-family residences which are not in large developments are on lots of ten or more acres to avoid subdivision regulations of the health department, not to avoid the requirements of Act 250. Nevertheless, it has been reported that many land developers have managed to circumvent the overall objectives of Act 250 by selling land in ten-acre parcels and "promoting the idea that landowners can sell off a few lots of land to help pay for the house."[44]

Regardless of whether individual homeowners or land developers were the primary cause of farmland conversion under the ten-acre exclusion of Act 250, it was a tremendous incentive to waste land. The ten-acre loophole results in small tracts of land which cannot be farmed economically, and sprawling urbanization, with its attendant problems of increased costs for roads and infrastructure.

In response to these problems, a bill which called for the elimination of the ten-acre exemption from Act 250 was introduced into the 1981–82 legislative session and finally passed in the 1983 session. In essence, the major provisions of the "Ten Acre Exclusion" bill closed the loophole by amending the definition of "lot" in Act 250. Developers are now required to obtain an Act 250 permit for any subdivision of land into ten or more parcels, regardless of the size of the parcels. A less controversial provision of the bill, which did not pass, would also have improved Act 250's effectiveness by authorizing state courts to assess a civil penalty for violations.[45]

The proposed legislation generated a major debate in Vermont, pitting prodevelopment and property-rights advocates against the state's strong environmental interests. Supporters of the amendment such as Don

Hooper, assistant director of the Vermont Natural Resources Council, maintained that:

> Closing the loophole is a minor adjustment of Act 250 that will remove one incentive to waste land....It does not affect a farmer, for example, who needs to sell off an occasional lot to pay his or her mortgage....With the farm economics picture as gloomy as it is, it may well be time to cheer even a little bit of progress [in farmland protection efforts]. The more massive structural changes that are needed in prices and marketing, not to mention new products, may be further away than ever.[46]

Strong and persistent lobbying efforts by environmental groups finally paid off with the closing of the ten-acre loophole in 1983, but it remains to be seen how much more protection Act 250 will provide for farmland.[47]

Definitional Problems of Criterion 9-B. There are other reasons why Act 250 has not been the mechanism of farmland protection that some advocates hoped it would be. One reason is that the District Environmental Commissions are unable to implement effectively the "primary agricultural soils" requirements of criterion 9-B. It is argued that the decentralization inherent in the system comprising nine District Environmental Commissions allows for too much interpretation of the Act 250 criteria and therefore leads to uncertainty in the application of the criteria. The uncertainty can be exploited by landowners seeking to convert agricultural land. The problem of decentralization is compounded by the lack of a statewide land use plan which could add strength and definition to the ten criteria of the Land Capability and Development Plan. Others point out, however, that decentralization is not necessarily a problem. It enables a DEC to base its decision partially on circumstances unique to the local area. Moreover, a certain degree of consistency is guaranteed by the process of appeal to the state-level Environmental Board, which oversees the implementation of Act 250 statewide.[48]

Critics of Act 250 feel that the DECs often are not aggressive when seeking evidence to determine whether a developer is complying with the Act 250 criteria. The Commissions must consider nine other criteria, and they tend to evaluate each application on the evidence that is presented to them. It is claimed that criterion 9-B, concerning prime agricultural soils, often is overlooked.[49] Another viewpoint is that the DECs are quasi-judicial bodies, not regulatory bodies, whose function is to render decisions based on the evidence brought before them, not to conduct independent investigations.

The most specific complaint regarding the DECs' evaluation of the agricultural lands criteria relates to the requisite inquiry into the financial return that can be made on the value of a parcel of land (criterion 9-B). Commissions have often equated "reasonable return on the fair market value of [the] land" with the return that could be realized through development, thus easing the burden of proof to developers.

The inability of some of the DECs to understand how "reasonable return" should be determined is symptomatic of two basic problems with the enforcement of Act 250. First, DECs are composed of lay people who generally have no legal training and may have difficulty with sophisticated analyses required by some of the arguments posed by developers.[50] Second, statewide regulations have never been adopted for the implementation of Act 250. In the absence of regulatory guidance from the professionally staffed State Environmental Board, the District Environmental Commissions are simply ill-equipped to interpret some of the Act 250 criteria. The Environmental Board has attempted to establish regulations to guide the DECs, but these efforts have been thwarted by the joint legislative committee, who by Vermont law must approve all regulations promulgated by state agencies.[51]

Despite the liberal interpretation of "reasonable return" which landowners have received from DECs, in recent years there have been further efforts to ease restraints on the development of agricultural land. Legislative proposals have been introduced which would require DECs to consider economic considerations as a criterion in determining whether agricultural lands were of "prime" quality or not. Although criterion 9-B allowed DECs to consider the "reasonable return on the fair market value" of the land in determining whether land uses could be reasonably restricted to farming, Act 250 generally defined "primary agricultural soils" as soils which have a potential for growing food and forage crops, are well drained with a slope of no more than 15 percent, and are capable of contributing to an economically viable farm.[52] In some cases, DECs have denied development permits because the land was of "prime" quality when in fact the land may not have been farmed for years because it was no longer economically viable. Thus, there have been several efforts to define prime agricultural soils more specifically and to include criteria which would require that agricultural land have easy access to markets, services, and supplies and be taxed appropriately. Land which does not meet these criteria, regardless of its physical characteristics, would not be considered prime and therefore would fall outside the primary agricultural considerations of Act 250. Given the difficult economic conditions of farming in Vermont today, proposals such as this could have the effect of defining all farmland in Vermont as less than prime and thus outside the agricultural review criteria of Act 250. Therefore, such proposals have been strongly opposed by farmland protection advocates.

A good illustration of some of the problems with criterion 9-B is the case of the Windsor Industrial Park application in the town of Windsor, Vermont. In October 1979, the Windsor Improvement Corporation, a public development entity, proposed to develop an industrial park on a parcel of land containing thirty-three acres of prime farmland in a forty-four-acre site. After numerous administrative hearings involving public agencies and private citizen groups, the DEC twice denied the Windsor Corporation a development permit. Windsor then moved the appeal to the Windsor Superior Court, which ruled in January 1983 that the land could be developed as planned. Supporters of the decision point out that the land was last used by a dairy farmer in 1945, and that future farming was unlikely because the land was somewhat hilly and divided or bordered by a river, a railroad, power lines, and highways.[53] Indeed, the entire parcel had been zoned as industrial for about twenty years, and only one farm was in operation in the vicinity. Development of the land, therefore, was a locally planned use that appeared to make economic sense.

Critics of the decision believe that conversion of the land reduces the critical mass needed for nearby farm operations to succeed and that the parcel itself was some of the best farmland in the region. They point to the rather bizarre process of decision making as indicative of the inadequacy of Act 250 to halt the decline of farmland use. Specifically, they point to the disagreement between the DEC and the Superior Court judge on whether the development would significantly reduce the agricultural potential of primary agricultural soils, whether a reasonable rate of return could be had without developing the land as planned, whether the plan would minimize the reduction of agricultural portential, and whether development would jeopardize agriculture on adjoining lands. These contrary opinions were the product of a lack of clarity in the criterion 9-B language, which are only partially resolved if the Environmental Board uses its rule-making powers. Notably, the phrase "adjoining lands" was interpreted by the court to include only abutting property.

As the primary feature of farmland protection under Act 250, criterion 9-B has met with poor success in protecting Vermont's agricultural lands.[54] Since the late 1970s, some Act 250 development permits have been denied solely because they conflicted with the primary agricultural soils criteria; however, these have been the exception. Richard Cowart, former executive director of the State Environmental Board, feels that there are three basic reasons why the agricultural soils criteria have not been implemented in a meaningful way. First, Vermont has a strong antibureaucratic tradition, and it was therefore difficult for those implementing the Act to tackle all the criteria at once. Second, there has been an understandable tendency to address the "bread and butter" environmental issues first, such as air and water pollution, and to look at the less pressing environmental concerns after the mechanisms for correcting the primary ones had been established. Third, declining agricultural acreage has only become an issue of national concern in recent years, and the Vermont citizenry has only recently adopted this concern along with the rest of the country.[55]

Despite its flaws, Act 250 does serve two functions that are important to farmland protection efforts in Vermont. First, the Act serves an educational function: the 9-B criterion is addressed at public hearings, and discussion in such a forum serves to alert citizens to the importance of the issue. Second, Act 250 actually has worked to protect a discrete number of acres of primary agricultural soils in a few cases. The basic problem, however, is that Act 250 can only operate on a case-by-case basis. Had the statewide Land Use Plan been adopted, matters may have been different. The areas of the state deemed most appropriate for agricultural production would have been identified and set aside. As the program currently stands, however, it is simply not capable of addressing long-range farmland resource planning on a short-range incremental basis.

Some farmland protection advocates are not convinced of the value of Act 250 as a vehicle of farmland protection even if the ten-acre loophole and criterion 9-B problems were solved. They feel that Act 250, with its case-by-case review and failure to identify specific lands worthy of protection, creates an "illusion of protection" and allows people to avoid dealing with the difficult issues of farmland protection. Ken Senecal, former Executive Officer of the State Environmental Board, wonders if Act 250 really is a useful weapon in farmland protection efforts. According to Senecal:

> A farmland program must have some teeth in it and I assume that is why so many people look to Act 250 for part of the solution to the problem. In my view, however, we have been far too quick to put teeth into the Act without any idea of what they were going to bite. The result has been turmoil and controversy with no identifiable benefit to farmland protection objectives.[56]

Indeed, because Act 250 operates on a case-by-case basis without reference to a general plan, it is probably even less effective in protecting farmland in a meaningful way than its supporters allege. It has been argued that it makes little sense to "examine the use of one piece of property without reference to the present and future uses of adjoining lands."[57] It makes little sense, for instance, to protect farmland if all the land surrounding it is used industrially. Thus, perhaps some of the acreages which have been denied development permits would have been put to some better use than farming.

Moreover, in response to the claim that Act 250 has served to educate the public on the farmland protection issue, Ken Senecal believes:

> There must be a better way to "educate" the public than by dashing the hopes and aspirations of private property owners because someone puts the farmland insignia on their property. It is not the private property owner who is to blame for the loss of Vermont farmland. More often than not, it is the tax dollars we spend for roads and sewage systems which create irresistible pressures for conversion of farmland to other uses by driving up land values and property taxes. These and other purely economic factors, such as the value of farm products and increased energy costs, cannot be addressed effectively in a case-by-case review under Act 250.[58]

In retrospect, it is not surprising that the statewide Land Use Plan mandated by Act 250 was never adopted. Substantial questions of equity are inevitably raised with such plans that restrict the use and thereby reduce the value of a farmland owner's property. Farmland owners in Vermont, like other states, are not supportive of regulatory programs and plans which suddenly reduce the value of their only asset and guaranteed hope for the future—their land—particularly at a time when farm economic conditions are most difficult. Without the support of farmers, it becomes increasingly difficult to develop a meaningful farmland protection program.

After a hopeful decade of experience with Act 250, most farmland protection advocates in Vermont are in agreement that Act 250 alone cannot have a significant impact on maintaining active agricultural land retention. The complexities of farmland protection stretch far beyond the capacity of a single comprehensive land use law. For a program such as Act 250 to have a significant impact on the retention of agricultural land, there must be a strong commitment among the citizens and legislature of the state that it do so. Local and state politics must place considerable importance on the issue, and towns and regional commissions must incorporate strong, clear statements relating to the protection of agricultural lands into local plans. Richard W. Carbin, Executive Director of the Ottaquechee Land Trust in Woodstock, Vermont, commented:

> Farmland loss have become one of the most serious environmental—not to mention economic and social—problems facing the State. Act 250 has not been effective in dealing with this issue. Act 250 alone cannot solve this problem, but it can be a strong tool to help, if—and this is a big if—Vermont has the courage to carry on effective land use planning at the local and regional levels.[59]

This sort of broad consensus for farmland protection, even in an environmentally conscious state such as Vermont, is difficult to achieve, particularly at a time of national agricultural surpluses when the economic viability of New England agriculture is in question.

The Act 250 experience in Vermont is also illustrative of the difficulties of relying on a single comprehensive land use law to assure farmland protection. Throughout the decade of the 1970s, Vermont's farmland protection efforts have been heavily focused on Act 250, a program which was not specifically established to protect agricultural land.

In the early 1980s farmland protection efforts began to shift away from such a heavy dependence on Act 250. As concern over the conversion of agricultural land has grown, farmland protection advocates in Vermont have sought to broaden the range of programs and make use of several other early conservation programs existing in the state when Act 250 was adopted and amended. The most important of these vehicles include:

1. The modification of existing or creation of new tax-incentive programs
2. Enabling legislation to allow the local purchase of development rights
3. Zoning innovations
4. The participation of private land trusts in farmland purchases

The following section describes the recent activities which have utilized these programs. Programs which were first established in the 1980s and directed solely at farmland protection will be discussed in a later section.

Tax Incentive Programs

Aside from the land use controls of Act 250, tax incentives have become an increasingly important component of Vermont's conservation and farmland protection efforts. In the 1970s, it became apparent that improved tax incentives to protect farmland and open space were needed because by then Vermont farmers were carrying tax burdens between two and four times as great as they were in 1960. Much of this increase was caused by the dramatic increase in the price of undeveloped land during the 1960s and 1970s, resulting in a shift of tax burden from buildings to open land. The problem of burdensome farmland taxation was exacerbated when local governments began to appraise land at 100 percent of its fair-market value.[60]

The Vermont legislature had sought to relieve the tax burden on farmers in the late 1960s through the use of two programs.[61] The first program allowed the establishment of development easements on farmland, which would restrict land uses for a limited period of time, in exchange for use-value assessment, thereby lowering the tax burden on the landowner for a specified period of time. The program was never utilized, however, partly because the concept of development easements was still in its infancy and misunderstood at both the state and local levels. The second program permitted owners of farmland, forest, and open space to enter into contracts with local governments fixing the land value and property tax for a period not to exceed ten years. This program also was not effective in preventing farmland conversion.

In 1973, Governor Salmon sought to reduce the farmer's tax burden further by recommending a plan that would, among other things, limit property taxes on farms to 6 percent of a farmer's gross receipts from farming.[62] The 1974 legislature practically ignored the recommendation, however, due to controversy which had erupted in the legislature over the ill-fated Land Use Plan of Act 250 which also was defeated in 1974.

A combination of farmer concern over the tax burden and a general concern on the part of the Vermont citizenry over the conversion of farm and forest land ultimately caused the legislature to adopt the Current-Use Tax Law in 1978. The specific goal of this law is to protect farm and timber land currently in active production. The details of Vermont's major tax programs affecting farmland are outlined below.

Tax Stabilization Contracts. In 1955, the Vermont legislature adopted several statutory provisions as part of a broad program of economic development. One of these new laws allowed local governments to enter into contracts with certain business enterprises allowing land values and taxes to be held constant.[63] The legislation, which was amended in 1967 and 1974 to include farmers and agricultural landowners among those who could enter into such contracts, currently allows a municipal corporation to contract with owners, lessees, bailees, or operators of agricultural property to fix or stabilize the tax on such property.[64] A majority of the town must vote to approve the contract, which cannot extend over ten years in duration.

In 1977, a separate statute was adopted allowing the legislative body of a municipality (i.e., the Board of Selectmen) to enter into a tax stabilization contract without voter approval.[65] Under the 1977 law, however, only farmland owners can enter into contracts with the town.

The other major requirements of the contract are that the land must be at least twenty-five acres and be both "actively and exclusively" devoted to farming. The Supreme Court of Vermont ruled in 1982 that under the "active and exclusive" use requirement of the 1977 law, a town can require that at least two-thirds of a property owner's income be derived from farming to be eligible for the program.[66] Under a penalty provision of the 1977 law, if the land use is converted from farming, a land use change (conversion) tax will be imposed to compensate for the back taxes (up to three years) that were lost under the stabilization program.

This scheme, allowing Vermont municipalities and individual landowners to enter into contracts to stabilize property taxes, has provided substantial property-tax reduction to those owning or leasing land for agricultural, forest, and open space uses. By 1981, thirty of Vermont's cities and towns employed the program, aiding 650 landowners of approximately 100,000 acres of farm, forest, and other undeveloped land. Total tax relief was approximately $330,000 or an average of $3.50 per acre.[67] Those communities that have adopted such programs generally contain less than the average proportion of farms with farmland generally assessed at above average value.[68] The program appears to be more popular in the relatively more urbanized areas of the state where the concern that farmland will be converted for development is greatest.

One potential drawback of the program is the negative impact on local governments which is caused by the loss of revenue from such contracts. Unlike the Current-Use Tax Stabilization Law, which was adopted by the legislature several years later, there is no specific provision for state reimbursement of lost local tax revenue. A 1977 amendment to the State Aid to Education legislation was intended to alleviate the added burden that tax stabilization programs for farms and forests placed on local governments. At present, however, the monies flowing from this amendment do not match the monies lost from the tax programs, and thus it has been suggested that a more equitable system should be implemented.[69] By 1985, no action had been taken, and therefore, municipalities are often reluctant to engage in or encourage such contracts.

Most landowners who were under the contract system have chosen to remain under the system rather than place their land under the Current-Use Tax program, when it was enacted in 1978. This was often because landowners receive a greater tax benefit under a specific contract than they would under the Current-Use Tax program. In addition, many landowners preferred to remain under the tax-stabilization agreements because they are administered locally and are of specified duration.

Capital-Gains Tax on Land Sales. In 1973, the Vermont legislature adopted a law that required increased taxation on the profit realized through a sale of land held for a short period of time. Known as the Land-Gains Tax,[70] this tax program aims to deter land speculation, raise revenues lost through other programs, reduce the subdivision of land, and slow down the escalation of land values.[71]

The Land-Gains Tax applies to any landowner who realizes a capital gain on the sale or exchange of land held for six years or less. The tax rate can be as high as 60 percent of the gain if the land is bought and sold within one year, 50 percent if it is bought and sold within two years, and so on, with the tax ceasing to be imposed when land is held for six or more years.[72] The tax applies only to the portion of the profit that is related to the value of the land, not to the portion of the purchase price attributable to buildings.

The law originally applied to primary, one-acre home sites, but real estate industry lobbyists successfully obtained amendments to these provisions of the bill in 1974, 1976, and 1978.[73] Today, the tax does not apply when the sale involves a primary residence incorporating up to ten acres of land, "whether it is the site for the seller's principal residence, the purchaser's residence, or a lot sold to a builder of a principal residence." The original law was upheld as constitutional in a challenge before the Vermont Supreme Court in 1974.[74] A further amendment in 1981 exempted certain charitable organizations whose purpose is to protect agricultural lands if the acquired land is used for agriculture for at least ten years.[75]

The tax had the significant effect of an increase of property bought by landowners for their personal use, rather than for speculation. This was largely because the out-of-state large parcel investors withdrew from the market. The tax has mainly affected Vermont residents, with only approximately 25 percent of the tax revenue from out-of-state investors. Moreover, the added cost of land incurred by the tax was marginal. The increased sale price of unimproved property was less than 6 percent and less than 1 percent on improved property. The tax produced only between $500,000 and one million dollars of revenue per year.[76]

In his 1986 report on the Vermont Land-Gains Tax, Thomas Daniels, assistant professor of Community and Regional Planning at Iowa State University (and native of Burlington, Vermont), made several policy recommendations designed to increase revenue, to control subdivision activity, and to evaluate the impact of the gains tax on land markets. To increase revenue, he suggested that the following adjustments be implemented:

1. Change the administrative rule allocating only 8 percent of the gain from condominium sales to the land element. A figure of 15 to 20 percent would tend to double revenues from condominium sales.
2. Extend the length of the liability period. Governor Salmon originally proposed that the gains tax apply to land held less than ten years.
3. Raise tax rates. However, higher rates might further discourage land turnover and produce even less revenue.

To control subdivision activity, Daniels proposed that extending the liability period and raising tax rates would effectively reduce the profitability of subdivisions. Moreover, the tax could be restructured to apply specifically to land divisions, regardless of ownership. Finally, to evalute the impact of the gains tax on the land market, he suggested monitoring land sales subject to the gains tax and organizaing property transfer records.[77]

Despite its limitations, the land-gains tax is viewed as a complementary land use control to Act 250, and a short-term measure aimed at a specific segment of land sellers. As such, it is an important indirect method of farmland protection.

Current-Use Tax Law. In 1978, the legislature enacted a law allowing use-value appraisals of land for tax purposes when the land is devoted to certain qualified uses. Like the Farmland Assessment Law in Massachusetts, the law is designed to provide an economic incentive to keep productive farmland in production by easing the tax burden of farmers who would otherwise be taxed at rates based on fair market value of their land.

When adopted in 1978 (effective 1980), the Current-Use Tax Law became the state's first program designed specifically to protect agricultural land, placing Vermont well behind its New England neighbors in adopting such a program. The expressed purpose of the Vermont legislation providing for current-use taxation of farm and forest land is:

> to encourage and assist the maintenance of Vermont's productive agricultural and forest land; to encourage and assist in their conservation and preservation for future productive use and for the protection of natural ecological systems; to prevent the accelerated conversion of these lands to more intensive use by the pressure of property taxation at values incompatible with the productive capacity of the land; to encourage and assist in the preservation and enhancement of Vermont's scenic natural resources; and to enable the citizens of Vermont to plan its orderly growth in the face of increasing development pressures in the interests of public health, safety and welfare.[78]

Like differential taxation programs in other states, under Vermont's current-use taxation law, owners of farmland or forest land are eligible to apply to their local property tax assessor for land appraisal on the actual use of the land rather than its potential "highest and best" use. To qualify for such an appraisal, a farmland parcel must:

1. Contain at least twenty-five acres and be devoted to active agricultural use
2. Have produced a gross income of $2,000 in one of the previous two years, or three of the previous five years, or
3. Be owned by a farmer and be part of an overall farm unit. A farmer is defined as a person who derives at least one-half of his or her gross annual income from farming.[79]

The eligibility of forest land is somewhat more stringent. The requirements generally are that the land be greater than twenty-five acres and managed under a ten-year forest management program approved by the Department of Forests, Parks, and Recreation. The owner of the land must submit an annual report indicating conformance with the management plan.

Use values, which are based on land productivity, are set by the state Current Use Advisory Board. Unlike in Massachusetts, productivity is primarily based upon soil quality, although other site conditions may be a consideration. A farm's value is then adjusted to the common appraisal level of its town.

A "land use change tax" is imposed when land which is or has been appraised under the Current Use Tax Law subsequently is developed. The tax is equal to 10 percent of the fair market value of the land as developed. A landowner may withdraw from the program at any time. The land use change tax is not due unless or until the land is developed.

The Vermont current-use taxation program includes a unique feature known as a use-tax reimbursement fund. This fund is generated both through general appropriations and through monies collected pursuant to the land use change tax and is used to reimburse local governments, up to a maximum amount per acre, for the revenues lost to them through the operation of the program. When the amount contained in the reimbursement fund is insufficient to make total reimbursement to the municipality, the fund makes prorated payments and the individuals being taxed under the program are required to account for the difference. In this event, property owners may elect to withdraw from the program and are relieved of any obligation to pay a land use change tax if the land is developed. As a result of the use-tax reimbursement program, therefore, municipalities lose little or no revenue and are consequently more likely to encourage farmland owners to participate.

Enrollment in the current-use tax stabilization program in the first two years of its operation was high, and it appears to have slowed the pace of agricultural land conversion somewhat. In 1980, 120,000 acres in over 500 parcels were enrolled in the program. In 1981, an additional 80,000 acres were bought under the program.[80] Of this total acreage, participating landowners paid an average of 70 percent less in taxes for land qualifying for the program, amounting to a total savings of about $850,000 or $4.25 per acre in taxes.[81] Despite this high enrollment many eligible landowners remained under the municipal tax stabilization contract system discussed previously, often on the advice of the State Current Use Advisory Board.[82]

By 1982, roughly 90 percent of the land enrolled under the program was in forest use and only 10 percent in agricultural use, roughly the same as the proportion of potentially eligible forest-to-farmland in the state as a whole. This has caused those who view current-use taxation as a "save-the-farmer" program to be very critical. The statute explicitly states, however, that it is to apply to land actively in production in both farms and forests. It is apparent that many people feel that farming is the more important land use to promote in the state and it is thus possible that future amendments of the current-use law will reflect this preference.

By mid-1983, many Vermont towns were ending their participation in the tax-stabilization program because of the reimbursement provisions of the current-use law. Farmland owners, therefore, are being encouraged to shift into the state program. This trend has continued, with about 550,000 acres of land enrolled in the current-use program by 1985.[83]

A deficiency of the program as originally adopted is the insufficient state appropriations for reimbursement for lost local tax revenues. In 1980, the reimbursement cost to the state was $400,000; in 1981, the cost was $800,000; in 1982, $1,100,000; in 1983, $1,500,000; and in 1984, $2,000,000. It has become obvious that to avoid limiting the benefits of the program, larger budgets will be required. Thus a budget request of $3.0 million was submitted to the legislature for 1985.[84] The increasing cost of the program is a major concern, particularly since the cost is increasing much more quickly than the average annual 6 percent total state budget increase. If future costs exceed budget allocations, it is hoped that an emergency appropriation would supply the difference. The danger of an insufficient appropriation is that all landowners who participate in the program would be required to make up the difference to the municipality and would then be given the option of withdrawing from the use-value appraisal program without paying a land use change tax. The result would be that many landowners who have enjoyed property-tax savings for several years would be free to develop their property.

Closely tied to the practical problem of how much money is available for reimbursement is the political question of how much the state should be willing to pay to keep a given piece of land in its current use. In one example, a parcel of land was taxed under the Current Use Tax Law at $2 per acre; if that same parcel were to be taxed based on its fair-market value, it would be $700 per acre. It has been suggested that it would be better if the large amount expended were to be used to promote the current use of a larger acreage elsewhere.

A related problem is how to handle agricultural or forest land in an area zoned for industrial use. Some officials say that it is where development pressures are the greatest that the greatest effort to protect farmland should be made. Others argue that such land simply costs too much to protect and may, in any case, conflict with a local government's determination that the farmland area is best suited for industrial or other developed use. Thus, the importance of local planning to farmland protection is underscored. Without local support, farmland protection programs cannot work well and are essentially in conflict with local pressures.

This quandary relates to another problem with the Current-Use Taxation Program as originally adopted, which is the possibility for towns to increase their revenues at state expense. Since the state reimburses towns for taxes lost, a town may generate extra revenue by inflating land value by a small amount in the assessment procedures (but not by so much that the value will be questioned). Additionally, the town can zone an area containing agricultural land for industry and then attract an industry into the town to build on the nonprime soil. The town thus receives additional revenue both from the new industry and from the state, which now must reimburse the town for the taxes on the development value of the land classified as industrial.[85]

To address this problem, the Vermont legislature amended the Current-Use Tax Law in 1982 to put a cap on the per-acre reimbursement to towns. This amendment provided that the maximum per-acre reimbursement could not exceed 5 percent of the highest use value established by the State Current Use Advisory Board for tillable agricultural land.

Another political problem in evaluating the Current Use program has been in identifying its fundamental aim. There is a question whether the law should benefit needy landowners, provide a tax benefit to all landowners who qualify regardless of need, or seek to tax land simply on the basis of its use value regardless of who owns it. The Current Use Tax Law's statement of purpose states that its intent is to tax land simply on the basis of its use value, regardless of who owns it. In practice, however, the present property-tax system is not structured so that land can be taxed directly on the basis of its use value. Rather, a tax bill is presently divided into the "state's share" and the "landowner's share." This practice creates the illusion that those qualifying for the program are receiving a positive benefit that other taxpayers do not realize and hence are receiving a form of tax break. If the situation is perceived as one in which only certain individuals are allowed to receive this break, the political issue of who deserves it is raised. This issue has been especially difficult in circumstances where some very wealthy individuals are the owners of Vermont forest land.

Despite these questions, the results described above indicate that Vermont's early experiences with the Current-Use Tax Law generally have been positive. Officials of the Current Use Advisory Board feel the law is helping to keep agriculturally productive land in production. Like many farmland-protection programs, the problems appear to be practical and political, and the primary issues are related to costs and process—how much it will cost to keep the land in its current use, the importance of local planning, and how to apply the program without creating public misperceptions while fulfilling its basic objectives.

Other Tax Programs. Several years ago, most Vermont farmers paid little in the way of estate and gift taxes when passing their land on to succeeding generations. With the incredible upward spiral of Vermont land values, however, even medium-sized farms have come to represent a significant value as part of a family estate. As estate taxes have increased, many family farm operations have been required to sell a portion of their land to meet this obligation when the farm is passed to a succeeding generation. Although it is unclear how much such estate taxes have actually encouraged the conversion of Vermont agricultural land to development, in recent years concern has grown to reduce taxes which might cause otherwise viable agricultural parcels to be divided.

At the federal level, this concern has grown as well, and in response, Congress passed the federal Tax Reform Act of 1976 and the Economic Recovery Act of 1981. These acts reduced estate, death, and gift tax burdens as they relate, among other things, to family farms. Moreover, Congress provided specifically for use-value appraisal under certain conditions,[86] for estate valuation, and allowed the tax liability to be paid over a number of years.[87] In 1980, the Vermont legislature followed the lead of the federal government and altered its estate tax laws to benefit farmers; the Vermont estate tax would be equal to the federal credit for state death taxes. The Vermont gift tax was also eliminated at that time. As a result of these estate tax changes, most farmland owners can achieve significant savings on their estate taxes, particularly on family-owned and -operated farms.[88]

The Vermont Homeowners Rebate Program[89] also helps farmers, although it does not specifically address them. Under the law, farmer-homeowners are entitled to an income tax rebate if their local property taxes exceed a certain amount of their adjusted gross annual income. Therefore, a farmer who holds valuable lands but has a low income for a given year is given an income tax reduction to help his operations remain economically viable. Vermont is the only state in New England with this so-called circuit-breaker legislation.

In summary, the various tax incentive programs are all of some benefit to farmers, although they achieve their effect through different means. The tax-stabilization contract program, as amended in 1977, has provided substantial property-tax reduction to those owning or leasing agricultural, forest, or open space land. The major drawback of this program, however, has been the loss of revenue to the local community. The land-gains tax, which does not directly aid farmers, serves as a short-term complementary land use control to Act 250 to deter land speculation. Considered to be Vermont's first program designed specifically to protect agricultural and forest land, the current-use tax permits farmland or forest landowners to apply to have their land appraised at its actual use value rather than its fair-market value. Unlike the tax stabilization program, there are reimbursement provisions to the Current-Use Tax Law; the amount appropriated for these reimbursements, however, has been insufficient to cover lost revenues. In general, this law has primarily benefited forest landowners, but state officials feel the law is also helping to maintain productive farmland. In addition to these programs, changes in estate and other gift tax laws have helped keep family farming economically viable. Finally, the Homeowners Rebate Program allows an income tax rebate for farmers whose local property taxes exceed a certain percentage of their adjusted gross annual income. The overall effect of this range of tax programs has been generally positive in farmland protection efforts even though Vermont officials believe that a more uniform taxation system would be more effective.

Purchase of Development Rights (1969)

In addition to the above mentioned tax incentive programs, farmland protection advocates sought to broaden the range of programs by initiating a local purchase-of-development-rights (PDR) program. In1969, the Vermont legislature adopted legislation providing for the purchase of development rights in order to protect agricultural, forest, and other undeveloped land. The law's express purpose is:

> to encourage and assist the maintenance of the present uses of Vermont's agricultural, forest, and other undeveloped land and to prevent the accelerated residential and commercial development thereof; to preserve and to enhance Vermont's scenic natural resources; to strengthen the base of the recreation industry and to increase employment, income, business, and investment; and to enable the citizens of Vermont to plan its orderly growth in the face of increasing development pressures in the interests of the public health, safety, and welfare.[90]

The law seeks to accomplish these objectives by allowing landowners to sell, donate, or otherwise transfer any or all interests in their property to a municipality or a selected state agency. Among the interests that may be acquired by the state or local government are: "Fee simple, subject to right of occupancy and use, which may be defined as full and complete title subject only to a right of occupancy and use of the subject real property or part thereof by the grantor for residential or agricultural purposes."[91] As originally enacted, the interests could be acquired for only the number of years specified in the individual agreement. A 1975 amendment, however, provides that the rights and interests in land may now be conveyed "in perpetuity."[92]

Unlike similar PDR programs in Massachusetts and Connecticut, the Vermont program was not specifically established to protect farmland. Indeed, when it was adopted in 1969, farmland protection was not the issue of great concern that it later became. Hence the 1969 legislature failed to include the Vermont Department of Agriculture as one of the state agencies which could accept rights and interests in land under the PDR program. In 1983, however, farmland advocates successfully convinced the legislature to amend the law to allow the Department of Agriculture to acquire agricultural land interests.

The major incentive for a landowner to sell development rights under the law is not so much the revenue derived from the sale of the rights as the savings realized in property taxes. The law provides that after public acquisition of a right of interest in the property, the owner of any remaining right or interest must be taxed only upon the value of those remaining rights and interests.[93] Thus, the legislation, which was adopted at the height of the so-called development crisis discussed earlier, allows farmers and others to sell the development rights that so drastically have increased land values in certain areas of the state. As with the PDR programs in the other New England states, owners of farmland and other open space land then pay a substantially reduced property tax on the basis of substantially reduced property value.

Despite the great concern generated by Vermont's development crisis in the early 1970s, the Vermont PDR program was seldom used by state agencies during the entire decade following its adoption. The central problem, as in other states, was cost. In 1980, a $2 million funding bill was introduced into Vermont's State Agriculture Committee, but strong support was lacking and the legislation was never adopted.[94] The bill was designed not only to fund development rights purchases by state agencies, but also to fund methods of enhancing farming productivity and marketing. Efforts to establish permanent state funding for the purchase of farmland development rights appear unlikely.

The law can also be used at the local level if a city or town wishes to purchase the development rights to farm or forest land. However, until 1982, no city or town had used the law to protect farmland. In March 1982, Vernon, a small agricultural community in the Connecticut River Valley near Brattleboro in southern Vermont, became the first town to establish a fund for the purchase of farmland development rights under the 1969 enabling legislation. This action made Vernon one of the few towns with a PDR program in the county.

Of Vermont's 247 cities and towns, Vernon is among the smallest in land area, with a population of less than 1,200 residents. The town possesses some of the most fertile agricultural land in the state, with a large percentage of the town's land area devoted to agricultural uses. Vernon is thus one of the most rural towns in Vermont. It is also the home of the Vermont Yankee nuclear power station, which generates in excess of $2,000,000 per year in tax revenues for the town. With a small local population demanding little in the way of public services, the Vermont Yankee power station generates a significant volume of excess discretionary revenue for the town. In March 1982, the town appropriated $50,000 of this revenue as a fund for the purchase of local farmland development rights.[95]

To implement the local program, the town sought the assistance of the private, nonprofit Ottaquechee Land Trust (OLT), based in Woodstock, Vermont. Due to the legal complexities of operating a PDR program, Vernon was unprepared to administer the program on its own at the local level. The OLT, which is the leading private land trust in the state, had gained considerable experience in the late 1970s and early 1980s in agricultural land protection efforts and was able to provide the necessary administrative assistance to the town.

In June 1983, the OLT assisted the town in its first purchase.[96] Eighty-six cropland acres of a 200-acre farm were purchased for $40,000, the land's full value being over $120,000. In four years the town will resell the property to a young farm couple at the purchase price with a development restriction on the land. In the interim, the town will receive $100 a month in rent from the couple, but no property taxes. The purchase is considered successful because the farm, formerly owned by an older farmer needing money to retire a bank loan, was under considerable development pressure. Assuming that farming remains economically viable, a major parcel of Vernon farmland will now be protected in perpetuity.

The Vernon example underscores the problems of high cost and limited applicability of PDR programs as a method of farmland protection. Given the high land values which the land development market can generate, PDR programs are nearly impossible to finance without a major funding source, which is usually outside the capabilities of farmers and farm communities. Another problem with the Vermont PDR program is that there is no agency or individual, at the state level, with the responsibility to promote, explain, and facilitate the acquisition of farmland development rights. In consequence, some feel that PDR participation has suffered because of a lack of knowledge by farmers of the PDR program or how it could protect farmland. Until a steady source of funding is available, and the administration of the program is facilitated, major participation in the PDR program will be unlikely.

Farmland protection advocates also began to initiate zoning innovations in the early 1980s in order to shift the emphasis away from Act 250. The Vermont Planning and Development Act, adopted in 1968, enabled such zoning mechanisms. There are conflicting views on the impact of zoning techniques on farmland protection in Vermont. Following is a description of the zoning practices in Vermont as they relate to farmland protection.

Zoning

The concern over the increasing loss of productive farmland has stimulated a new interest in zoning as a means of slowing this trend. The Vermont Planning and Development Act authorized planning at the local and regional levels and empowered localities to create exclusively agricultural or rural zones. The purpose of the Act, in part, is to provide for sufficient open space in appropriate locations, to enable property-tax mitigation, and to protect agricultural uses from high-density uses.[97]

Before a zoning regulation can be adopted, the municipality must first adopt a master plan approved by the community. The zoning regulation is the bylaw that carries out the plan. Over 200 of Vermont's 246 cities and towns currently have planning and about 190 have zoning bylaws.[98] However, some towns have plans but have not adopted zoning regulations because they rely primarily on Act 250 to control out-of-state development.

At this time, no Vermont towns have adopted exclusive agricultural zoning regulations. This is largely true because effective agricultural zoning needs the support of both farming and nonfarming citizens. Although local proposals to create exclusive agricultural zones have been considered lately, farm owners are suspicious of restrictive regulations that could depreciate their land value or prohibit its future sale and accordingly do not support such proposals.

In Vermont, the agricultural and rural zones that were created are nonexclusive and allow residential uses on lots of twenty-five acres or more. A town can make the zone more permissive,[99] however, by amending its zoning regulation and revising its master plan. Moreover, individual landowners can apply for higher-density development in such nonexclusive agricultural zones.

The nonexclusive zoning regulations have been criticized for not being coercive enough. A publication of the USDA Extension Service at the University of Vermont pointed out:

> Because zoning regulations can be amended as growth pressure builds or a change of personnel serving on Planning Commissions or Town Boards can shift priorities within a community, a zoning regulation cannot be considered a permanent solution to protecting the land from development.[100]

The Municipal Planning and Development Act was later amended (1983) to accommodate the objectives of the Agricultural Lands Task Force Report. The Task Force identified the need to inject agricultural considerations into the local public planning process and to establish a system whereby farm landowners can voluntarily commit their land to agricultural uses. The Report and its recommendations are discussed more fully in a later section of this chapter.

Agricultural zoning is a relatively new concept, and indeed there are only a handful of exclusive agricultural zones in the country. At this time, however, zoning in Vermont seems to have an uncertain future because of the contrasting views on its merits. As a result, certain other local efforts—

such as the use of private land trusts—are becoming more popular in Vermont.

Private Efforts—Land Trusts

Land trusts, as private, nonprofit corporations, can take advantage of existing tax deductions allowed for charitable donations of land (or land interests) to achieve farmland protection objectives while satisfying the financial objectives of a landowner. Using the same techniques available to the private sector, a trust can gain control of the land (or its rights) from a landowner who might otherwise choose to sell for development. Hence, a landowner can donate, or sell at a reduced price, land or the development rights through the activities of a trust. In this way, the land can remain in agriculture permanently.

Private land trusts are proving quite effective in farmland protection efforts in many states, and Vermont is no exception. Indeed, some observers in Vermont believe that the efforts of a few private land trusts in the state have shown the most effective results in farmland protection in recent years, given the ineffectiveness of Act 250 and the subsequent short experience with other state programs.

One of the most active land trusts in Vermont is the previously mentioned OLT, which uses private action combined with various public programs to protect Vermont farm and timberland. According to Richard Carbin, executive director of the OLT, the trust began in 1977 with the voluntary efforts of members of the Ottaquechee Regional Planning Commission, which was seeking to find alternative approaches to the protection of land, particularly farmland:

> The Commission was particularly concerned about the failure of the zoning process and other regulatory methods to protect farmland and potentially productive timber land. From their explorations of private approaches to land use, the Commissioners concluded that no matter how distasteful the idea might be to conservationists, land in the American economic and social system is treated not as a resource but as a commodity. The tradition of private property rights in this country is so strong that even when the argument "of the public good" is used to affect individual land use decisions, there will always be strong emotional opposition. Any attempt to conserve land as a resource must recognize and respond to this basic American ethic.[101]

The OLT was born in response to this land ethic. To accomplish its objectives, the trust uses the same techniques available to the private sector to gain control over land to be protected from development, including the right of first refusal, traditional financing, and purchase option agreements, among others. The Trust is a private, nonprofit corporation and can therefore utilize tax laws governing charitable contributions of property to achieve its land conservation objectives and also meet the financial objectives of a landowner who might otherwise choose to sell the land for development. Bargain sales, charitable contributions of land or development rights, donations of conservation easements, or other less than fee-simple land transactions have all been accomplished through the activities of the OLT to the benefit of landowners and conservationists alike.[102]

During the first two and one-half years of its existence, until December 1980, the Trust operated on a total annual budget of $1,000. A separate, voluntary board of directors had been established, but the trust relied heavily on volunteer labor and assistance from the Ottaquechee Regional Planning and Development Commission. Initially, the activities of the Trust focused on the protection of any farmland; efforts soon became more specifically focused on *productive* farm and forest land.

By 1980, the workload of the Trust had grown to such a point that a full-time professional staff was required. By the end of 1982, the OLT had completed twenty-five farmland protection projects totalling approximately 8,000 acres of land. Roughly 30 percent of this acreage is cropland or pastureland, with most of the remainder in forest. Most of this forest land is contained in existing farm units, as the trust has not focused specifically on the protection of forest land. At the beginning of 1983, the OLT had twenty to thirty additional projects in progress, including the previously mentioned efforts in the town of Vernon. These projects successfully placed an additional 8,000 to 10,000 acres of farm, forest, and other critical resource lands under permanent protection.

Although the Trust operates on a statewide basis, most of its activity and support have been focused in the more heavily urbanized areas of the state around Burlington, the Champlain Valley, and southern Windham County, where development pressures on farmland have been the greatest. The Trust has been least active in the northeast corner of the state. This is largely attributed to the lower pressures for development and more conservative local attitudes which often view such innovative financial and legal mechanisms with misunderstanding and distrust.

The OLT is now the principal private land trust in Vermont, focusing specifically on the protection of agricultural land. Two other active trusts in Vermont are the Lake Champlain Islands Trust and the Vermont Chapter of the Nature Conservancy. Both have supported farmland protection efforts, but their emphasis is different: the former focuses on the general protection of Lake Champlain's islands, and the latter on natural and scenic lands. Thus, their results in farmland protection are very limited.

The experience of the OLT in Vermont illustrates how a well-planned private program can utilize existing public incentives to achieve significant results in farmland protection efforts. It is noteworthy to compare the results of the OLT with the Massachusetts APR program over the same period. As discussed in Chapter 3, by December 1984 the state-run Massachusetts APR program had placed restrictions on 9,800 acres at a total cost to the Commonwealth of over $15 million. In contrast, over the same period the OLT had succeeded in protecting 20,000 acres of Vermont farm and forest land without requiring new laws or legislative appropriations. The techniques utilized by the OLT and other private land trusts do involve costs to the public in lost tax revenues, but the true costs are never calculated, and such hidden costs appear to be much more acceptable to the legislature of the state (and to Congress) than the direct appropriation of large sums for outright purchase by the state. Such private efforts, combined with careful local planning, are proving to be effective tools of farmland protection in Vermont.

Recent Agricultural Land Protection Programs

As the concern for farmland protection has grown in Vermont, a range of new legislative programs has been adopted in an effort not only to protect agricultural land directly but also to address the problem of farming's economic viability. In the 1970s, the Vermont legislature established two programs, among others, which had only a marginal effect on the retention of active agriculture. First, in 1975, the legislature proposed an Agricultural Development Division within the Department of Agriculture to provide marketing assistance for native producers. Designed to promote a permanent and integrated agricultural economy in Vermont, the program fell short with respect to implementation guidelines.[103] Then in 1977, enabling legislation for Conservation Commissions was enacted.[104] This statute was similar to those passed in the other New England states in the 1960s. Of greater significance to farmland protection and agriculture, however, are the programs established at the beginning of the 1980s.

The three major initiatives of the 1980s include the Governor's Executive Order 52 and two farmland-related pieces of legislation. In 1980, Governor Richard A. Snelling took specific action to protect farmland by signing Executive Order 52, which mandates that all state agencies do not take actions that needlessly cause the conversion of important agricultural land in the state if there is a reasonable alternative. Moreover, each agency is required to specify policies and actions to minimize the impact of development on agriculture.

In the 1981 legislative session, two important new pieces of farmland-related legislation adopted were a right-to-farm law, which protects reasonable farming activities from nuisance lawsuits, and an Institutional Marketing Law, which requires state agencies to purchase Vermont farm products if there is a dependable supply and if they are of comparable quality and price to those produced outside the state. These two programs are not specific land protection measures, and yet both laws were adopted in response to the ever-growing concern over the conversion of agricultural land and a desire to improve the economic viability of farming to slow conversion.

Executive Order 52

On September 25, 1980, Governor Snelling issued Executive Order 52.[105] The order directs all state agencies not to "eliminate or significantly interfere with or jeopardize the continuation of agriculture on productive agricultural lands or reduce the agricultural potential on primary agricultural soils unless there is no feasible and prudent alternative and the facility or service has been planned to minimize its effect on such lands."[106] Although the governors of several states recently have issued similar executive orders, the Vermont order is unique in describing a detailed planning process for agencies to follow to meet this general goal. Under the Order, all state agencies whose actions may affect farmland were required to consult with the State Planning Office and prepare a report stating their policy on farmland protection. This task was completed by January 1, 1981.

The Executive Order also created an Agricultural Lands Review Board. The Board is composed of the heads of five major state agencies, including the Commissioner of Agriculture, who is the chair; the Director of the State Planning office; the Secretary of Environmental Conservation; the Secretary of Transportation; and the Secretary of the Agency of Development and Community Affairs. The State Planning Office is directed to provide staff and administrative support to the Review Board. The Board meets only at the request of the governor to review any proposed actions by state agencies that have a significant impact on productive agricultural lands or primary agricultural soils. The Board makes recommendations within fifteen days of the governor's request.

By 1983, the Agricultural Review Board had met twice,[107] once to review an industrial development project and once to review a new hydroelectric project. In the first case, a Canadian company sought to build an industrial development which would require financial assistance from the state. Following review of the project by the Board, the developer agreed to develop only a portion of the 113-acre site, building primarily on the nonagricultural soils. A five-year lease to a farmer would be negotiated on the remaining agricultural land. In this case, the Board felt that these mitigating measures were adequate.

In the second instance, the Board met to review the Saxtons River Hydroelectric Project. The case was particularly complex because recent state policy supports small-scale hydroelectric projects as alternative sources of energy. The Saxtons River project would result in the flooding of approximately ninety acres of prime farmland located within the hundred-year floodplain of the Saxtons River. The project would require a farmer to move from the floodplain, which has virtually no other pressures for conversion since urban development in the hundred-year floodplain is not allowed. Unlike the industrial development, no on-site mitigating measures were possible since the entire area would be flooded. As an alternative, the developers of the project agreed to purchase a nearby farm with an equal number of similar-quality acres and move the farmer at their expense. The Board found this arrangement satisfactory even though farmland protection was not the end result. Indeed, because the land was high-quality prime farmland in a floodplain area where conversion pressures are virtually nonexistent, the net loss of farmland was doubly troublesome.

Despite the marginal benefits of these examples, most observers agree that Executive Order 52 has been beneficial in farmland protection efforts. According to Robert Wagner, Land Use Consultant of the Vermont Department of Agriculture, the order has been a success because consideration of agricultural lands is now a part of state agency planning processes.[108] As a result, several ad hoc planning and mediation processes have taken place between the Department of Agriculture and various state agencies in an attempt to negotiate and resolve agricultural conflicts without convening the Board. In two counties, this has resulted in comprehensive Industrial Site Surveys, designed to address agricultural land protection. This type of preplanning will help assure farmland protection before expensive development conflicts arise.

To make Executive Order 52 more effective, legislation (S.B. 13) was introduced in 1983 to make it a state law. An important provision of the proposed legislation would have allowed the Commissioner of Agriculture as well as the governor to convene the Board. This would have allowed the Board to evaluate state agency projects even if a governor chose not to use the review procedure. Although the legislation was not adopted in 1983, it was expected to be reintroduced in 1986.

Right-to-Farm Law (1981)

Following a growing trend among the New England states, in May 1981, the Vermont legislature adopted a Right-to-Farm Law.[109] Farmers in the state supported the legislation because, as spreading urbanization approached farmland areas, farmers were increasingly under threat of nuisance suits. Noise, dust, odors, and other externalities of farming often conflict with the quiet residential communities desired by many city and suburban dwellers moving to the countryside. Prior to the Right-to-Farm Law in Vermont, newcomers could sue farmers for routine farming operations, sometimes forcing the farmers to stop or alter their farming practices. Accordingly, the Vermont legislature said that nuisance suits can "encourage and even force the premature removal of the lands from agricultural use."[110]

The legislative purpose of Vermont's Right-to-Farm Law is to "protect reasonable agricultural activities conducted on farmland from nuisance lawsuits."[111] The statute accomplishes this goal by creating a rebuttable presumption that agricultural activities conducted on farmland are not a nuisance if they are consistent with good agricultural practices and were established before other, more recent activities. If the farming activity is conducted in accordance with current federal, state, and local laws and regulations, there is a presumption that it is a proper agricultural activity. Farming activity can be considered a nuisance if it can be proven that the agricultural activity has a "substantial adverse effect on the public health and safety."[112]

What constitutes reasonable farming activity is not explicitly defined in the statute. The Vermont legislation leaves this issue for the courts to decide, as other states with similar laws have also done. Also, as in other states, another potential problem with the Vermont statute is that state and local health boards still have the power to abate nuisances.[113] Therefore, what may be a reasonable farming practice in a rural area can become a public nuisance (i.e. public health hazard) simply through urbanization. Nevertheless, the law does provide an important incentive for farming by providing public policy protection for farmers and the recognition that farming operations often involve undesirable activities that should be allowed to continue.

Institutional Marketing Law (1981)

Like the Right-to-Farm Law, other recent programs which help to protect agricultural land in Vermont seek to do so by making farming more profitable either by removing difficult restrictions on farming or by easing financing and marketing problems. The Institutional Marketing Law, adopted in 1981, is such a program. It seeks to assist the direct marketing of Vermont-produced agricultural products by requiring the purchasing director of any state agency to purchase products grown or produced in Vermont when available and when they meet quality standards established by the Commissioner of Agriculture,[114] assuming other considerations, such as price, are equal. In this respect, the Institutional Marketing Law in Vermont differs from those in Massachusetts and Maine where state agencies are authorized to spend more for local products.

Although this program by itself can do little to improve the economic viability of farming in Vermont, like similar efforts in other New England states, it is viewed as one part of a wide range of direct interventions by state agencies to assist in maintaining a viable farm economy. Farmland protection advocates are investigating other methods of direct and indirect state assistance which encourage the economic viability of farming.[115] Legislation utilizing a variety of methods for improving the viability of farming in Vermont have been introduced into every legislative session since 1980. However, despite great concern over the problems facing Vermont farmers, no major new proposal specifically aimed at farmland protection had been adopted by the end of the 1985 legislative session.

Pending Agricultural Land Protection Programs

In recent legislative sessions, a broad array of agricultural development programs have been considered. Nearly all of the proposed legislation addresses the underlying economic difficulties of farming in Vermont rather than specifically addressing the protection of farmland. To date, most of the proposals have not had the necessary broad support of farm organizations, the Agriculture Department, environmental groups, and general citizenry necessary for implementation. The most important of the proposals are outlined below.

Vermont Agricultural Development Authority

This proposal seeks to create an Agricultural Development Authority, similar to the Vermont Industrial Development Authority, with the power to issue bonds for the purpose of providing grants and loans for agricultural development projects. The new Authority also would be authorized to acquire and resell or lease farmland and facilities which are threatened by development to nonfarming uses. A related proposal would make agricultural enterprises, excluding the acquisition of land, eligible for financing by industrial development revenue bonds which would be issued by the Vermont Industrial Development Authority. This proposal effectively could achieve the same result as the Agricultural Development Authority without requiring the creation of a new agricultural authority.

Family-Farm Security Act

This proposal was modelled after a program established by Minnesota's Family Farm Security Act. If adopted, the program would allow credit-worthy individuals who have net total assets of less than $50,000 and who wish to own and operate a farm to obtain low-interest, state-guaranteed loans. The program would provide state funds to subsidize interest payments and guarantee repayment of the loan if the borrower defaulted. The loans would be approved by the Commissioner of Agriculture and a five-person advisory council consisting of farmers, nonfarmers, agricultural economists, and officers from participating lending institutions.

Agricultural Districts

As in nearly all of the New England states, agriculture protection proponents in Vermont have sought in recent years to adopt a statewide program of agricultural districts similar to legislation adopted in New York state in the early 1970s. Like the New York program, proposals in Vermont are intended to give farmers sufficient incentive to voluntarily identify areas where farming would be the favored activity to be protected against economic considerations which might make it impracticable, unprofitable, or undesirable to farm.[116] Despite the popularity of this approach among farmland protection advocates, however, passing legislation to create agricultural districts in Vermont has proven difficult, since it is easily perceived to be a type of exclusive agricultural zoning which is not favored by landowners. As with the similar Agriculture Incentive Areas Act in Massachusetts, Vermont officials believe this is a misconception.

Conservation of Farmland Resolution

During the 1981 legislative session, all of the above proposals, as well as legislation to close Act 250's ten-acre loophole, were debated in an effort to deal with the farmland-conversion issue. Only two, the Right-to-Farm Law and the Institutional Marketing Law, were adopted. Confusion over which way to proceed caused a bill to be introduced into the 1981 legislative session requesting that a task force be formed to study the farmland conversion problem comprehensively.

Recognizing the complexity of the issue and the difficulty of finding a solution, Agriculture Commissioner George Dunsmore appointed an Agricultural Lands Task Force (ALTF) and a technical backup committee. The ALTF's objective was to assemble a comprehensive agricultural enhancement package for consideration in the 1983 Vermont legislative session. It had forty members, including agricultural leaders, farmers, legislators, and members of the planning, development, and financial community. A ten-member Technical Committee did much of the legwork and prepared many of the proposals for review and adoption. As an initial step, in January 1982, the ALTF submitted to the legislature, and won approval of, a widely supported Conservation of Farmland Resolution,[117] known as Joint Resolution 43, which formalized their legislative assignment, calling for a full report on the farmland protection issue.

The resolution begins by identifying Vermont agriculture as the "major contributor to the economy of the state and the region" and declares the state's farmland to be a "unique and irreplaceable resource, whose conservation is essential to present and future sustained agricultural activities and of great benefit to the welfare of the people of the state of Vermont." To this end, the resolution has the following general provisions:

1. Encourages the efforts of both public agencies and private organizations to protect and maintain open agricultural land within Vermont for present and future use.
2. Requests that the ALTF develop a report describing the loss of farmland in Vermont and possible responses for consideration by the governor and 1983 General Assembly.
3. Encourages federal, state, and local cooperation in efforts to conserve Vermont farmland.[118]

Following approval by the legislature of Joint Resolution 43, the ALTF proceeded to collect and analyze all available data on the location, extent, past and present uses, and reasons for change in the use of Vermont's agricultural land. As a result of this effort, two interim reports were published which analyzed the extent of primary agricultural soils in Vermont and changes in the Vermont landscape.[119] The ALTF also reviewed existing farmland protection programs in Vermont as well as the experience of other states in farmland protection.

In February 1983, the ALTF submitted its recommendations to the Vermont legislature in a short and concise report entitled *The Farmland Issue in Vermont—Findings and Recommendations.* The recommendations of the ALTF were simple and to the point, with the objective of creating a package of existing state programs and new legislation that would provide a "comprehensive approach to farmland protection at the local level."[120] To accomplish this goal, two important steps were identified: "the need to inject agricultural considerations into the local public planning process and the establishment of a system by which farmer-landowners can, by their own choice, commit land to agricultural uses." The approach to these two steps recommended by the ALTF combines initiatives to encourage local town planning to protect farmland with state-supported incentives, technical assistance, and guidelines to encourage farmland protection by individual landowners.

To encourage local town planning to protect farmland, the ALTF recommended legislation that would require or strongly encourage towns to develop local farmland protection plans. This would be mandated by an amendment to the Vermont Planning and Development Act, such that each town in the state would be required to:

1. Review the agricultural resources of the town and identify those lands considered to be of local agricultural importance.
2. Develop a land protection program to provide for the present and future use of these agricultural lands.
3. Prepare a plan that outlines areas of the town that may qualify as important agricultural areas and details the development plans and a protection program for these lands. This plan will be submitted to a newly created Agricultural Lands Protection Board (ALPB) for review and approval.[121]

Following approval of the plan, farm landowners may enter into land-commitment contracts in accordance with the land-protection program. The signing of such a contract between the farmer-landowner and the local board of selectmen "triggers" eligibility for various economic incentives which would be provided to the farmer-landowner by the state. The incentives recommended by the ALTF were considered to be the very least necessary to ensure farmer-landowner involvement in a protection program. Indeed, there was concern among many members of the ALTF that the recommended incentives were too weak to be appealing to Vermont farmers, particularly considering the cost-price squeeze which dairy farmers have felt in recent years.

Incentives for landowners were recommended at two levels, depending upon the degree of commitment of farmer-landowners to the program. At a minimum, differential tax assessment, such as use-value appraisal, would be provided to farmland owners entering into a temporary agreement. More permanent land-protection commitments, such as allowances for cluster development, granting of conservation restrictions, or property-rights transfers, would allow participating landowners to receive additional financial and technical benefits, including farm financing assistance and assistance from a newly created Vermont Farmland Protection Fund. Farm financing assistance would be available through two recommended new programs, a Guaranteed Low Interest Loan Program and a new Agricultural Development Authority. These recommended new programs are similar in their provisions to the Family-Farm Security Act and the Vermont Agricultural Development Authority, which had been proposed but not adopted in the 1981 legislative session. The ALTF recommendations thus seek to make these two programs part of a coordinated local land protection package.

The ALTF also recommended a special new fund. The Vermont Farmland Protection Fund would provide flexible financial assistance to farmers and farmland conservation organizations by providing a revolving loan program to assist, for example, in quick purchases of farmland and financing for bargain sales. In addition, in certain cases, the program would allow for direct purchase of farmland or rights in farmland at times when conversion to other than agricultural uses is likely. The Vermont Farmland Protection Fund thus could be useful in assisting the efforts of land trusts such as the OLT. The ALTF recommended the establishment of such a fund for use only on farmland considered to be of statewide importance by the Agricultural Land Protection Board, who would also be responsible for administration of the fund.

The ALTF thus has recommended a coordinated range of programs to encourage farm landowners to sign local land protection contracts. The two levels of landowner incentives are summarized in Table 4.3.

The ALTF also recommended that incentives be provided to encourage towns to plan carefully for agriculture. The primary incentive would be reimbursement of the cost of the agricultural planning effort (i.e., "seed money"), provided by the Agricultural Lands Protection Board on completion and approval of the required local agricultural protection plan. In addition, state agencies would provide technical assistance to towns in the preparation of the plan.

In February 1983 the ALTF's recommendations were submitted to the legislature as required by the Conservation of Farmland Resolution of 1981. It is unlikely, however, that the entire package of recommendations will be adopted in a single legislative session. Legislative members of the ALTF believed that some of the programs, such as the provisions for town agricultural planning, could be introduced and adopted quickly and inexpensively. Some of the more costly incentive programs, however, probably will be a subject of considerable debate, particularly in light of huge budget deficits facing the state in the mid-1980s. The incentive programs thus will probably require refinement and consideration in future legislative sessions.

If the Vermont legislature does amend the Vermont Planning and Development Act to require local agriculture protection plans, perhaps the most important of the ALTF's recommendations will be in place. Local planning for farmland protection will then be mandatory. Although this does not guarantee that farmers will enroll in protection plans proposed by the towns, at the minimum a process of local consideration of agriculture will begin.

As the farmland protection experience from Vermont and other New England states indicates, the most effective farmland protection programs appear to be those that have been initiated and developed at the local level. The Vermont experience is particularly illustrative of the difficulty of

Table 4.3
Individual (Farmer-Landowner) Incentives Recommended by the Vermont Agricultural Lands Task Force (February 1983)

Level I Contract (minimum)

1. *Differential assessment for qualifying land*
For land retained in agriculture, under a first-level contract, assessment or appraisal for property tax purposes will be based on the value of such land for production of food or fiber crops. The intent of this incentive is to change the basis of property tax appraisal or assessment to reflect the continued agricultural use of such land, and to effect a similar property tax burden on the farmland owner.

Level II Contract (additional conditions)

1. *Differential assessment for qualifying land*
The method of land appraisal, administration procedures, length of commitment, shift of local tax (revenue) burden, and comprehensive effect of such programs may be structured to increase the impact of providing property tax relief. At least, such statutes and programs should be clarified to local communities and landowners, to identify or detail statewide program objectives and reduce existing deficiencies.
2. *Farm financing assistance programs*
A. Guaranteed Low Interest Loan Program. This program is aimed to aid eligible farmers (particularly young, beginning farmers) in the acquisition of farm real estate. This aid will be in the form of state money in guarantee of loans.
B. Agricultural Development Authority. The Agricultural Development Authority will provide low interest loans and grants for farmers and/or operators of agricultural enterprises. The purpose of this element would be to provide increased opportunities for the production, marketing, and processing of agricultural products.
i. Operating Loan Fund. This operatng fund will be used by the authority as a nonlapsing fund for assisting start-up farmers or operators of agricultural enterprises that demonstrate certain needs.
ii. Development Fund. This development loan or grant fund would be available to farmers or operators of agricultural enterprises to finance the establishment or expansion of agricultural marketing, processing, or reprocessing facilities.
3. *Farmland Protection Fund*
The purpose of the Fund would be to provide flexible financial assistance to farmers and farmland conservation organizations to assure the permanent protection of Vermont's most important agricultural soils. The Fund would be available to assist in the purchase of farmland or rights in farmland at such times when conversion to other than agricultural uses is most likely: when a farm is for sale, when a farm has failed, upon the death of a farmer, etc. The Fund would not, however, be strictly limited to a revolving loan program. There should be enough flexibility built into the program to allow for purchase of rights of first refusal, options, outright acquisition, and purchase of development rights.

relying primarily on comprehensive statewide regulatory programs to achieve farmland protection. After relying on Act 250 as a statewide vehicle of farmland protection for over a decade, land conservation advocates in Vermont have begun to realize that such comprehensive planning is nearly impossible to accomplish without the support of the towns. The Vermont Agricultural Land Task Force recommendations thus acknowledge the primary importance of local planning for farmland protection and seek to focus future legislative efforts at this level.

Future of Agricultural Land Preservation in Vermont

For years, the Vermont farmer—like farmers throughout New England—has found it increasingly difficult to compete with the farms of the western United States. This trend was becoming apparent even as early as 1937, a time when Vermont could still grow certain crops relatively efficiently. Charles Crane, a writer of the time, noted this efficiency:

> Though the so-called corn belt of the country is in the West, Vermont is far ahead of most of the country in the amount it can grow to the acre. In the past decade, Vermont has averaged better than forty bushels of corn to the acre, with only two states ahead. When Vermont grew practically all of its own wheat, as it did a century ago, the state stood third in the crop. Though wheat is not a minor crop in Vermont, I understand we can grow an average of twenty bushels to the acre (one year an average of twenty-nine), the average for the United States being only 12.9.[122]

Technology and efficiency of the midwestern farms were increasing, however, and economic conditions for the hill farmer of Vermont were deteriorating. For a while the slack was absorbed by the state's dairy industry, which was able to serve the urban centers of the East and thus was able to compete with the West by virtue more of proximity than of efficiency. Reporting on the situation in 1932, Charles Crane referred to the "dominance of the 'milk train'" and proudly declared that "[e]very night while we Vermonters sleep, some eighty carloads of Vermont milk and cream are carried to Boston and New York consumers."[123] By the early 1960s, however, technology again had overtaken the Vermont farmer. New developments in the storage and transportation of dairy products removed the edge that had helped support the industry in Vermont, as dairy products from large dairy regions began to influence the entire U.S. market. Declining farm economic conditions thus began to result in the conversion of thousands of acres of Vermont farmland to forests or urban development.

Whether the farmland protection efforts discussed in this chapter will be successful in Vermont is perhaps more difficult to determine than for any other New England state. Determinants far outside the control of Vermont farmers or the Vermont legislature may continue to have more significant implications for Vermont farming than any of the state-initiated programs. The difficult economic conditions nationwide facing the dairy industry do not bode well for farming in Vermont, which depended upon dairy operations for 82 percent of all farm cash receipts in 1981. Because Vermont is so heavily dependent upon a single farm commodity, recent huge surpluses of dairy prices nationwide and resulting cuts in federal price suports can have a significant impact on Vermont farm economics. This cycle, in turn, will make it more difficult than ever before to maintain Vermont farmland in active production.

Understandably, pressures will be placed on farm landowners to sell their land for development or conversion to other uses if the opportunity is presented. In such times as these, Vermont farmers will be the last to support farmland protection programs if they find the programs reduce the value of their land, which may be their only source of income for the future. Thus, the future of agricultural land protection efforts in Vermont looks uncertain at best. As Don Hooper, assistant director of the Vermont Natural Resources Council and member of the advisory Technical Committee of the ATLF, commented:

> Farmland preservation programs are going to be harder to get in Vermont in coming years. In the past you could look at the rest of New England with its heavy urbanization and loss of farmland and say "we don't want to be like them." But with dairy prices going down as they are, nobody knows quite what to do about it.[124]

Despite uncertainty about how to deal with the farmland conversion problem in Vermont, important advances have been made. At first glance, Vermont's comprehensive approach to land use control embodied in Act 250 would appear to be the most logical means of assuring a balance of land uses in the state and thereby protecting agricultural lands. Indeed, at the time of its adoption, Vermont was hailed as a leader in positive land use controls to protect the quality of its physical environment. The protection of farmland was seen as one of the future benefits that would be derived from Act 250's comprehensive approach.

Experience has illustrated otherwise, however. Although such a comprehensive approach to land use planning may be the most ideal method of assuring the protection of agricultural lands, as the Vermont experience indicates, such an approach proved to be politically impossible. By depending so heavily upon Act 250 as a farmland protection device, Vermont found itself several years behind leading states such as Massachusetts in farmland protection efforts. This is not to say that Vermont citizens and legislature were slower to express their concern over dwindling farmland resources, but only that Vermont probably to date has pursued the less effective course in addressing those concerns.

Despite the poor results through the 1970s, farmland protection efforts have made a tremendous advance in Vermont in the early 1980s. Two facts may soon make Vermont a leader in farmland protection among New England states. First, Vermont is actively seeking to initiate a broad range of programs including regulations and financial incentives to protect its farmland resources. In a relatively short period since 1980, several new programs and private actions have developed. Second, Vermont is the first New England state to seek legislation for farmland protection planning actively at the local level, where it will do the most good. If such local planning legislation is adopted, and the prognosis is good, many of the political difficulties faced by the statewide Act 250 may be overcome and farmland protection may develop a more grassroots, home-rule approach.

Notes

1. V.S.A. 10, §600. et seq.
2. *The Farmland Issue in Vermont: Findings and Recommendations, Report of the Agricultural Lands Task Force* (Montpelier: Department of Agriculture, February 1983).
3. *1978 Census of Agriculture*, 481.
4. *1880 Census of Agriculture*. Figures include grand total of all land in farms including improved cropland and unimproved woodland.
5. *Primary Agricultural Soils and Vermont Agriculture* (Burlington: Soil Conservation Service, USDA; Montpelier: Vermont Department of Agriculture, 1982): 8.
6. *The Changing Vermont Landscape: A Resource Inventory Report* (Burlington: Soil Conservation Service, USDA; Montpelier: Vermont Department of Agriculture, October 1982): 9.
7. Ibid., 23.
8. Ibid., 9, 23.
9. *AIP Study*, 47.
10. *1978 Census of Agriculture*, 504.
11. Steve Kerr, *Farmland: Keeping Developers Off It versus Keeping Farmers On It, Vermont Environmental Report* (Montpelier: Vermont Natural Resources Council, November/December 1980): 3.
12. Peirce, *The New England States*, 244.
13. Phyllis Myers, *So Goes Vermont* (Washington, D.C.: The Conservation Foundation, February 1974): 9.
14. *AIP Study*, 47.
15. "Vermont Leads N.E. in Self-Employment," *Boston Globe*, November 21, 1982, x.
16. *1978 Census of Agriculture*, 123. Twenty-one percent of farms were owned by white-collar workers, 11 percent by retired persons, and 7 percent by blue-collar workers. Corporate and foreign owners controlled less than 2 percent of the land, but recent stories of foreign owners buying Green Mountain farms have fueled, in part, a concern to control development.
17. David G. Heeter, "Almost Getting It Together in Vermont," in Daniel R. Mandelker, *Environmental and Land Controls Legislation* (New York: The Bobbs-Merrill Company, Inc., 1976): 325.
18. U.S. Census. It should be noted that most of the population growth was due to a natural population increase. Estimates are that immigration accounted for an increase of only 15,000 persons during the ten-year period from 1960 to 1970. See also Heeter, "Almost Getting it Together in Vermont," 326.

19. Robert G. Healy, *Land Use and the States* (Baltimore: The Johns Hopkins University Press, 1976): 36.

20. Ibid., 37.

21. Meyers, *So Goes Vermont*, 10–11.

22. Concern over proper sewage disposal eventually played a central role in transforming Haystack Corporation, which had received over $9 million in mortgage rights to Wilmington property that was to support 2,200 housing units, into an insolvent shell. This concern was also directly responsible for halting the development contemplated by International Paper mentioned earlier: the 20,000 acres on which the proposed development was to be built were located on land with a thin layer of topsoil over bedrock that was insufficient for septic systems. Governor Deane G. Davis eventually personally persuaded the corporation not to proceed with its plans. Davis's actions at this time were instrumental in stimulating public demand for the 1970 legislation to enact land use controls. *The Collapse of a Development*, 48–55.

23. Meyers, *So Goes Vermont*, vii.

24. Healy, *Land Use and the States*, 37.

25. Officials of the Vermont Division of Property Valuation and Review point out that increases in the fair-market value of land will not automatically result in a property tax increase. Property taxes increase only if (a) the town must raise more money, or (b) the value of one type of property increased at a faster rate than the value of another type of property (e.g., if the value of land increased faster than that of buildings, then owners of land would pay a larger percentage of taxes in the town than they had previously).

26. Healy, *Land Use and the States*, 36.

27. The state legislature expressed some of these concerns in Vermont's 1981 Right-to-Farm Law: "Agricultural production is a major contributor to the state's economy; agricultural lands constitute unique and irreplaceable resources of statewide importance, that the continuation of agricultural activities preserves the landscape and environmental resources of the state, contributes to the increase in tourism, and furthers the economic self-sufficiency of the people of the state; and that the encouragement, development, improvement, and preservation of agriculture will result in a general benefit to the health and welfare of the people of the state (V.S.A. 12, §5751).

28. V.S.A. 10, §600 et seq.

29. Heeter, "Almost Getting it Together in Vermont," 329.

30. Ibid., 331.

31. Under the provisions of Act 250, proposed developments falling into the following specified categories must apply to the appropriate local district commission for a permit to develop. Each commission consists of three laypeople appointed by the Governor. Developments requiring a permit are those that fit at least one of the following descriptions:

   a. Developments involving the construction of housing projects and the construction or maintenance of mobile homes or parks of ten or more units;
   b. Developments involving the construction of improvements for commercial or industrial purposes on a tract of more than one acre in towns without permanent zoning and subdivision bylaws and on a tract of more than ten acres in towns with such controls;
   c. Developments involving the construction of improvements for state or municipal purposes of a size of more then ten acres;
   d. All developments above an elevation of 2,500 feet. V.S.A. 10, §6001(3).

32. A 1973 amendment allowed a first appeal to go to a county court rather than the environmental board. V.S.A. 10, §6089.

33. Act 250's ten original regulatory criteria require that an application:

   1. Will not result in undue water or air pollution...
   2. Does have sufficient water available for the reasonably foreseeable needs of the subdivision or development;
   3. Will not cause an unreasonable burden on an existing water supply, if one is to be utilized;
   4. Will not cause unreasonable soil erosion or reduction in the capacity of the land to hold water so that a dangerous or unhealthy condition may result;

5. Will not cause unreasonable congestion or unsafe conditions with respect to the use of the highways, waterways, railways, airports and airways, and other means of transportation existing or proposed;
6. Will not cause an unreasonable burden on the ability of a municipality to provide educational services;
7. Will not place an unreasonable burden on the ability of the local governments to provide municipal or governmental services;
8. Will not have an undue adverse effect on the scenic or natural beauty of the area, aesthetics, historic sites or rare and irreplaceable natural areas...
9. is in conformance with a duly adopted development plan, land use plan or land capability plan...
10. is in conformance with any duly adopted local or regional plan or capital program... V.S.A. 10, §6086(a).

34. The Land Capability and Development Plan supplemented the original ten criteria under Act 250 with several additional criteria and requirements. Under Criterion 9, which requied that a development must be in conformance with a duly adopted land capability plan, several specific natural resource, energy conservation, and capital planning criteria were added. Criterion 9-B identifies primary agricultural soils for consideration in the granting of a developing permit. Criterion 9-B states:

Primary Agricultural soils. A permit will be granted for the development or subdivision of primary agricultural soils only when it is demonstrated by the applicant that, in addition to all other applicable criteria, either, the subdivision or development will not significantly reduce the agricultural potential of the primary agricultural soils; or,

1. the applicant can realize a reasonable return on the fair market value of his land only by devoting the primary agricultural soils to uses which will significantly reduce their agricultural potential; and
2. there are no nonagricultural or secondary agricultural soils owned or controlled by the applicant which are reasonably suited to the purpose; and
3. the subdivision or development has been planned to minimize the reduction of agricultural potential by providing for reasonable rates of growth, and the use of cluster planning and new community planning designed to economize on the cost of roads, utilities and land usages; and
4. the development or subdivision will not significantly interfere with or jeopardize the continuation of agriculture or forestry on adjoining lands or reduce their agricultural or forestry potential. V.S.A. 10, §6086(a) (9) (B).

The criteria for secondary and forest lands are essentially the same as the first three of the above critiera. The burden of proof varies depending on the criteria. The burden is on the applicant to establish that the development will conform to the criteria pertaining to agricultural lands if it is shown that the proposed development involves primary agricultural soils and the development significantly will reduce agricultural potential in the soils.

35. V.S.A. 10, §6043.

36. Healy, *Land Use and the States*, 51.

37. Ibid., 54.

38. Ibid., 50 (quoting Senator Arthur Gibb).

39. Vermont's Land Use Plan and Act 250 (Montpelier: State Planning Office, January 1974): 10.

40. Don Hooper, *Looking Through the Loophole, Vermont Environmental Report* (Montpelier: Vermont Natural Resources Council, March–April 1982): 2.

41. Indeed, this practice can most easily occur in those areas where, arguably, the state most wants to keep land in agriculture. In regions where agriculture is common and economically feasible, land values have not been affected by development pressures to as great an extent as have those in more highly urbanized areas where the value of one large piece of land is much less than the value of that same land when it is subdivided into smaller plots. Consequently, a developer in agricultural regions has relatively little to lose and much to gain by building on large acreages to evade the Act 250 requirements.

42. "Loophole Causes Rash of Vermont Developments," *Boston Sunday Globe*, September 17, 1972, 45.

43. Hooper, *Looking Through the Loophole*, 2.

44. "Loophole Causes Rash of Vermont Developments," 45.

45. *Vermont Natural Resources Council Bulletin*, 1.

46. Hooper, *Looking Through the Loophole*, 2.

47. Efforts to close the loophole in the 1981–1982 legislative session were not successful and amendments were made to delete the ten-acre exclusion provision to achieve passage of the civil penalties provision of the bill. Members of the House Natural Resources Committee felt that the provisions for civil penalties were more important and would not be approved by the full House unless the ten-acre exemption were deleted. The ten-acre exclusion provisions and the civil penalties provisions were reintroduced into the 1983 legislative sessions as two separate bills; the ten-acre exclusion provision passed, while the civil penalties provision did not. House Bill 81 (1983 session); House Bill 82 (1983 session).

48. Richard Cowart, comment, August 7, 1981.

49. Driebeek and Lenz, *Act 250 and the Preservation of Agricultural Soils*, 11.

50. Some developers, for instance, have employed a "bootstrapping" agrument at the commission level: they have argued that a fair return on the value of their land is the value that would be realized through development, when arguably their land has no development value at all until the district environmental commission has granted the Act 250 permit. The commissions have had difficulty dealing with this and similar arguments. Richard Cowart, comment, August 7, 1981.

51. Richard Cowart, comment, August 7, 1981.

52. V.S.A. 10, §6001(15):

> "Primary Agricultural Soils" means soils which have a potential for growing food and forage crops, are sufficiently well drained to allow sowing and harvesting with mechanized equipment, are well supplied with plant nutrients or highly responsive to the use of fertilizer, and have few limitations for cultivation or limitations which may be easily overcome. In order to qualify as primary agricultural soils, the average slope of the land containing such soils does not exceed 15 percent, and such land is of a size capable of supporting or contributing to an economic agricultural operation. If a tract of land includes other than primary agricultural soils, only the primary agricultural soils shall be affected by criteria relating specifically to such soils.

53. John Howland, state senator, remarks, Vermont Law School's Environmental Law Center Act 250 Conference (May 13–14, 1983), in response to Richard W. Carbin, *Act 250 and Agriculture, The Windsor File*.

54. Richard Cowart, comment, August 7, 1981. The Environmental Board recently conducted a quick, informal survey of all of the permit requests that had been considered during a six-month period from late 1979 through early 1980, at around the time when agricultural criteria were just beginning to be applied on a consistent basis. It found that Criterion 9-B was a consideration in from 10 to 15 percent of the cases. The Board divided those cases in which the agricultural land criteria were important to the decision into four categories: "avoidance," "denial," "hope," and "mitigation." There were roughly ten cases falling into each of the first three categories. The avoidance category was established for those cases in which the district commissions were still failing to apply properly the 9-B criterion. There were good policy reasons in some of these cases for the commission decisions, but nonetheless, they were not permissible under the statute. The denial category was established for those cases in which permits had actually been refused on the basis of the agricultural land criteria. The hope category included those cases in which, strictly speaking, the developer's permit might have been denied under Act 250, but the district commission was willing to accept the developer's representation that most of the agricultural land at issue would remain in farming.

The largest percentage of cases fell into the mitigation category. These cases were those in which the district commission attempted to accommodate the interests of both the developers and those who desired to preserve agricultural land. One mitigating factor that the district commissions found especially persuasive was the establishment of homeowners' associations. These associations are bound by covenants, one term of which requires that a specific acreage of common land be kept in active agriculture. One basic problem with ordinary subdivisions is that, even if a number of adjacent lots come together still have an acreage large enough to render farming economically feasible, high transaction costs prevent a farmer from entering into lease arrangements with each individual landowner. The association eliminates this problem: the farmer can bargain with a single body. Thus, the homeowners' association facilitates both development and the continued agricultural use of productive soil.

55. Richard Cowart, comment, August 7, 1981.

56. Ken Senecal, *The Act 250 Farmland Protection Myth, Vermont Environmental Report* (Montpelier: Vermont Natural Resources Council, March/April 1981): 5.

57. Ibid.

58. Ibid.

59. Comments by John Howland, state senator, and Richard W. Carbin, executive director, ORLT, Vermont Law School's Environmental Law Center Act 250 Conference (May 13–14, 1983).

60. Thomas Vickery, comment, Lincoln Seminar 1.

61. *State Programs*, 60–61.

62. Healy, *Land Use and the States*, 56.

63. V.S.A. 24, §2741.

64. These contracts can allow: (a) fixing and maintaining the valuation in the grand list; (b) fixing and maintaining the rate or rates of tax applicable; (c) fixing the amount of annual tax; or (d) fixing the tax applicable as a percentage of the annual tax. V.S.A. 24, §2741(a).

65. V.S.A. 32, §3846.

66. *Town of Cambridge* v. *Bassett*, 142 Vt. 171 (1982).

67. *Farmland Retention Techniques*, Brieflet 1332-B, 1.

68. Robert L. Townsend, *Tax Stabilization of Farm, Forest, and Open Space Property in Vermont* (Burlington: University of Vermont, The Extension Service, 1980): 21.

69. Ibid., 1.

70. V.S.A. 32, §10001 et seq.

71. For a more complete discussion of the purposes of this law, see R. Lisle Baker, *Taxing Speculative Land Gains: The Vermont Experience* (Cambridge, Mass.: Lincoln Institute of Land Policy, Tax Policy Roundtable No. 5, 1980): 4; Thomas L. Daniels, *Occasional Paper No. Ten*, "Land Gains Taxation: The Vermont Case" (Burlington: University of Vermont, Center for Research on Vermont, 1986).

72. V.S.A. 32, §10003.

73. Baker, *Taxing Speculative Land Gains*, 5.

74. *Andrews* v. *Lathrop*, 132 Vt. 256, 259, 315 A.2d 860 (1974). The Court found that "the tax places a burden on short-term ownership and on high profits in the resale of lands, two attributes of property closely linked to the holding of lands for speculative purposes."

75. V.S.A. 32, §10002(i).

76. This discussion abstracted from Baker, *Taxing Speculative Land Gains*, 7–10.

77. V.S.A. 124, §3751.

79. V.S.A. 124, §7352.

80. Deborah Brighton, comment, August 10, 1981.

81. *Farmland Retention Techniques*, Brieflet 1332-B, 1.

82. Deborah Brighton, Vermont Current Use Advisory Board, comment, Lincoln Seminar 2.

83. Deborah Brighton, comments, August 10, 1981, January 13, 1983, and August 9, 1985.

84. Ibid.

85. Deborah Brighton, comment, Lincoln Seminar 2.

86. 26 United States Code (U.S.C.) §2032A.

87. 26 U.S.C. §6166.

88. *Farmland Retention Techniques*, Brieflet 1332-E, 2.

89. V.S.A. 32, §5961 et seq.

90. V.S.A. 10, §6301.

91. V.S.A. 10, §6303(a)(2).

92. V.S.A. 10, §6308(b).

93. V.S.A. 10, §6306(b).

94. "Agriculture Fund Rejected," *Herald Day Morning,* March 12, 1980, x; "State Could Buy Development Rights Under Bill Approved by Senate Panel," *Herald Day Morning,* February 28, 1980), x.

95. Robert Wagner, Agricultural Land Resource Consultant, Vermont Department of Agriculture, comment, February 1983.

96. "Vermont Town Discovers a Way to Save the Farm," *Boston Sunday Globe,* July 3, 1983, 19.

97. V.S.A. 24, §4302(a)(1), (3), and (5).

98. *Farmland Retention Techniques,* Brieflet 1332-B, 2.

99. *In re Zoning Permit of Patch,* 140 Vt. 158, 174 (1981) (involving Town of Wallingford).

100. *Farmland Retention Techniques,* Brieflet, 2.

101. Richard W. Carbin, *The Ottauquechee Regional Land Trust, Vermont Environmental Report* (Montpelier: Vermont Natural Resources Council, November/December 1980): 5.

102. Prior to the ORLT efforts to help establish the Vermont PDR program, probably one of the most important Trust projects was the protection of the Shelburne Farm in Shelburne, Vermont, one of the most unique farm properties in the state. Established in 1886 by William Seward Webb and Lila Vanderbilt Webb, the original farm estate contained approximately 3,800 acres of high-quality farmland on the shores of Lake Champlain. A landscape master plan and forest management plan were prepared by Frederick Law Olmsted and Gifford Pinchot, and state-of-the-art agricultural technology was employed. Noted architects of the period were consulted in the design of several notable buildings on the property. Indeed, for fifteen years, between 1890 and 1905, Shelburne Farm was considered to be a model of the agricultural estates.

In the 1940s, Derick Webb, grandson of the farm's founder, inherited 1,700 acres of cropland and woodland containing the core facilities of Shelburne House, the farm barn, and the coach barn. As farming in Vermont became less and less economically viable and land values began to rise, Shelburne Farm began to experience financial pressures for subdivision and development. To avoid this conclusion, in 1972, Shelburne Farms Resources (SFR) was established as a nonprofit educational corporation. It was funded by the sale of an adjacent parcel to the Nature Conservancy, and Derick Webb conveyed to SFR the three major buildings remaining on the property. Today, SFR conducts a wide array of educational and cultural activites and promotes an integrated use of the farm's assets by teaching land management techniques, experimenting with new crops, and creative forms of crop marketing such as direct milk and cheese sales, an on-farm bakery, etc. Although SFR controlled the buildings, the agricultural and forest land remained in private family ownership.

By 1979, however, debt obligations on the land threatened the continued existence of the farm. The owners of the property decided it was time to retire the existing mortgage on the property and make long-range plans for the permanent protection of the farmland resources at Shelburne Farms. Negotiations began with the Ottaquechee Regional Land Trust (ORLT), which was instrumental in these efforts. The owners of the 1,700-acre farm decided to sell approximately 600 acres on the periphery of the property that were not essential to the farming operation. However, because the land was adjacent to the farm and highly visible, the ORLT was asked to help find a buyer whose plans for the property would be consistent with the farm's setting and objective.

In March 1982, an agreement was signed with a group of five partners who planned to construct eight to sixteen permanent and second home dwellings on approximately 100 acres of the 600-acre property. As part of the agreement, permanent conservation restrictions on the remaining 500 acres would then be conveyed by the five partners to the ORLT, who will resell the restricted land at agricultural use value to one or more farmers who could not otherwise afford the high price of farmland in Shelburne. The remaining 1,100 acres of land and existing farm operations will be owned by a new corporation, Shelburne Farms, Inc. The stock in this new corporation will gradually be transferred to the ORLT to be held in perpetuity. A long-term lease agreement will allow the nonprofit SFR continued access to the property for its educational and cultural programs.

Darby Bradley, legal counsel to the ORLT, outlined the important role which the private trust was able to play in preserving the agricultural resources of Shelburne Farms: "Future tax savings from the charitable gift of conservation restrictions and the sale of land for agriculture and limited development allowed the buyers to pay a higher initial price for the 600 acres than a conventional developer could justify. This, in turn, has given Shelburne Farms more capital with which to continue its operations. . . .Through imaginative use of land planning concepts and income and estate tax planning, the Ottanquechee Trust helped find a solution which meets the needs of the seller, the objectives of the buyers, and the concerns of the public." Darby Bradley, *Conservation and Limited Development Will Preserve Shelburne Farms, Vermont Environmental Report* (Montpelier: Vermont Natural Resources Council, March/April 1982): 6. For a complete history of Shelburne Farms see William C. Lipke, ed., *Shelburne Farms: The History of an Agricultural Estate* (Burlington: Robert Hull Fleming Museum, University of Vermont, 1979).

103. V.S.A. 6, §2961 et seq., Natural Organic Farms Association Legislative Committee to Speaker of the House, February 5, 1979.

104. V.S.A. 24, §4501 et seq.

105. Executive Order 52 of Governor Richard A. Snelling, September 25, 1980.

106. Ibid., 1.

107. Robert Wagner, comment, February 1983.

108. Ibid.

109. V.S.A. 12, §5751 et seq.

110. V.S.A. 12, §5751.

111. Ibid.

112. Ibid.

113. V.S.A. 12, §5753(b).

114. V.S.A. 6, §4601.

115. Steve Kerr, Director of Agricultural Development for the Vermont Department of Agriculture, commented on the role of state institutions in assisting these efforts:

> The goals of agricultural development are (1) to increase opportunities for farmers, (2) to increase farmers' incomes, and (3) to increase the number of variety of jobs in agriculture. There are a number of ways to do this. The Agriculture Department is looking into cooperative storage, regional processing facilities, and cooperative marketing and transportation to make farming more efficient and to reduce costs. If Vermont farmers could store large enough quantitites of grain, they could take advantage of lower rates for large volume shipments. Piggy-back milk hauling could reduce transportation costs, but we also need to improve transportation networks. We need to decide which roads and bridges are most critical and concentrate on maintaining and improving them.

Kerr, *Farmland: Keeping Developers Off It Versus Keeping Farms On It*, 3.

116. Although the final version of an agricultural district program is probably years away, the following provisions have been considered in Vermont:

District Definition. The definition of a district has caused some confusion and controversy in Vermont. Since the state is so largely agricultural, the Vermont Farm Bureau believes the entire state should be designated as agricultural district. Even farmland advocates point out this is impractical in the development-prone areas of the state. More feasible proposals specify that for agricultural districts to be formed, landowners within the district must own at least 250 acres or 10 percent of the land proposed for the district, whichever is greater. In addition to size, such factors as the viability of agriculture in the district, soil types, surrounding land uses, and the availability of surrounding markets and related services must also be considered.

Benefits to Farm Landowners. Incentives to be provided landowners for voluntarily establishing agricultural districts are also a source of controversy. Don Hooper, assistant director of the Vermont Natural Resources Council, believes that the major flaw in recent agricultural district proposals is a lack of incentives to the landowner. Certain provisions common to agricultural legislation in most states, such as freedom from nuisance suits or limitation on development projects undertaken with state funds, are already in place in Vermont or are under consideration as separate programs. Provisions which limit special assessments and other tax incentives of the program are not considered strong enough to induce widespread use of agricultural districts considering the range of tax programs already available to farmers.

Despite the lack of consensus on agricultural districting in Vermont, most proponents of farmland preservation in the state see a well-designed program of agricultural districts playing an important role in encouraging viable agriculture in the state. As in several of the New England states, agricultural districts in Vermont are likely to be one of the more important farmland protection proposals on the agenda of legislative sessions in the early 1980s.

117. Joint Resolution of the House 43 (1982 session).

118. This clause has been referred to as the "Land Trust Clause," giving legislative blessing to the farmland protection efforts of private land trusts in the state.

119. *Primary Agricultural Soils and Vermont Agriculture* and *The Changing Vermont Landscape: A Resource Inventory Report* (Burlington: USDA Soil Conservation Service and Vermont Department of Agriculture, University of Vermont, 1982).

120. *The Farmland Issue in Vermont: Findings and Recommendations, Report of the Agricultural Lands Task Force* (Montpelier: Department of Agriculture, February, 1983): 7 (emphasis added).

121. The Agricultural Lands Protection Board should be composed of "Representatives from the Department of Agriculture, the farming community, the Office of State Planning and financial interest. In addition..., individual communities should be represented by the Supervisor of the local Natural Resources Conservation District and the Chairperson of the local Regional Planning Commission for the specific region of the State" under consideration. Ibid., 11.

122. Charles E. Crane, *Let Me Show You Vermont* (New York: Alfred A. Knopf, 1937): 144.

123. Ibid., 143.

124. Don Hooper, Vermont Natural Resources Council, comment, April 12, 1982.

5

Connecticut Programs for the Protection of Agricultural Land

For most travelers, Connecticut is thought to be the state that bridges New England with the rest of the nation. In fact, the state is both a bridge and a destination. Although New York City encroaches from the southwest, numerous old colonial towns and seafaring communities remain largely untouched by modern development. Indeed, most of the state is wooded, with the Berkshire Hills and the varied seacoast an unmistakable part of New England. Severing the state is the Connecticut River, which winds through Hartford and enriches thousands of acres of the most productive farmland in New England.

This state has always maintained a commitment to agriculture. During the Revolutionary War, Connecticut was named the "provision state."[1] The state still operates the oldest agricultural experiment station in the country. Like its southern New England neighbors, Connecticut became heavily industrialized during the nineteenth and early twentieth centuries, but its citizens long have sought to maintain a balance in its urban-rural landscape. In fact, Connecticut has long been one of the states in the forefront of conservation efforts specifically targeted to agricultural lands.

Food production has been the basis for Connecticut's agricultural land protection programs. Such a tangible focus is easy for the public to support because it presents an objective basis for determining how much land is needed for agriculture. The core of Connecticut's program has been a purchase of development rights (PDR) program, which combines effective local planning with the state role. Under such a program, municipalities identify and establish zones of fertile agricultural land and the state then selects and purchases the development rights on certain of these parcels. The PDR program is expensive to operate, however, and other incentives and devices recently have been employed to delay the conversion of agricultural land to other uses until more appropriations become available for PDR programs. As a complement to the PDR program, these incentives and devices form the basis for a more coordinated range of agricultural land protection programs.

Public Concern to Protect Agricultural Land

Public awareness of the importance of agricultural land in Connecticut arose initially in response to the post–World War II decline in farming. Spreading urbanization resulted in an undesirable decrease in local food production, often had a negative effect on the physical environment, and began to erode the small-town, rural lifestyle which was valued so highly by many of the state's citizens.

In 1976, journalist Neal Peirce described the allure of the Connecticut countryside and captured the feelings of many people in the state who lament the recent conversion of rural and farmland acreage:

> Connecticut farmland is some of the most pleasing in America to behold. One thinks of the rich bottomlands of the Connecticut River Valley and, in the hillier sections, farms with their stone walls and barns that provide such a pleasing backdrop to the wooded areas and bring natural old New England charm to the center of the East Coast megalopolis. Thus it is sad to report that the number of Connecticut farms has been dropping rapidly....In large part, of course, this reflected the pressure on farmland from commercial and industrial developers. Partly because of the decline in farming Connecticut has twice as much wooded area as in the late 19th century.[2]

Decline of Farmland Acreage

As in the other New England states, Connecticut citizens have witnessed a substantial decline in the number of farms and in farmland acreage since the late nineteenth century. According to the U.S. Census of Agriculture, in 1880, 78 percent of the state's acreage (2,453,541 acres) was farmland. By 1940 this figure had dropped to 49 percent (1,512,151 acres), and by 1969 it had dropped to 17 percent (541,372 acres). During the thirty-year period following World War II, farmland acreage underwent the most dramatic decrease. In 1945, Connecticut had approximately 22,000 farms on about 1,593,000 acres, while by 1974, there were only 3,400 farms on 440,000 acres.[3] Farmland conversion has continued in recent years such that by 1982

Table 5.1
Statewide Change in Acreage of Farmland and Number of Farms in Connecticut 1880–1982

Year	*Acres*	*Number of Farms*	*Change*	*% Change*	*% Change from 1880*	*Average % Change per Year*	*Average Farm Size (acres)*
1880	2,453,541	30,598	—	—	—	—	80.2
1910	2,186,000	26,815	–267,541	–10.9	–10.9	–0.4	81.5
1940	1,512,151	21,163	–673,849	–30.8	–38.3	–1.0	71.5
1950	1,272,352	15,615	–239,800	–15.9	–48.1	–1.6	81.5
1959	884,443	8,292	–387,909	–30.5	–64.0	–3.4	106.7
1969	541,372	4,490	–343,071	–38.8	–77.9	–3.9	120.6
1978*	455,731	3,519	–85,641	–15.8	–81.4	–1.8	129.5
1982	444,242	3,754	–11,489	–2.5	–81.9	–0.6	118.3

*The 1978 Agricultural Statistics were revised in 1982 to compare more accurately to the 1982 and previous U.S. Agricultural statistics.
Source: U.S. Bureau of the Census, Census of Agriculture.

only 14 percent of the state's land area was devoted to agriculture. Table 5.1 summarizes the decline in farmland acreage from 1880 to 1982.

Despite the decline in acreage, as with farming throughout New England, average per farm output increased dramatically over this period because of scientific advances. In the dairy industry, for example, ten times more milk was produced per Connecticut farm in 1980 than in 1940.[4] During the decades of the 1960s and 1970s, however, total statewide production of many farm products fell with the decline in farmland acreage despite continued advances in technology. Between 1967 and 1975, the overall production of vegetables fell 31 percent (45,250 to 31,350 tons); hay fell 18 percent (239 to 195 tons); milk fell 9 percent and milk fat fell 8 percent (cows from 67,000 to 55,000 and milk from 677 to 613 million pounds); and tobacco fell 29 percent (9,020 to 6,387 pounds). Conversely, corn for silage production was one of the few commodities to increase during this period, rising 11 percent (752 to 832 tons).[5]

Although most of what was formerly farmland is now in forest use, much farmland became developed in response to urban development pressures. Many farmers with small, inefficient farms ceased operating and sought off-farm employment, using their farms as a residence only. Frequently, the Connecticut farmland was sold for development, and this urbanization became especially common in the lower Connecticut River Valley and in coastal counties.[6] In other areas, the best cropland was rented by neighboring farmers while the remaining acreage reverted to forest.

With respect to the farming which remains, dairying accounts for 80 percent of the state's farmland in either forage-crop production or cropland pasture. Broilers and tobacco (largely for cigars), although once common in Connecticut agriculture, have been declining. Greenhouse and nursery operations, however, which greatly expanded in the 1970s, are more secure. Also secure is Hartford County's truck farming and fruit production industry. Although the state produces a high proportion of the milk, milk products, eggs, and sweet corn its citizens require,[7] in 1980, 78 percent of total food needs were supplied by other states.[8]

Official statistics for Connecticut agriculture for 1982 reveal that agriculture remains a significant industry in the state. The total annual cash sale of crops, livestock, and livestock products on 4,300 farms with 490,000 acres was $309,092,000.[9] By comparison, Connecticut placed behind Maine, Vermont, and Massachusetts in total marketing cash receipts. Approximately $200 million in related processing, supply, equipment, and marketing industries were generated by this production.[10] Figure 5.1 and Table 5.2 depict total production and cash receipts from farm marketing for 1982.

At the turn of the 1980s, some observers believed that the agricultural industry was making a modest comeback. Both small or part-time and large farms had grown in number and size.[11] Modern farm operators produced only the most profitable products, used the best lands, used advanced technology, and learned sound management skills to enable them to compete with Western farms.[12] One product that was not flourishing, however, was tobacco, a product which makes Connecticut agriculture somewhat unique. Although the state ranked ninth nationally in tobacco cash receipts in 1982,[13] "[f]uture production is uncertain because cigar sales are declining, production costs are up and the land the tobacco grows on is flat and nicely located for garden apartments, industrial parks and shopping malls."[14]

Many people were becoming concerned about the agricultural industry in general. Adding to this concern, Commissioner of Agriculture Vincent Majchier reported in October 1983 that "farming accounts for 13% of the land use. Indeed, up to 1975 Connecticut had been losing farmland at a faster rate than any other state."[15]

As in most areas of New England, public concern over the decline of farmland acreage is a relatively recent phenomenon. Prior to World War II, market forces determined the desired mix of land uses in Connecticut. With the tremendous economic growth of the postwar period, however, there emerged a growing concern among the state's citizens that intense development of the land and "urban sprawl" posed a genuine threat to agriculture and other "desirable" land uses. The Whyte Report, by William H. Whyte, discussed later in this chapter, was one of the first efforts which responded

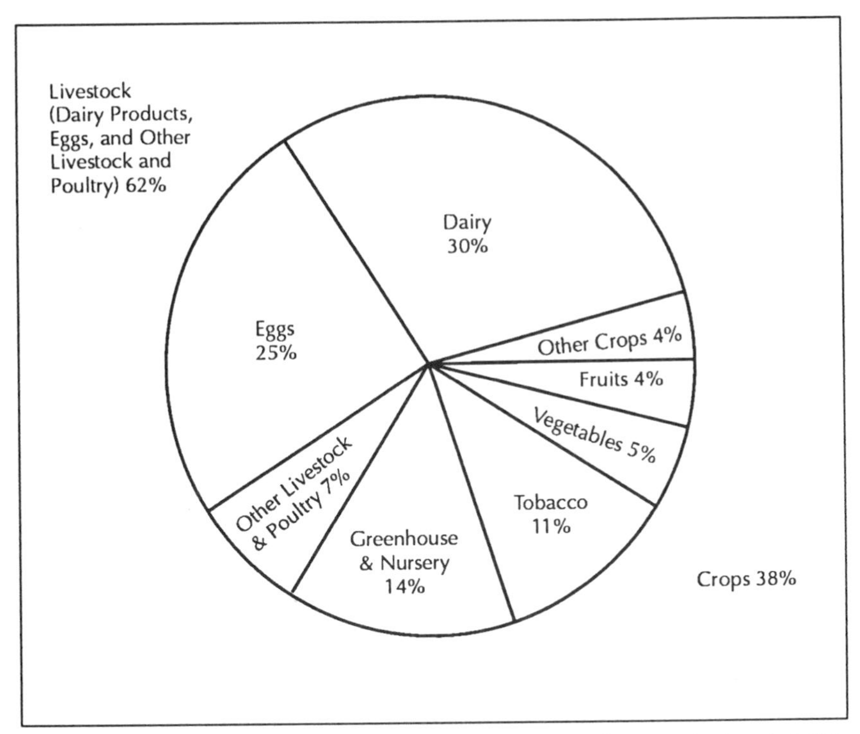

Figure 5.1. Distribution of cash receipts from Connecticut farm marketings in 1982. *Source:* New England Crop and Livestock Reporting Service.

to the problem. On the other hand, there were many people who did not view the development of agricultural land as undesirable. In 1975, for example, analysts Aubrey Birkelbach and Gregory Wassall at the University of Hartford responded to the Governor's 1974 Task Force Report on the Preservation of Agricultural Land by questioning whether Connecticut should try to protect agricultural land at all:

> The Task Force report assumes that free market forces cannot be allowed to determine the amount and location of acreage for food production. Specifically if we allow developers to continue acquisition of farm land, "...when we need the land to feed us, it will be irretrievably subdivided into lots and criss-crossed into paving." To be fair, they might have added that this land will also be converted into sites where our manufactured goods and services are produced and sold, and that these provide the bulk of income to our residents. Gross cash receipts from farm output per productive acre in 1973 were $700.00. Compare this to values added per acre in manufacturing in Connecticut, which is $342,703.00. Gross sales per acre from a regional shopping center average about $900,000.00, and these types of activities are often located on what was formerly farm land.[16]

Table 5.2
Connecticut Crops: Acreage, Yield, Production, and Value (1982)

Crops	*Acres Harvested*	*Yield per Acre*	*Total Production (1,000)*	*Value of Production*** *(Thousands of $)*
Corn for Silage	48,000	16.5	891 tons	24,681
Hay, all	89,000	2.27	202 tons	18,988
Potatoes	1,800	235	423 Cwt.	1,988
Tobacco, all	2,670	1,590	4,244 pounds	21,759
Apples, com'l*	—	—	1,310 42-lb crates	9,700
Peaches	—	—	48 48-lb crates	1,035
Pears	—	—	1.55 tons	775
Sweet Corn	3,200	63	202 Cwt.	2,727
State Total	144,670			81,653

Notes:
*Production is the quantity sold or utilized.
**Value relates to marketing season or crop year.
Source: New England Crop and Livestock Reporting Service, USDA, Concord, N.H, 1983.

Yet according to Irving Fellows, an agricultural economist at the University of Connecticut, little conflict between development and agricultural land protection need exist. There are over one million acres of open land on which development could take place, often with more satisfying results than the development of cropland areas. In fact, prime agricultural land constitutes less than 300,000 acres or 10 percent of the total land area in the state, and less than 100,000 acres of this land needs to be protected to meet recommended food production levels. Generally, therefore, there prevails a strong public concern that the better farmland areas should not be converted to development uses.[17] This concern over conversion of farmland evolved into a more serious public concern that the state's food production would become inadequate.

Concern over Decline in Food Production Potential

As agricultural acreage and the agricultural industry has declined in Connecticut, public concern has grown over the impaired ability of the state to supply its own food needs. As a result, numerous agricultural land protection programs adopted in the state consistently have recognized the importance of native food production. In fact, this concern was the major impetus for most of the agricultural land protection programs.

One of the state's earliest programs, the 1963 differential taxation law, carried with it an expression of purpose that read, in part, "it is in the public interest to encourage the preservation of farm land, forest land, and open space in order to maintain a readily available source of food and farm products close to the metropolitan areas of the state."[18] Similarly, in 1974, the charge to the Governor's Task Force for the Preservation of Agricultural Land explained that part of the desire to maintain agricultural land was attributable to a desire to provide for a nearby source of food to help meet the demands of a growing population.[19] Probably the most comprehensive statement indicating that the state legislature believed the protection of agricultural land would promote the production of food within the state is found in a draft bill entitled "An Act Concerning the Preservation of Agricultural Lands" introduced by the Committee on the Environment to the General Assembly during the January 1975 session:

> The general assembly finds and declares that the preservation of agricultural lands for the production of food is of paramount importance to the well-being of the people of Connecticut; that there is increasing national and worldwide demand for food and increasing potential for shortages; that Connecticut farmers annually produce and could continue to produce approximately one-third, by average, of the milk, eggs, poultry, fresh vegetables and apples consumed within the state; that the disappearance of agricultural land is the single largest threat to the preservation of food production in Connecticut, largely due to the disparity between the use value of land for agriculture and its sale value of other uses and to the impetus to convert farmland to other uses imposed by inheritance taxes; and that agricultural lands represent an integral part of Connecticut's economy and culture.[20]

In 1978, the legislature reiterated this position and the Purchase of Development Rights (PDR) Program was adopted with the intent "to maintain and preserve agricultural land for farming and food production."[21]

Efforts to help make farming economically viable, thereby encouraging agricultural production, were important to the public and their legislators because they feared that food would be unavailable and that prices would skyrocket. Dr. Irving Fellows and Patrick Cody, authors of the state's Food Plan, observed in 1979 that "there is an intuitive belief that complete dependency upon other areas for all the food consumed in Connecticut places the citizens in a dangerous and costly position." Local production was viewed as economically viable because Connecticut has a comparative advantage over western states in producing certain products. Iowa's corn and hog producers, for example, can produce milk cheaper than can Connecticut's dairy farmers, but since corn and hog production is more profitable in Iowa and because transportation costs have risen sharply in recent years, Connecticut farmers can compete for some markets.[22]

Some people, however, such as analysts Birckelbach and Wassell, believed that the effort to preserve the level of agricultural production in Connecticut was ill-advised, claiming that other parts of the country are better suited for farming, while "Connecticut has skilled manpower, a locational advantage and large markets for its industrial goods and its commerce."[23]

This was not the view adopted by the Connecticut legislature, however, as fostering the continued in-state production of a substantial portion of Connecticut's food needs has become one of the explicit aims of the state's agricultural land protection programs. This aim has been incorporated into the state's PDR program, which provides for the protection of some farmland in order for the state to produce a certain percentage of its food needs. Connecticut, therefore, has taken firm pragmatic steps to reach its food production goals.

Impact on the Physical Environment

Public concern over the need to protect environmental quality also has been a reason for the adoption of farmland protection programs. Perhaps more than any of the New England states, Connecticut specifically has identified the protection of basic environmental processes as a goal of farmland protection. Farms, forests, and open space lands are considered an essential part of the hydrologic cycle. Awareness has grown that these areas must exist to receive and store precipitation, to moderate evaporation and runoff, and to stimulate infiltration and the recharge of aquifers. As the population increases, agricultural areas will assume greater significance in the water cycle.

The realization that undeveloped open land is valuable for water management is another important reason behind the implementation of legislation and the protection of certain Connecticut open lands from development. The Connecticut Conservation Association observed in 1974 that agricultural land can play an important role in sewage treatment systems:

> As more sewage treatment plants are built, the disposal of their nutrient-laden effluent can be expected to add additional nitrogen and phosphorous to our lakes and streams, some already choked with weeds and algae. Tertiary treatment to remove these nutrients is expensive, but purification by filtration through a soil-crop filter removes nutrients cheaply, grows a useful crop, and recharges the ground water.[24]

Agricultural land is not, of course, the only important part of a water management system. It is estimated that half of the remaining undeveloped forest land in the highly urbanized southwest area of the state, stretching from Greenwich to New Haven, is owned by utilities who acquired these open watershed lands to protect the purity of the region's drinking water supply.[25] Indeed, the natural filtration provided by undeveloped land is seen as even more efficacious than that afforded by the systems that current technology can provide.

Although probably not as important a consideration as water management in the decision to implement farmland conservation or protection programs, the maintenance of air quality is regarded by many Connecticut citizens as a valuable contribution of such programs. The Connecticut Conservation Association has noted that the vegetation occupying undeveloped land can, for instance, remove such pollutants as ozone and sulfur dioxide from the air.[26] Moreover, crops or forests utilize carbon dioxide and release oxygen. Thus, an important goal of farmland protection programs is to provide and maintain a healthy environment.

Concern over Loss of Open Space and Community Character

Aside from the tangible aspects of environmental quality, the visual character of the state is a public concern that contributed to legislative action. As in the other New England states, the so-called amenity value of open space was probably one of the most important concerns of the Connecticut citizenry leading to the institution of programs to protect farmland.

With the postwar suburbanization boom, many citizens were concerned that unbridled development ultimately would destroy the beauty of the remaining undeveloped areas of the state. For example, in the mid-1970s when the New Haven Water Company announced its intention to sell some of its open space land surrounding water supplies for development, the residents of many communities were outraged—not so much out of concern for the drinking water supplies but rather out of a desire to protect the character of their surrounding landscape.[27] Like these forest lands, agricultural tracts were considered an essential component of the landscape. Farmland acreage decline has coincided with public efforts to protect farmland.

Agriculture's visual character also has encouraged greater tourism, an important state industry. In 1974, the Connecticut Conservation Association estimated that about 400,000 people will use the state's open space for sightseeing by the year 2000.[28] Much of the desire to tour the countryside is derived from the quality of farmland. Former Commissioner of Agriculture Leonard Krogh has summarized some of the aesthetic concerns significant to the state:

> Farmland provides an important contribution to the quality of life and environment. Agriculture assists in keeping Connecticut green. It provides breaks in the pattern of settlement from city to suburb, and from suburb to shore. It provides thousands of acres of cultivated, privately owned, tax-paying, open space.[29]

In summary, Connecticut has witnessed a substantial decline in its farmland resource base, with most one-time agricultural parcels now lying in forest lands. This decline prompted a major concern over the availability of foodstuffs in times of emergency and over the state's present and future self-sufficiency. In addition, the citizenry realized that well-located farmland helps improve air and water quality. Finally, the amenity value of agriculture, with its important contribution to tourism, also served to garner support for the implementation of Connecticut's farmland protection programs.

Conservation Programs and Early Efforts to Protect Agricultural Land

The first efforts to protect agricultural lands in Connecticut were initiated in the 1950s in response to the encroachment of urbanization on the countryside. Many groups helped foster support for governmental action. As a result, in 1955 the legislature passed legislation allowing localities to define farming broadly so that local zoning and tax efforts could protect agricultural land.[30] The same legislation also established a commission to study farmland retention in Connecticut.[31] The primary recommendation of the commission, a differential taxation law, was not immediately adopted by the legislature. Nevertheless, agricultural activity was viewed in the late 1950s as an important and distinct activity largely for aesthetic and environmental reasons. Throughout the 1960s, efforts grew to protect agricultural lands as one type of open space land use. Only recently, however, has agricultural land come to be viewed as an open space priority in the legislature.

As another part of the early efforts to retain land for agriculture, Conservation Commissions have come to play a vital role in gaining public support through educational efforts.

Conservation Commissions (1961)

Like the other New England states, Connecticut followed Massachusetts's lead in adopting Conservation Commission enabling legislation in 1961.[32] By 1983, these bodies were active in most of the state's 169 municipalities. Although not as active as in Massachusetts, these commissions have been vital in developing effective mechanisms to retain land for agriculture and in gaining support for agricultural land protection laws through educational efforts.[33] These commissions also have bolstered support for specific open space projects and are involved in the 1980s in the state's farmland purchase of development rights program.

The Whyte Report (1962)

At the executive level of state government, in 1962, Governor John M. Dempsey requested from the director of the Connecticut Development Commission and the Commissioner of Agriculture and Natural Resources a report with plans to protect the state's open space and natural resources. Author and conservationist William H. Whyte prepared the report, entitled "Connecticut's Natural Resources—A Proposal for Action."[34] This report proposed state-level acquisition of open space lands, grants to localities for such acquisitions, a use-value tax assessment law, and various water pollution control programs.[35] All of these recommendations resulted in legislation within a few years.

The Green Acres acquisition program has received several million dollars over the past twenty years. This funding allowed numerous parcels of open space to be purchased for active and passive recreation and for wildlife sanctuaries. It is important to note that many local Conservation Commissions were instrumental in these purchases.

It was hoped that these land acquisitions would encourage communities to consider the use of a variety of land use controls which would include the conservation of prime farmland.[36] Certainly, the acquisition program has had that effect indirectly, and it introduced the concept of state purchase which is employed in the purchase of farmland development rights.

More critical to early farmland protection efforts, however, was the state's use-value assessment law. Following is a discussion of that law's evolution.

Public Act 490 (1963)

The first major attempt by the legislature to slow the rate of agricultural land conversion was the consideration in 1956 of a bill which would have provided property tax relief for farmers. This bill initially was defeated when it became embroiled in partisan politics, but the 1963 legislature passed "An Act Concerning the Taxation and Preservation of Farm, Forest and Open Space Land," or Public Act 490 as it is more commonly known. This differential tax law was the first of its kind in New England.

This tax law was designed to prevent premature development of farmlands[37] and to delay the conversion or abandonment of farmland before a state land acquisition can occur. Act 490, like its sister statutes subsequently passed in other New England states, reduces the property tax burden on qualifying farmland owners by taxing their lands at agricultural use value rather than at potential development value. These lower assessments help make farming more economically viable. Supporters of Act 490 believe the program can make a substantial contribution to Connecticut's farmland protection efforts by providing an economic incentive to the farmland owner.

To receive tax benefits under the program, farmland owners must apply to their local assessor. Connecticut assessors have the greatest discretion among their counterparts in New England in determining what constitutes qualifying farmland. The statutory definition of farmland is "any tract or tracts of land, including woodland and wasteland, constituting a farm unit." The statute requires local assessors to consider several factors in determining whether a tract constitutes farmland,[38] including acreage, extent of farming operations, productivity, income derived from the land, type of farming equipment used, and whether farmland tracts are contiguous. In addition, extensive interpretive guidelines were recommended by the University of Connecticut Department of Agricultural Economics.

Despite these guidelines, as one might expect, wide differences in classification among towns exist. Some towns allow farmland classification for large gardens, while others require tracts to be large commercial farms.[39] The failure of a local assessor to classify land under the program entitles the landowner to an appeal, but as a practical matter, these appeals occur rarely because local assessors have wide discretion in administering the program. In disputes involving the statutory definition of farmland, the state courts have held that tilled nursery land is farmland, but land used for loam and gravel removal is not.[40]

Once a farmland tract is classified under the program, its valuation is also subject to wide local variations. Despite this discretion, local assessors do not operate in a vacuum: they have available the average use values of farmland throughout the state for various crops. Categories of major current crops indicate average agricultural productivity. Initially, these use values were issued by the Department of Agricultural Economics at the University of Connecticut, but they are now developed by the State Board of Assessment Advisors of the Office of Policy and Management. The recommended use values are obtained in one of two ways: (a) capitalization of the annual net rental value of a given parcel of land in a given commodity use; or (b) capitalization of annual net returns for a given parcel of land in a given commodity use. Of the two methods, the former is more highly recommended to assessors when sufficient market information is available, as it represents exactly that amount that a farmer is willing to pay for the opportunity to put land to a given use.[41]

Local assessors are under no obligation to apply state-established commodity-use values strictly, and some variation is expected. It is consistent with the aims of Act 490 that values are flexible to allow for adaptation to local conditions rather than to reflect rigidly the statweide averages. Accordingly, valuation criteria and assessments vary widely.[42] While actual use values assigned to a parcel can be high or low, depending on where the land in question is located, local assessors are encouraged not to vary the relationship among the recommended use values for the various categories of land established by the state. Dr. Irving Fellows believes that such a practice, which would subject agriculture in the state to a "major artificial disturbance," has not occurred:

> If you work in Connecticut you soon learn that the assessor at the town level is the king as far as setting assessments and he is the only one by law that can assess property. Therefore, we had to tread very easily toward him by saying "These are recommended values and this is how we got them. If you want to change them you may, but at least have them consistent within the framework of use value assessment." We have been quite successful and there has been little or no arbitrariness in valuation.[43]

The penalty provision of Connecticut's use-value assessment law also is unique among the New England states. As originally enacted in 1963, Act 490 did not have a penalty provision to deter removing the land from agriculture, although the legislature had considered such a conveyance tax. Not until 1972 was legislation for such a tax adopted to discourage land speculation.[44] The amendment requires that the conveyor of the land pay a penalty equal to 10 percent of the purchase price of the land if he sells within one year of placing it under Act 490. The tax decreases by 1 percent for each year from the time that the current owner was enrolled in the program, so that a person who has owned a given parcel of agricultural land for ten years is not subject to any conveyance tax. In a test of the validity of the provision, a state court held in 1977 that the variations in the penalty based on the length of enrollment in the program does not violate the landowners' constitutional right to equal protection under the law.[45]

Since its inception, the conveyance tax has rarely been used. Most existing farmland changes ownership infrequently.[46] Thus, after ten years of enrollment a farmland owner can convert the use of the land and suffer no penalty. Developers, of course, plan ten years ahead to avoid the tax.[47] Moreover, because the conveyance tax is set on a sliding scale, the penalty becomes marginal in comparison to the profit to be realized upon sale for development in those cases where the land is converted. Act 490, therefore, has been criticized as a haven for nonfarm tax avoiders, since it is often abused by speculators and developers. Indeed, at least in the case of forest lands, several Fortune 500 companies receive a tax break while waiting to develop the land.

There have been several proposals in recent years to establish a more effective penalty provision, but none has been adopted. Although supporters of such proposals believe that a stronger disincentive is needed to prevent the sale or conversion of farmland enrolled in the program, opposition to such proposals has been strong for two reasons.[48] First, there is a fear that changing the law at this time would encourage a huge number of sales and conversions before the stronger penalty becomes effective. Many people believe farmland conversion should not be risked where the public currently receives the benefits of active agriculture. Second, there is a fear that farmland owners no longer will join the program because participation could be too costly later if the owner wanted to convert. This latter argument appeals to those people who are sensitive to the precarious economic existence of many farmers. As a result of these concerns, the penalty provision has not been changed.

Partly because of the comparatively weak penalty provisions in Act 490, enrollment has been high. At first, Act 490 was predominantly used in Fairfield County, which borders New York City, since development pressures were strongest there.[49] By 1971, Dr. Irving Fellows reported that in "most towns of the central [Connecticut River] valley, all qualifying farm and forestland ha[d] been placed under Act 490 and appear[ed] to be holding land from development."[50] Throughout the 1970s, use of this program rose because of the rising value of farmlands. Between 1971 and 1977, the average value of farmland in Connecticut jumped from $1,036 to $2,024 per acre, a 96 percent increase. (The average for New England jumped from $540 to $1,013.)[51] Actual market value of farmland, however, is not always reflected in property tax assessments. Thus, Act 490 applications have been most frequent after decennial property tax revaluation.[52] This conclusion is illustrated by specific examples of the program. In 1969, an eleventh-generation farmer reported that his real estate taxes jumped from $2,400 to $28,000 and that only Act 490 prevented him from selling his land for development to pay his taxes.[53] In 1976, a sixty-three-acre farm in Wilton had a market value of $436,590 and a use value of $35,000. Under Act 490, the landowner received almost a $13,800 reduction in taxes, paying only $1,200. On numerous other occasions, assessment reductions have been greater than 70 percent under the program.[54] It is not surprising, therefore, that by mid-1983, almost all of the state's farmland was enrolled in the program.

Despite its widespread use, Act 490 does not hold land permanently in agriculture. In some cases the value of land in a certain area is so high that the farmland owner, even with the aid of use-value assessment, still cannot afford the property taxes because of high building and equipment assessments and other farming costs. The Connecticut Conservation Association observed:

> There is little question that P.A. 490 for many farmers, is the difference between staying in business and not. However, the land tax burden, even at reduced levels, continues to grow. The simple fact that thousands of farms have gone under since 1963 seems to illustrate that for all P.A. 490 is worth—it is not enough.[55]

It was such a trend of spiraling tax rates that appealed to the New Haven Water Company in the 1970s and made the sale of forest land attractive. Despite the fact that its watershed lands were taxed at the lowest possible rate under Act 490, the company still paid over $2 million in taxes on this land to the City of New Haven in 1975.[56]

Additionally, because the conveyance tax provision does not penalize long-term owners, most parcels of farmland have no conversion disincentive under the Act. There are, moreover, two critical times at which the Act can cease to have real impact, namely, upon the retirement or death of the farmland owner. In the former case, the owner often is willing to sell to the highest bidder, often a developer; in the latter case, high inheritance taxes can force the devisee(s) to sell at least part of the land to pay the taxes.

Commissioner of Agriculture Vincent Majchier reported in October 1983 that "[f]armland values in Connecticut are third in the nation. Because of these high values which are caused primarily by development pressures, very few farms are transferred to the next generation."[57] The problem of high inheritance taxes has been addressed by legislation adopted in 1978 and will be discussed later in this chapter.

Aside from its limited effect in certain cases, the program also has been criticized as aiding landowners who are not committed to furthering the goals of farmland protection. Land speculators and developers often receive a tax benefit on many parcels.

Although Act 490 is not trouble-free, most observers feel the program is a success. Act 490 was designed only to prevent premature development and to provide sufficient time for state land acquisitions. The Legislative Program Review and Investigations Committee concluded that the program "significantly has delayed the conversion of farmlands."[58] In addition, because the program has helped gain support for other farmland programs,[59] it has helped agricultural protection efforts.

As with any program, these benefits must be weighed against the cost of the differential taxation law. In this regard, the primary drawback is a shift of tax burden among landowners, but this shift is relatively small. In 1968, all of the land eligible for valuation under the Act represented less than 3 percent of taxable property in the state, and use-value assessments were close to market value assessment levels in rural and unrevalued areas. With revaluation, however, the shift in tax burden today could be significant in some urban fringe communities. Yet, according to Dr. Irving Fellows, the development that probably would take place if Act 490 were not in effect would require expensive support services such as schools, roads, and sewer lines. Thus, there may be minimal additional tax burden to nonfarm properties with this program, as opposed to the increased taxes needed to service the land if developed.

In the future, Act 490 may be integrated more directly with other farmland programs. For example, Connecticut's purchase of development rights program (PDR), discussed later, can make use of Act 490 assessments when valuing such rights. Moreover, uniform recordkeeping among local assessors could prove helpful. Currently, recordkeeping varies widely because of local discretion. While some towns have mapped lands participating in the program, most have not. This data could be used not only for local tax planning purposes, but also for the PDR mapping system. Whether this responsibility will be placed on localities remains uncertain.

Development Commission Report (1966)

The farmland conversion problem which Public Act 490 was designed to address was symptomatic of a larger land use problem facing Connecticut during the 1960s and early 1970s, namely urban sprawl, or the tendency of development to chew up undeveloped land surrounding urban areas in a semmingly haphazard fashion. The Connecticut legislature found the problem sufficiently acute that in June 1965 it passed Special Act 249 directing the Connecticut Development Commission to undertake a comprehensive study of the extant planning and zoning statutes and to recommend changes. The resulting report, prepared by the American Society of Planning Officials, described the problem as follows:

> Advances in transportation and in technology, a growing complexity of organization and demands on local government, increases in personal wealth, rising standards of consumer taste and a steady growth in population are all important factors changing the landscape of Connecticut. No longer can communities easily escape or even slow down the tide of urbanization. The incessant and unyielding pressures of urbanization require the State to take stock of its resources and capacities to deal with future development problems.[60]

One of the major recommendations of the report was that planning should be undertaken at the state level to ensure that local efforts be coordinated and that the goals of all the state's citizens considered. Connecticut, in 1917, had been one of the first states to allow local governments to plan and zone land. Prior to World War II, however, few communities took advantage of this opportunity because there was little adverse growth pressure. When the economic development boom began in earnest following World War II, the results were predictable: the first communities to be plagued with the problems of uncontrolled urban development were the first to use the planning and zoning legislation.[61] Local governments struggling to cope with the pressures of development rarely considered how their actions would affect other communities. While some communities tried zoning to prevent the negative effects of development, other communities facilitated development through zoning to increase their tax base. The latter policy often tended to sabotage statewide goals for protecting agricultural land, because it increased the already considerable pressure on the farmer to sell for development. In 1974, the Connecticut Conservation Association noted:

> This pressure becomes acute when the farmer sees his land placed on the "zoned for residential" or "zoned for commercial" development list. One farmer, who had just put up a for sale sign, told CCA [Connecticut Conservation Association] that the last straw came when his local assessor and other town officials let it be known that they were "anxiously awaiting" the day when his farmland could be turned over to a developer. "When fellow townsmen and neighbors don't want farming in the community it's time to get out," he said.[62]

Thus, although the Development Commission's report of 1966 recommended that considerable planning and zoning power remain with the cities and towns, it did not suggest "uncontrolled delegation of authority to localities as now exists under the present enabling act. The State must decide, in some measure, what statewide goals of development should be achieved so that there is some standard of determining when the combination of State and local actions is falling short of the goal." The report, therefore, focused on the need for coordination of land use controls.

The recommendations of the Development Commission specifically addressed the issue of agricultural land protection. First and foremost, local agricultural zoning was suggested. The report recommended that municipalities with community development programs approved by the state be given the authority to designate holding zones of open space which could only be developed according to a specified local timetable. Agriculture, forestry, and large-lot residences would be the only uses permitted in these areas. Rezoning would need to be limited to minimize the opportunities of favoritism or corruption. The report further recommended that the state purchase the development rights of agricultural parcels in floodplain areas.[63] As discussed later in the chapter, this second proposal was expanded to include the purchase of agricultural land in all areas and is currently a key component of Connecticut's agricultural land protection efforts.

Proposed Conservation and Development Plan

The Connecticut legislature adopted the basic recommendation of the Connecticut Develoment Commission in 1969 by creating the Office of State Planning within the Department of Finance and Control, and by directing this agency both to prepare statewide land use plans and to coordinate regional and state economic, transportation, recreation, and other planning activities. The legislature renewed this charge in 1971 through a Joint Resolution seeking an intensified effort on the part of the Office of State Planning to prepare a land and water resource Plan of Conservation and Development. The Office responded by producing, in January 1973, its Proposed Conservation and Development Plan.[64] On September 27, 1974 the Plan became the official policy of the executive branch of state government through Executive Order No. 28. In 1976, the general assembly adopted legislation allowing intensive review and consideration of the Plan,[65] but it was never adopted by that body.

The Plan acknowledged the land use challenge facing Connecticut, which in 1972 was the fourth most densely populated state in the nation. Although the gross population density of the state had risen during the 1960s, the size of the average residential lot had doubled and the amount of space devoted to urban areas had risen from 10 percent to 16 percent of Connecticut's total land area in the ten-year period. The Office of State Planning felt that "[c]ontinued development at these relatively low densities would eventually reduce Connecticut's hills and valleys to a vast suburbia—destroying both the natural environment and urban life—and creating a dispersed society impractical to serve with utility, health, education, and other public services."

To counteract this problem, the Office recommended that the state assume a leadership role in allocating land and water resources to accommodate the interests of all citizens. To further this policy, the Office prepared maps illustrating the existing and recommended future land use for the entire state. The Plan sought not to favor one land use over another but rather to ensure that each land use be confined to appropriate areas in the state. The Plan expressly acknowledged the importance of farming:

> Agricultural and forested areas are important resources that need preserving in Connecticut. Agriculture plays a significant role in the State's economy and food markets. The state's forest lands provide recreation in the form of hiking, swimming, boating, fishing, hunting and an increasing amount of winter sports activity and are also critical elements in the natural water cycle. In addition, both agricultural and forested areas act as giant air purifiers, without which our urban centers would have substantially less clean and more harmful air. The continued sprawls of development in the form of large lot single-family housing presents a significant threat to resources. If the trend toward townhouse and multi-family housing construction were reinforced by a conscious government policy for higher density development within areas suitable for Urban Development, the protection of large forest and farming areas would not present a major obstacle to needed housing and job opportunities.

The Plan, therefore, recommended that state and local governments "[e]ncourage continuation of major agricultural and forest areas of the

State in their present use in accordance with the Land Use Policy map."[66] Despite this policy of preserving farming, the Plan was criticized by farmland-protection advocates, such as the Connecticut Conservation Association, for failing to offer "any hard guidelines for maintenance of the industry."[67] Public criticism of the Plan became so widespread that the protection of agricultural land became a public priority in the 1970s.

Recent Agricultural Land Protection Programs

Early Connecticut efforts to protect agricultural land, leading to the Development Commission Report and the Proposed Conservation and Development Plan, viewed the importance of agriculture as one of many desirable land uses. When the Governor's Task Force Report for the Preservation of Agricultural Land was issued, however, the protection of farmlands emerged as an open space priority. Central to this goal was a purchase of development rights program. Playing a lesser role in the network of Connecticut's programs are tax programs, a right-to-farm law, zoning enabling legislation, and a bond review program.

Governor's Task Force Report (1974)

Because the efforts to institute comprehensive state land use planning were unsuccessful in halting the conversion in farmland, in April 1974, Governor Thomas J. Meskill established a Task Force "to study and then recommend a land policy to maintain agriculture." This Task Force of twenty-five people was composed primarily of farmers and members of state and private agricultural groups. The Task Force found farmland to be diminishing at such an alarming and persistent rate that it published its report four months early, in December of the same year.

The chief recommendation of the Governor's Task Force was that municipalities identify and establish zones of fertile agricultural land within their jurisdiction and the state then purchase the development rights of land in these reserves.[68] This concept of state land acquisition, of course, had been proposed several times in Connecticut, including the Whyte Report, the Connecticut Development Commission Report, and the Proposed Conservation and Development Plan. A program for the purchase of agricultural development rights, however, was not given clarity and a sense of urgency until the Task Force Report.

The scheme is simple: by paying a farmland owner the difference between the value of his land for producing food and its value for development, the state would become the owner of the right to develop, and the land, even if conveyed, could be put to no more intensive use than agriculture. The Task Force estimated that to preserve the level of agricultural enterprise then existing in the state (which was at one-third self-sufficiency), Connecticut should plan to reserve at least 325,000 of the 500,000 acres devoted to farming. These acreages would be located in town-designated reserves, and development rights purchases by the state would be at the landowner's option. It was calculated that the development rights would cost the state about $1,500 per acre and that the total cost of the program to the public would be around $500 million. The Task Force recommended financing through bond authorizations, and funding to retire the bonds through a 1 percent tax on all real estate transfers. Such a tax was expected to yield $30 million a year.

One of the goals of the Task Force, of the Connecticut legislature, and of the people who acted upon the Task Force's recommendation, was to promote farming as an economically viable industry in the state. The Task Force realized, of course, that many factors other than the land use policy of the state affected the viability of the farm industry. It noted, for instance, the significant impact that the high price of imported feed had on the price of dairy products. The Task Force also realized that merely setting aside sufficient acreage to yield one-third of Connecticut's present food needs would not be enough in itself to ensure the continued production of this proportion of food in the future: the steady increase in the population might well place increased demands on the agricultural industry in the future. The Task Force felt, however, that a necessary first step in preserving agriculture as a viable industry in the state was to assure farmers currently working the land that this same land would be in production ten or twenty years from now and that measures they took to maintain or increase productivity would not be wasted. It was the Task Force's belief that "[i]f we make farm ownership secure and work hard at our Experiment Stations, we can reasonably hope to increase our yields as our population increases and thus continue growing a third of our food from the acres we use today."

The Task Force's recommendation, of course, was not supported universally. Aubrey Bickelbach and Gregory Wassall, who had challenged the premise that Connecticut should attempt to protect its agricultural land at all, also challenged the conclusion that if farmland should be protected, the purchase of development rights was the best means possible. These critics felt that if the Task Force recommendations were adopted, the following results would be ensured:

1. The transfer of development rights would only temporarily forestall the free agricultural market mechanisms which have caused agriculture's demise.
2. The loss of 325,000 acres of land to the potential development market would result in increases in developable land values, particularly on the fringes of our urban areas.

3. Such a plan would provide the state with only 11 percent or so of its food requirements at the very most [the authors estimated that the state was at best 11 percent self-sufficient in 1973].[69]

Despite such controversy, a purchase of development rights bill was introduced into the Connecticut General Assembly in 1975, shortly after the Governor's Task Force released its report. The legislature failed to adopt the proposal largely for reasons not related to its merits: the 1 percent conveyance tax proved politically unacceptable.[70] The Connecticut Association of Realtors and other real estate development groups successfully argued that the tax was overly burdensome because it was too high, overly broad, and affected people totally unconnected with agricultural lands.

In 1978, however, a revised purchase of development rights pilot program without the conveyance tax provision was approved by the legislature. This law had no regular funding provision and no local zoning requirements. Nevertheless, the program operates with Act 490 and several more recent initiatives to provide Connecticut with a modestly effective range of programs designed to stop agricultural land conversion in the state.

Purchase of Development Rights Program (1978)

The comprehensive PDR pilot program adopted by the legislature in 1978 has become the focus of agricultural protection activity in Connecticut. The objectives of the program were simply stated by Commissioner of Agriculture Vincent Majchier in his status report, released in December 1983:

1. Retain the best and most productive agricultural land;
2. Provide an opportunity for farmers to purchase farmland at affordable prices;
3. Help farmland owners overcome estate planning problems which often result in farmland loss;
4. Provide working capital to enable farm operations to become more financially stable;
5. Address other personal ownership problems such as age and health, which contribute to the likelihood of the land being converted to nonagricultural use.[71]

The program is funded out of general state revenues and enables the state to purchase a deed restriction from any farmland owner eligible under the program, thereby ensuring that the land will remain available for agricultural use.

The program adopted by the legislature has three parts. First, a Food Plan study was mandated by the Act to determine what types of foodstuffs the state needs to meet its agricultural land protection goals. Second, mapping of farmlands was required to provide the needed technical assistance in acquisition determinations. Third and most importantly, an acquisition process was established through which development rights could be purchased by the state.

The Food Plan. The PDR legislation required the Commissioner of Agriculture to work with the College of Agriculture at the University of Connecticut to develop a food plan which would include an evaluation of the following:

1. Prospective essential food requirements in Connecticut at intervals of ten and twenty years from the present [1978], assuming population growth in the State similar to that experienced since 1975;
2. Prospective sources of supply for such food requirements, including estimates as to agricultural land requirements for that portion of such food requirements to be produced in Connecticut; and
3. Recommendations of priorities to be established with respect to agricultural production and agricultural land requirements, considering such future food requirements in the state and the land available for such production.[72]

These requirements were met, and the Food Plan Report by Dr. Irving Fellows and Patrick Cody was submitted to the Governor and General Assembly on December 15, 1979.

The Food Plan set forth three plans. Plan One simply analyzed the farmland acreage requirements to maintain current production levels without the full benefits of an aggressive agricultural preservation policy. Plan Two, the recommended plan or "Food Plan," called for greater production of certain farm products. Plan Three discussed the acreage needed for annual or seasonal consumption of certain crops, in the event of a national food-related calamity which would seriously diminish Connecticut food supplies.

The Food Plan carefully evaluated the expected growth and food consumption patterns in the state's population and the nature of the farm economy. This Plan focused on the problems which result from food dependency on other states and put forward the notion that Connecticut has a comparative economic advantage over other states in producing certain products. The authors of the Food Plan also felt that it "accomplishes the policy objectives of an improved supply of local food products and of a broad spectrum of environmental benefits." Agricultural production would be assured under this plan because the PDR program would provide "greater stability, lower capital requirements, and greater profit potential" for farmers willing to keep their lands active.

The Food Plan projects that 83,500 acres of cropland would be needed in the year 2000 to provide for 33 percent of the state's food needs. Some 80 percent of the cropland would be used for dairying, providing for 60 percent of the state's needs. Fresh vegetables and fruits would account for most of the remaining cropland, shifting prime agricultural land out of tobacco and nursery production, to produce almost all seasonal needs. The Food Plan excludes the very limited cropland needs of poultry, eggs, and meat production, as well as that of tobacco farms, sod farms, nurseries and greenhouse, and livestock farms.[73] A total of about 300,000 acres would be required, including acreage for uses such as pasture, forest, drainage, and open space.[74]

The Commissioner of Agriculture must consider the Food Plan's goals when applications for the PDR program are evaluated. The Plan, therefore, focuses on the land-intensive farming activities which the PDR program is designed to protect. The goals are intended to be consistent with the general concerns of the public in protecting farmland, as opposed to farm activities.

Land Use Maps. The PDR legislation also required the Secretary of the Office of Policy and Management (OPM) to work with other public officials and agencies to develop maps of the state's existing land uses.[75] At a minimum, the maps were to describe soil types, active and inactive farmland, type of crops being produced, relevant local zoning, planned and existing sewer and water lines, and forest and open space lands. These maps were to be delivered to the Commissioner of Agriculture to assist him in deciding which lands to purchase under the program.

By mid-1985 it appeared that the maps would never be completed. Because of fiscal constraints, the OPM has completed other maps utilizing current data, such as from the Soil Conservation Service, supplemented with new data, such as from county agents.[76] There is doubt whether the maps are sufficiently detailed to be useful in the PDR evaluation process.

Initial Selection Process. The selection of land for the pilot program was comprised of a process involving application, screening by various agencies, appraisal, price negotiation, bonding approval, and purchase. To ensure that the best lands would be considered under application, media announcements were made and public discussions held by the Commissioner of Agriculture and his appointed Advisory Committee. In response, about one hundred applications were submitted by the November 1978 deadline. From these applications, the Commissioner ultimately selected thirty which represented landholdings "in severe jeopardy" of being put to nonagricultural development.[77]

To decide the finalists in the program, the Commissioner and his Advisory Committee developed a scoring system based on eighteen factors. These factors were based on the statutory requirement that the major factor in evaluating applications be jeopardy ("the probability that the land will be sold for nonagricultural purposes"). The other factors include current and future productivity, soil classification, the land's physical characteristics, the value of acquisition to the preservation of the state's agricultural potential, the nature of any encumbrances, and cost.[79] The eighteen factors became part of the regulations for state agencies.

After scoring the thirty second-round applicants, the commissioner and his advisory committee selected ten finalists. Appraisals were made on the properties, and the Commissioner negotiated a purchase price with each landowner. Surveys and title searches were conducted. Eight farms eventually were selected in the pilot program from six different counties. Five of the parcels were dairy farms, two were corn and/or hay farms, and one was a fruit and vegetable farm. The total acreage was 2,321 acres. The total cost of development rights was over $3.5 million at an average cost of $1,620 per acre. By 1980, two additional landowners had completed negotiations with the state. These parcels added 264 acres to the program at a cost of about $510,000.

Statutory and Regulatory Amendments. There have been several changes in the program since 1978, and many more changes have been suggested. Most of these changes are refinements of the original legislation rather than major amendments. These revisions seek to improve the program and make it less subject to possible abuse by landowners who are not committed to cropland, dairy, or other forms of traditional land intensive farming.

Two amendments were adopted by the legislature in 1979. The first Act simply required the Commissioner of Agriculture to adopt regulations for the program because the original legislation only gave him permission to do so and firm guidelines were needed. Another procedural Act set forth various acquisition requirements for the Commissioner such as appraisal procedures. These revisions were undertaken to help the Commissioner run the program more smoothly.

Aside from further procedural changes, the 1980 legislature took the step of reestablishing the program, which was scheduled to terminate on July 1, 1980. Five more years were added to the program, since it was no longer considered a "pilot" type. In December 1980, then–Commissioner of Agriculture Leonard Krogh also changed the regulations relating to scoring values.[80] The original regulations favored dairy farmers over small vegetable farmers[81] because the point totals given for production were substantial. The regulations were revised to give greater emphasis to the suitability of the land for farming, a criterion Krogh felt should be given the greatest weight.[82]

The 1981 legislature took further steps to improve the program by amending the valuation requirements and limiting farm-related construction. Originally, the law required the Commissioner to pay the difference between fair-market value and farmland-use value. Yet given the initial flood of applicants, it became obvious that the state could pay less than the statutory formula, thereby enabling the Commissioner to make more purchases with a fixed sum. In the words of the Commissioner, by the end of 1980, "[t]he department has received over 200 applications from all eight counties in the state, totaling over 26,000 acres, . . . many applicants have such a strong desire to see their land remain in agriculture, they would be receptive to a much smaller payment to guarantee this assurance."

There was also a concern that developers wished to take advantage of the program because the housing market was more risky than development rights purchased by the state. If the state bought these, or other landowners' development rights, there existed an incentive to resell the land to people committed to farming at a value well beyond its restricted value. The Commissioner of Agriculture found that would-be farmers "are willing to place a value on this agricultural land much higher than that established by the department." Thus, the program had the potential of subsidizing land speculators.

Finally, the resources for the program were becoming depleted because inflation raised the value of farmland's highest and best use value about 12 percent each year.[83] Under the pilot program, about 95 percent of this value was being paid for the development rights.[84] In response, the Commissioner recommended, and the General Assembly adopted, a provision allowing the state to pay less for development rights than was allowed under the original law. The Commissioner can now bargain with an applicant and pay only a reasonable amount for the land. The highest and best use value is lessened by the agricultural value or the agricultural market value. The legislature also authorized the acceptance of development rights as gifts.

Another abuse of the program involved a farmer who received funds in exchange for his development rights and then proceeded to construct large farm buildings for such uses as egg and poultry production. Again, the landowner was considered to be circumventing the intent of the PDR legislation which was directed at preserving traditional, land-intensive farming activities. Dairy or crop production, for example, were desired activities because they kept the land open. The 1981 legislature, therefore, required that construction on restricted parcels not materially reduce acreage and productivity, and that it be limited to no more than 5 percent of a farm's total prime farmland.

The 1982 legislature addressed the problem of the heavy weight assigned to the jeopardy criterion in scoring applications, which effectively precludes the purchase of many important parcels. One parcel of land was purchased by a nonprofit organization that intended to sell the development rights to the state. When such an organization purchases a parcel in jeopardy, that parcel is no longer in jeopardy because of the new owner. The original law gave low scores to such lands, and the Commissioner therefore was unable to purchase the development rights on those parcels. This was a significant obstacle for land trusts because, as the Massachusetts experience reveals, nonprofit organizations often can be useful in protecting valuable farmlands in jeopardy but rely on the state to reimburse their working capital through a development rights purchase.

The new law requires the Commissioner to disregard nonprofit ownership if the parcel formerly qualified under the program. Effectively, this keeps the parcel in jeopardy and allows nonprofit, private organizations to purchase a parcel without threatening the development rights. This is a prudent amendment because private organizations usually can move more quickly than the state to save farmland that is truly in jeopardy.

Another 1982 amendment requires the farmland owner entering the program to sell the development rights to his entire farm parcel. Although it is less costly to purchase only part of a given parcel, purchasing less than the whole farm can have the effect of subsidizing development. Excluded farm parcels generally would increase in land value once the development rights are purchased because nearby open space is assured.

This problem was especially acute in the pilot program because the application process initially allowed the applicant to list what portion of his land he wished the state to take an interest in, although the Commissioner often changed that determination.[85] Again, the legislature responded with appropriate statutory amendments to fine-tune the PDR program to avoid its abuse.

Former Commissioner H. Earl Waterman effected a change in the scoring regulations, applicable after March 26, 1984, which further reduced the advantage of large producers, such as dairy operations, in the scoring process. The regulations now give higher scores to vegetable and fruit operations.[86]

Finally, acting on 1984 state enabling legislation for adopting local PDR programs, the town of Easton set up a municipal PDR fund to which donations were made in 1985.[87]

Current Selection Process. Beginning in the fall of 1983, the Commissioner, pursuant to his statutory authority, processes applicants through a nine-step acquisition process:

1. Applications are received.
2. Farmland is evaluated for agricultural potential by the Commission with the help of the Soil Conservation Service (USDA) and county agricultural agents (UConn).

3. Documentation is collected to examine the likelihood of sale for a nonagricultural use.
4. Commissioner consults with the Advisory Committee and approves the farm for appraisal.
5. Department of Environmental Protection Land Acquisition Unit assigns appraisals and reviews values which are discussed with the Agricultural Lands Preservation Advisory Council.
6. Negotiations are conducted between the landowner and the Commissioner.
7. The State Property Review Board must approve the application based on whether the Commissioner has complied with state statutes and regulations relative to the Farmland Preservation Act.
8. Funds are requested from the State Bond Commission for the cost of development rights, survey, and title search.
9. The Department of Environmental Protection secures contracts with surveyors to establish boundary lines of property on which restriction will be placed. This selection process is a slightly more streamlined version of the process used under the pilot program.[88]

Results to Date. As of January 1986, with $22.75 million, twenty-eight farms (6,870 acres) had been preserved; in addition, the state negotiated PDRs on eight more farms (1,200 acres) and has appraised another thirty-five farms (4,100 acres). The average acquisition price is $1,650 per acre.[89] The farms chosen came from a pool of 265 applicants, comprising 36,810 acres.[90] Thus, the program is moving slowly toward its goal of protecting 83,500 acres of cropland by the year 2000.

Release. A reversion of development rights process is provided for in the Act in the event development is desired by the landowner.[91] The process is so restrictive, however, that few landowners will be able to buy back their development rights after purchase by the state. A petition for release must be submitted to the Commissioner of Agriculture either by the landowner or by the town as approved by town resolution with the landowner's consent. The Commissioner must then consult the Commissioner of Environmental Protection and any advisory board he chooses to appoint. At least one public hearing must be held. If the Commissioner finds that there is an overriding public necessity to allow the release of the development rights, a local referendum will be held. The public hearings and the referendum are held at the expense of the petitioner. If the referendum is approved, the landowner must pay the full market value of the development rights to the state at the time of the sale.

Prognosis for the Future. Like the other New England PDR programs, the Connecticut program's major problem is cost and its major weakness is funding. The estimated total cost is from $300 to $400 million, but the sum could be as great as $500 million.[93] The state had not been generous with the program after its initial $5 million bond issue in 1978. From 1979 to 1982, only $4.75 million was appropriated. In 1983, however, the legislature both appropriated an additional $5 million and established a joint legislative study committee to formulate funding proposals. As a result, a further $8 million was appropriated for the PDR program during the 1984 and 1985 legislative sessions, for a total funding thus far of $22.75 million. Program proponents hope that greater revenues will continue to be appropriated as the economy improves and as the Department of Agriculture continues its purchases. Yet as a result of cost and funding problems there is no certainty that the Food Plan goals will be met.

One indication that the program may not warrant greater funding is the nature of the applicants. Although the Commissioner has received over 250 applications, relatively few of these represent land holdings in the rich Connecticut River Valley area. Because of urban growth presure in this area, the value of development is considered so great that many farmers in the Valley do not wish to enter the program. While the purchases thus far involve high-quality land, future acquisitions may not be as easily justified if they involve non-Valley farmland in a less critical resource area. Nevertheless, some analysts of the program believe that landowners in the Valley may make more applications to the program as it becomes more established.

A central problem with most PDR programs, including Connecticut's, is that market conditions may make a given parcel of land unprofitable for farming despite PDR. Moreover, a landowner can receive use-value taxation even if he does not produce agricultural commodities as long as the land is kept open. Therefore, there is only an incentive to produce food if it is profitable. The Connecticut Department of Agriculture and the Commissioner's Advisory Committee believe that farming will continue on lands in the program because the farms selected have outstanding profit-making potential when operated by skilled farmers.[94] Although under the pilot program it was deemed politically expedient to purchase the rights for farms in a number of different counties, since 1980 the state, like Massachusetts, has sought to protect blocks of farms so that marketing and other critical mass costs are reduced for a group of farmers.[95]

Furthermore, revisions in the scoring criteria of applicants could enhance the likelihood of future production. The scoring criteria used by the Commissioner, as mandated by the legislature, require jeopardy, not land quality, to be of primary importance in an acquisition. Assessment of the true conversion risk, however, is difficult and can be misleading. High scores are given if the landowner lists the property with a real estate agent or if the property is in probate court. Former Commissioner Krogh, therefore, requested that the legislature alter the criteria to focus on land quality.[96] Yet such requests were ignored, and so applicants will continue to manipulate the jeopardy criterion. Moreover, the scoring criteria still give great weight to current production. Future generations may question the virtue of this emphasis if land quality on highly productive lands declines. Future generations also may question why the state's deed restriction does not, as suggested by some, require active agriculture.

In the Legislative Program Review and Investigations Committee Report of January 1, 1980, it was concluded that the pilot program's benefits, at current funding levels, outweighed the costs. The Report found that the certain availability of farmland at farmland values will cut the cost of production and increase the volume of local produce. The Report also found that the protection of the natural environment, protection of rural aesthetics, and the "perpetuation of the state's agricultural tradition" were benefits of the program. The costs associated with the PDR program are the bonding sums and the increased value of developable lands as a result of a reduction in total available land supply. This latter concern, voiced by the Connecticut Home Builders Association among others, may be a cost future generations will not wish to bear. Yet there is no evidence that land costs have increased as a result of the program.[97]

The assessment today appears even more positive. Some funding appears certain for the future as the program is considered the cornerstone of farmland protection in Connecticut. The state's other programs are designed as a complement.

Succession and Transfer Taxes (1978)

Although the PDR program is the focus of Connecticut's farmland protection efforts, the state has determined that a range of programs is needed to protect adequately the agricultural resource base. In one response to this need, the 1978 General Assembly altered the state's inheritance tax law to ease the problem of farm estate taxation. Often the heirs to a farmer's estate were forced to sell a farm to pay the taxes due even though they intended to keep the land in agriculture. In one case the farmland estate heirs were forced to sell a farm with a market value of $350,000 and taxes due of $44,000 because the farmer had left little in other assets.[98] To avoid passing on such problems, an aged farmer may find it prudent to sell the farm before his death to avoid high taxes. This problem was acute in the 1970s because many of the state's farmers were near retirement. Indeed, in 1969 the average age of Connecticut's farm operators was 52.9 years.[99]

The law now allows for assessment of estate farms at their use value if the immediate family agrees to farm the land for ten years.[100] If the heirs fail to farm the land for ten consecutive years, then the property is revalued at its market value and tax penalties become due. Supporters of this law hope this provision will help keep many farms in agriculture, and continued production is especially likely if the development rights are purchased within the ten-year period.

Right-to-Farm Law (1981)

Like most of the other New England states, the Connecticut legislature has adopted a Right-to-Farm Law allowing farmers greater freedom in the type of operations they can conduct.[101] Passed in May 1981, the law's provisions are very specific and somewhat limited, making it less permissive for the farmer than most of the region's sister statutes.

Under the statute, a farming operation does not constitute a nuisance even if various types of odors, noise, dust, chemicals, and water pollution result from such activities, if the farming practice was in operation for at least a year without substantial change. The operation must also adhere to generally accepted agricultural practices, which can be established by inspection and approval by the Commissioner of Agriculture or his agents. Furthermore, the use of chemicals and the existence of water pollution are not immune from nuisance suits unless their existence is in accordance with certain other agency regulations and if drinking water is not polluted. Finally, negligent, willful, or reckless misoperation does not invalidate a farmer's legal liability. Thus, the statute is designed to allow normal, nonhazardous farming activities to continue, even if newcomers to an area adopt a municipal ordinance or promulgate a regulation to the contrary.

Despite the goal of protecting farmers engaged in reasonable farm practices from lawsuits, litigation can be expected to arise. First, a newcomer could successfully characterize a particular nuisance as falling outside the statute. For example, a noxious odor not emanating from livestock, manure, fertilizer, or feed would not be an immune nuisance. There is no catch-all clause to cover new situations. Second, requiring the farming operation to be unchanged substantially for at least one year could create hardships. For example, a long-term dairy farmer could decide to begin an egg and poultry operation or to expand his operation. A newcomer could then sue for nuisance if the new operation interfered sufficiently with the surrounding area.

These potential problems may not arise if the courts broadly construe the wording of the statute to allow modern or new agricultural activities. That such a construction may be given to this statute was suggested by a 1982 trial court decision in New Haven County concerning a new residential development constructed next to a dairy farm.[102] An incoming abutter resident sued the farmer under nuisance law because manure spreading generated odor and flies. An injunction to prevent this activity, as well as monetary damages, was requested. The judge found for the farmer because of the right-to-farm defense. The judge noted that the farm was in operation before the interlopers took up nearby residence. Moreover, the dairy farm had been in operation for over a year, had not changed substantially, and constituted a generally accepted farm practice. This case met the specific requirements of the statute, and the court appeared to have no difficulty making a determination that the farm activity was generally accepted.

Other local courts may not rule in favor of the farmer when the activity in question is not well established or is not conventional. The drafting of the statute, therefore, reflects a concern for the traditional farmer, but its provisions leave some farmers subject to nuisance suits. Farmers seeking to avoid nuisance suits may take advantage of certain pollution control programs, such as a 1974 law allowing partial reimbursement of waste management systems by the Commissioner of Agriculture.[103] An effective system can prevent the occurrence of some nuisance claims.

Upper Connecticut River Conservation Zone (1982)

In 1979, a Connecticut River Assembly was established temporarily by the legislature to protect and preserve the River floodplain for various purposes, including agricultural use.[104] The Assembly found that the upper river area north of Hartford possessed unique agricultural and other values that should be respected in the development process.[105] The Assembly recommended that the protection and expansion of agricultural land uses should be encouraged. In light of limited fiscal resources, the Assembly proposed that a permanent assembly regulate land use.

Following this recommendation, in 1982, the legislature established a permanent Connecticut River Assembly and a permanent conservation zone covering fifteen municipalities.[106] The Assembly, composed of the Governor, a member of the Capitol Region Council of Governments, a member of the Mid-State Regional Planning Agency, and a member from each of sixty-five municipalities in the Connecticut River Valley, is empowered to establish minimum standards for regulatory land use to be adopted in local zoning and subdivision regulations and to comment upon large land developments in the area. Negative comment by the assembly can be overridden by a two-thirds vote of the local land use agency considering the application for development. A municipality can withdraw from the Assembly one year after its legislative body votes to do so. Among the proposed land uses the Assembly must consider are residential uses of twenty-five acres, or more than fifty units; soil and earth removal projects involving 15,000 cubic yards or five acres of land; and hazardous waste facilities.

The conservation zone concept is important to the state's farmland protection efforts. As public fiscal resources remain limited for programs such as the PDR and because the PDR program suffered from a lack of Connecticut River Valley applicants, direct regulatory controls have become more important. The zone's effectiveness has yet to be determined, however, because the recession of the early 1980s has eased development pressure in the Valley. Moreover, the Assembly was not empowered to implement the land use standards until October 1, 1983. Important to its success will be the way in which the Assembly exercises its authority to affect farmlands.

Bond Review Program (1983)

Recognizing that state as well as private development projects can affect farmlands adversely, the General Assembly passed a bond review program on May 13, 1983.[107] Although this program is similar to the Executive Orders on public developments issued in Vermont and Massachusetts, its operation is unique. The salient differences are that this program only affects bond financed projects and that the Commissioner of Agriculture appears to have absolute veto power over many of these projects.

Under this program, any state bonds, except refunding bonds, that will be used to finance projects which convert twenty-five or more acres of prime farmland to a nonagricultural use require a statement of approval from the Commissioner of Agriculture. The Commissioner can give approval only if the development project promotes agriculture, promotes the goal of agricultural preservation, or is located on a site that has no reasonable alternative for the project. This law, therefore, appears to invest the Commissioner with a great degree of discretionary power.

The effect of the program on proposed capital projects is not yet clear. Certainly public planners will be forced to take the law into account. These planners may also need to consider several pending agricultural protection proposals as these proposals become legislation.

Zoning

The 1966 Development Commission Report and the 1974 Task Force Report, the 1980 Legislative Program Review, and the Investigations Committee's "Sunset Review" of the Agricultural Preservation pilot program all recommended agricultural zoning be implemented across the state. Although the state has regulated inland and tidal wetlands,[108] the concept of statewide zoning is as unpopular in Connecticut as it is elsewhere in New England.[109] The legislature has considered enabling statutes allowing municipalities to implement agricultural districts, which would restrict the use of a tract of at least ten to eighty acres, and agricultural zones, which would allow more uses on smaller minimum tracts.

As discussed above, some localities have zoned farmland for a nonfarm use in an attempt to force out farmers, often for tax base reasons. For example, the town of Fairfield, with zoning first enacted in 1925, classified some farmland as residential. One family farm which fell under this classification constituted a prior nonconforming use when the zoning ordinance was enacted and continued its operations despite the prevalence of newcomers. When the family started processing soft drinks in the 1970s, the town issued a cease-and-desist order. The Supreme Court of Connecticut upheld the order because of the farmer's change of prior use.[110] Thus, because of economic problems, farmland owners in a hostile area could be forced to convert their farmland to a higher use.

Nevertheless, localities may wish to adopt local farmland zoning ordinances. Although the state's general enabling legislation does not expressly grant this authority, towns are able to implement cluster development and performance zoning. One such regulation in Windsor allows farmland owners to develop certain tracts at unusually high densities in exchange for the zoning use restriction. Moreover, the differential taxation law, Public Act 490, allows municipal planning commissions to designate land as open space if approved by the majority of the local legislative body. Both the USDA Agricultural Extension Service and the Soil Conservation Service are providing soil data to help zoning boards direct development away from prime farmlands. Despite such technical assistance, many localities remain unaware of available zoning techniques or are uncertain of their ability to designate agricultural zones under the state's enabling legislation.[111]

One zoning innovation found in the enabling legislation of some states but not yet in Connecticut is a transfer of development rights (TDR) program.[112] The concept embodies the use of conservation and transfer zones. A transfer zone owner must purchase the development rights of a conservation zone owner to develop his land fully. There are, therefore, no direct public outlays to landowners. The concept presumes that undeveloped areas with public services now exist in the state to support greater development at a profit. In 1980, the Legislative Program Review and Investigations Committee found that this program deserves consideration in Connecticut towns in the urbanized state.[113]

Land Acquisition

One approach to farmland protection, which has not been accepted widely, is the outright purchase of lands by the state. The state could then operate marginal agricultural operations or lease the land to committed farmers. Although the original high financial cost and strong potential of an annual financial loss have been considered the major reasons for not utilizing this approach, sales of produce or the lease of the land could recover the purchase price in some cases. The main political problem with the program is a genuine fear that government would aggrandize its power through land ownership.[114] Nevertheless, the town of Farmington has established such a leaseback program along the Farmington River floodplain. Other muncipalities could utilize this approach.

The Future of Agricultural Land Protection Programs

Connecticut has maintained a long tradition of conservation activity related to farmlands, and, like Massachusetts, generally has been among those states in the spearhead. The state legislature first became concerned about the conversion and abandonment of agricultural land in the 1950s and passed the first current-use tax law in New England in 1963. Periodic legislative inquiries into the problem have produced a variety of programs, all of which have centered around a PDR comprehensive program.

The Governor's Task Force Report (1974) gave the existing state land acquisition program clarity and urgency. Based on a Food Plan, the state's PDR program is targeted toward specific food goals—to preserve existing agriculture for present and projected food supply needs. To this end, local planning is coordinated with the state's role. Municipalities must first identify and establish zones of fertile agricultural land within their jurisdiction, and the state will then purchase development rights on certain of these reserves. Even though food production is the basic aim of the program, the legislature has tailored the program to benefit traditional, open space, land-intensive farm operations as opposed to unsightly industrial operations.

Connecticut PDR legislation also required detailed maps of existing land uses to be completed by the Secretary of the Office of Policy and Management in conjunction with other officials and agencies. The purpose of these extensive maps was to assist the Commissioner of Agriculture to evaluate which land should be purchased under the program. However, fiscal restraints have prevented the completion of these maps in the suggested detail, and the experience indicates that these planning goals were not realistic.

The PDR program has evolved from a pilot program, with a number of refinements in its operation which were made by the legislature and the Commissioner of Agriculture. Connecticut's experience suggests that other states might use a pilot program to test the waters with innovations designed to protect agricultural lands.

The PDR program is costly and suffers from a lack of Connecticut River Valley applicants. Consequently, the state has employed other devices to protect its agricultural resource bases. Such devices operate with the program to provide Connecticut with a modestly effective range of agricultural land protection programs—a direction in which Connecticut is clearly heading. Such devices and initiatives include: tax incentives, a right-to-farm law, a bond review program, zoning techniques, and a purchase-and-leaseback program. Furthermore, regulatory controls, such as the Upper Connecticut River Zone, offer a low-cost alternative to the PDR program. Connecticut experience suggests, however, that such a local program could become more useful if all the landowners perceive that they were treated fairly by such bodies.

In conclusion, Connecticut appears to be moving in the direction of a range of programs to protect agricultural land from conversion and abandonment. The range of recent initiatives is designed to complement its comprehensive PDR program, which is based primarily on food supply needs. Such an objective basis renders public support, an essential component in agricultural land protection programs. Moreover, Connecticut's PDR experience underscores the importance of realistic local planning in conjunction with the state's selection and purchasing role. Generally, Connecticut appears committed to continue the battle against encroaching urbanization and declining farmland use through its PDR program and whatever other range of innovations become available.

Notes

1. Ronald F. Aronson and Irving F. Fellows, *Connecticut Agriculture* (Storrs: University of Connecticut, Cooperative Extension Service, 1981): 1.
2. Peirce, *The New England States*, 228.
3. One local authority puts the mid-1970s figure at 565,000 acres. *Land for Growing Food in Connecticut*, Bulletin 769 (Storrs: Connecticut Agricultural Experiment Station, 1976), reprinted in Lawrence K. Furbish, "Memorandum on the Preservation of Agricultural Land to the Honorable Audrey Beck" (Hartford: Office of Legislative Research, 1977): 4.
4. George A. Ecker and Irving F. Fellows, "Connecticut Agricultural Trends and Status, 1981" (Storrs: University of Connecticut, Cooperative Extension Service, 1981).
5. Lawrence K. Furbish, "Memorandum on the Preservation of Agricultural Land to the Honorable Audrey Beck," 1, 5–7. This memorandum was based largely on USDA statistics.
6. Mark B. Lapping, "The Land Base for Agriculture in New England" (Paper prepared for the National Agricultural Lands Study, October 1979): 25.
7. Aronson and Fellows, *Connecticut Agriculture*, 1.
8. Leonard E. Krogh, "Report of the Commissioner of Agriculture, December 15, 1980" (Hartford: Connecticut Department of Agriculture, 1980): 3.
9. *1982 Connecticut Agricultural Statistics* (Hartford: Connecticut Department of Agriculture, November 1983): 1. Due to different methods of data collection, the state agricultural statistics often differ from statistics from the U.S. Census of Agriculture.
10. George A. Ecker and Irving F. Fellows, "Connecticut Agriculture Trends and Status, 1982" (Storrs: University of Connecticut Extension Service, November 1983): 3.

11. Aronson and Fellows, *Connecticut Agriculture*, 2; *New York Times*, July 8, 1984.

12. Irving F. Fellows and Patrick H. Cody, *A Food Production Plan for Connecticut, 1980–2000*, Bulletin 454 (Storrs: Connecticut Agricultural Experiment Station, March 1980): 1.

13. *1982 Connecticut Agricultural Statistics*, 1.

14. "Today's Tobacco Road Is Not Easy," *Boston Sunday Globe*, July 11, 1982, 2.

15. *Status Report, Program for the Preservation of Agricultural Lands* (Hartford: Department of Agriculture, October 1983): 4.

16. Aubrey W. Birkelbach, Jr., and Gregory H. Wassall, "The Case Against the Sale of Development Rights of Connecticut's Agricultural Land" (Hartford: University of Hartford, March 1975): 8.

17. Irving F. Fellows to authors, February 4, 1983.

18. C.G.S. §12–107a.

19. Irving F. Fellows, comment, Lincoln Seminar 1.

20. Committee Bill No. 7598, §1(b) (January 1975 session).

21. C.G.S. §22–26aa.

22. Fellows and Cody, *A Food Production Plan, Connecticut*, 2.

23. Birkelbach and Wassall, "The Case Against the Sale of Development Rights of Connecticut's Agricultural Land," 3 .

24. "The Vanishing Land," *Connecticut Conservation Reporter* (Bridgewater: Connecticut Conservation Association, August–September 1974): 2.

25. Hallie Black, "Connecticut Puts a Damper on Water Company Land Sales," *Planning* (Chicago: American Society of Planning Officials, January 1977): 13.

26. "The Vanishing Land," 2.

27. Black, "Connecticut Puts a Damper," 12.

28. "The Vanishing Land," 1.

29. Leonard E. Krogh, "Report of the Commissioner of Agriculture" (December 15, 1979): 5.

30. C.G.S. §1-1.

31. "The Vanishing Land," 5.

32. C.G.S. §7-131a.

33. William J. Duddleson, *Supplementary Report* to Scheffey, *Conservation Commissions in Massachusetts* (Washington, D.C.: The Conservation Foundation, 1969): 178–79.

34. Duddleson, *Supplementary Report*, 178.

35. Irving F. Fellows, *The Impact of Public Act 490 on Agriculture and Open Space in Connecticut, Proceedings of the Seminar on Taxation of Agricultural and Other Open Land* (East Lansing, Michigan: Michigan State University, 1971): 49.

36. Duddleson, *Supplementary Report*, 179.

37. "New Directions in Connecticut Planning Legislation" (Chicago, Ill.: American Society of Planning Officials, February 1966): 93.

38. C.G.S. §12-107b(a), 107c(a), and 107c(d).

39. *The Agricultural Lands Preservation Pilot Program: A "Sunset" Review*, Vol. I-21 (Hartford: Legislative Program Review and Investigations Committee, Connecticut General Assembly, Jan. 1, 1980): 22.

40. *Johnson* v. *Board of Tax Review of the Town of Fairfield*, 160 Conn. 71 (1970). *Holloway Brothers, Inc.* v. *Town of Avon*, 26 *Conn. Sup.* 160, 162 (1965).

41. Irving F. Fellows, comment, Lincoln Seminar 1.

42. *A Sunset Review*, 22.

43. Irving F. Fellows, comment, Lincoln Seminar 1.

44. C.G.S. §12-504a(b).

45. *Curry* v. *Planning & Zoning Commission of the Town of Guilford*, 34 *Conn. Sup.* 52, 60 (1977).

46. *A "Sunset" Review*, 22.

47. Leonard E. Krogh, July 19, 1982.

48. *A "Sunset" Review*, 32.

49. Ibid., 20.

50. Fellows, *Impact of Public Act 490*, 52.

51. Furbish, "Memorandum on the Preservation of Agricultural Land to the Honorable

Audrey Beck," 1–4. These figures are based on data from USDA Farm Real Estate Market Development Report of March 1977.

52. *A "Sunset" Review*, 20. Decennial property tax revaluation is required by C.G.S. §12–62.
53. Fellows, *Impact of Public Act 490*, 52.
54. Betty Cochran, "A Practical Guide to Connecticut's Use Value Assessment Law: P.A. 490—The Open Spaces Act" (unpublished, May 1978): 121, reprinted in *A "Sunset" Review*.
55. "The Vanishing Land," 4.
56. Black, "Connecticut Puts a Damper," 11–12.
57. *Status Report, Program for the Preservation of Agricultural Lands*, 4.
58. *A "Sunset" Review*, 23.
59. Fellows, *Impact of Public Act 490*, 52.
60. "New Directions in Connecticut Planning Legislation," 9, 10.
61. Ibid.
62. *The Vanishing Land*, 4.
63. "New Directions in Connecticut Planning Legislation," 27, 49, 62–65, 71–72.
64. "Proposed: A Plan of Conservation and Development for Connecticut," Joint Resolution No. 40 (Hartford: Office of State Planning, January 1973): iii.
65. C.G.S. §16a–24.
66. "Proposed: A Plan of Conservation," iii, 8, 16.
67. *The Vanishing Land*, 7.
68. "Report of the Governor's Task Force for the Preservation of Agricultural Land" (December 20, 1974): 1, 3, 4.
69. Birkelbach and Wassall, *The Case Against Development Rights*, 38.
70. Leonard E. Krogh, July 19, 1982.
71. *Status Report, Program for the Preservation of Agricultural Lands*, 5.
72. C.G.S. §22-26ee(a).
73. Fellows and Cody, *A Food Production Plan*, 9, 11.
74. *A "Sunset" Review*, 11.
75. C.G.S. §22-26dd.
76. *A "Sunset" Review*, 12.
77. Ibid., 6.
78. *Connecticut Law Journal* (October 16, 1979): 7–8.
79. C.G.S. 22-26cc(a).
80. *Connecticut Law Journal* (January 6, 1981): 10–11.
81. *A "Sunset" Review*, 10.
82. Leonard E. Krogh, July 19, 1982.
83. Krogh, "Report of the Commissioner of Agriculture" (December 15, 1980): 6, 7.
84. Krogh, July 19, 1982.
85. *A "Sunset" Review*, 25.
86. *Connecticut Law Journal* (December 6, 1983): 50–70.
87. *Farmland Notes* (Washington, D.C.: NASDA Research Foundation Project, January 1986): 1.
88. *Status Report, Program for the Preservation of Agricultural Lands*, Table 2.
89. *Farmland Notes*, 1.
90. *Status Report, Program for the Preservation of Agricultural Lands*, 6.
91. C.G.S. §22-26cc(a).
92. Irving F. Fellows, comment, Lincoln Seminar 1.
93. *A "Sunset" Review*, 33.
94. "Connecticut PDR Program Revived with $5 Million Boost," *Farmland Notes* (Wasington, D.C.: NASDA Research Foundation Farmland Project, August, 1983): 2; "Farmland Preservation Moves Ahead in Connecticut," *New England Environmental Network News* (Summer 1983): 6, 15; Leonard E. Krogh, "Report of the Commissioner of Agriculture" (December 15, 1979): Appendix X.
95. Irving F. Fellows, comment, Lincoln Seminar 2.
96. Krogh, "Report of the Commissioner of Agriculture" (December 15, 1979): 6.
97. *A "Sunset" Review*, 16.

98. Cochran, "A Practical Guide to Connecticut's Use Value Assessment Law: P.A. 490—the Open Space Act," 66.

99. Birkelbach and Wassall, *The Case Against Development Rights*, 3.

100. C.G.S. §322-6c.

101. C.G.S. §19a-341.

102. *Farmland Notes* (Washington, D.C.: NASDA Research Foundation Farmland Project, December 1982): 1, 3.

103. C.G.S. §322-6c.

104. Special Act. No. 79-77.

105. "Recommendations and Findings of the Connecticut River Assembly: Final Report to the Connecticut General Assembly" (Capitol Region Council of Governments and the Midstate Regional Planning Agency, December 1981): 2.

106. C.G.S. §25-102 et seq.

107. P.A. 83-102, amending C.G.S. §3-20(g) and 22-6.

108. C.G.S. §22a-36-45. C.G.S. §22a-28-35.

109. *A "Sunset" Review*, 28.

110. *Wade's Dairy, Inc.* v. *Town of Fairfield*, 181 *Conn.* 556 (1980).

111. C.G.S. §8-2; C.G.S. §12-107(e). This power was held not to be an unconstitutional taking of property without just compensation in *Curry* v. *Zoning Commission of the Town of Guildford*, 34 *Conn. Sup.* 52, 61 (1977).

112. *A "Sunset" Review*, 19, referring to Frank Schnidman, "TDR: A Tool for More Equitable Land Management?" *Management and Control of Growth*, Vol. 4 (Washington, D.C.: Urban Land Institute, 1978): 52.

113. *A "Sunset" Review*, 29.

114. Ibid., 26.

6

Rhode Island Programs for the Protection of Agricultural Land

Rhode Island, the nation's smallest state, has a diverse landscape and a unique cultural history. Narragansett Bay severs much of the state, affording seagoing vessels a tranquil harbor. Historically, this large inlet supported an active mercantile industry and was actively engaged in whaling and the slave trade. Today, aside from commercial uses, the islands and beaches which dot the Bay enhance the tourist industry. The mansions of Newport remind visitors of how Rhode Island's rugged coastline lured the social elite at the turn of the twentieth century.

Narragansett Bay was responsible for Rhode Island's early development into a heavily industrialized state. Also important was the development of hydropower on the Blackstone, Pawtuxet, and other rivers. Inland cities such as Pawtucket and Woonsocket once were thriving textile centers. Many of New England's early enterprises, however, have left the region. Today, Rhode Island's economy has diversified, but many of the state's industries continue to face economic problems.

Despite Rhode Island's industrialization and small size, the state has not become urbanized from border to border. Indeed, today most of the land mass is either woodland or farmland. This geographical diversity stems, in part, from the localized nature of government: many of the thirty-nine cities and towns have resisted uncontrolled development. Generally, Rhode Islanders have sought to preserve a nostalgic New England lifestyle by resisting the pressures of urbanization, and, since the late 1970s, the state has made a substantial effort to protect the active farmland that remains.

Public Concern to Protect Agricultural Land

Rhode Island's interest in farmland protection programs, as in the other New England states, grew out of the increasing public awareness of the conversion of active agricultural land coupled with the knowledge of the specific benefits of continued agricultural activity. Although the desire to control urbanization was the most important reason farmland protection measures were adopted, concerns over food production and the quality of environment also were considerations in the broad consensus for legislative action.

Decline in Farmland Acreage

Since the 1960s, the public desire to protect agricultural lands has grown with the realization that such lands were rapidly converted to other uses. In 1966, the *Providence Sunday Journal* declared that based upon U.S. Census of Agriculture statistics from the five-year period 1959 to 1964: "Open space in the form of farms disappeared at the rate of one percent a year and will vanish completely in about fifteen years if the present trend continues."[1]

While many Rhode Island towns still have active farms, the state has suffered an enormous decline in its farmland resource, especially since World War II. In 1981, the Task Force on Agricultural Preservation, created by Governor J. Joseph Garrahy to study Rhode Island farming, concluded in its final report that:

> The role of agriculture in the Rhode Island economy and on the Rhode Island landscape has steadily and dramatically declined. This trend has been the inverse of the pattern of industrialization of the state....
>
> To say that agriculture was once a mainstay of the Rhode Island economy, and no longer is, should not be surprising—it is a logical outcome of this state's size, its geographic position on the northeast coast, the features of its land surface, and especially of the history of industrialization in North America and the revolution in agriculture which permits 4 percent of the population to feed and clothe the rest of us. What is of concern is that the linear continuation of the historical trend of farmland loss would mean that Rhode Island, along with several other states, would have no remaining farmland by the end of the twentieth century.
>
> A balance of equilibrium between farms and other land uses does not appear to have emerged in the natural course of events; the encroachment of industry, roads, and suburban development is a one-way process. The implications of this process for the Rhode Island economy, for the way of life for Rhode Islanders, and for the distribution of resources in the nation is beginning to receive public and governmental attention.[2]

In addition to describing how Rhode Island's agricultural land base has been reduced by industrialization, the Task Force reported isolated specific

conditions responsible for this trend.[3] The expansion of interstate highways, the development of produce refrigeration for transported goods, the invention of efficient farm machinery suitable only for large, flat tracts of land, and the superior irrigation methods and longer growing season of western farms, all made Rhode Island farming economically noncompetitive. Moreover, as population and industrial uses increased, more suburban, commercial, and recreational land was needed. Developers leapfrogged into more remote areas of the state, once used for farming, because of low land costs. To provide services to these new communities, local governments encouraged greater development to increase their tax base. As development spread, land values grew and real estate taxes increased. Many Rhode Island farmers were induced to abandon or convert their land to other uses because of the high tax rates. Indeed, in 1978 farm real estate taxes in Rhode Island averaged $25.96 per acre, the highest of any state in the country; the national average was $3.40 per acre.[4] Thus, both highly competitive modern farm practices in the West and the need for land to absorb urbanization in Rhode Island contributed to the decline of active agriculture.

Statistics compiled over the years by the U.S. Census of Agriculture illustrate the magnitude of the decline in Rhode Island farmland acreage. In 1880, almost 77 percent, or nearly 515,000 acres, of all land in the state was devoted to agriculture. By 1982, the figure was only 11 percent, or 62,466 acres, although the more recent data compiled from non-Census sources differ slightly.[5] Since World War II, the conversion of Rhode Island farmland to other uses has been particularly dramatic. In fact, between 1940 and 1970 the acreage of all farmland was cut roughly by two-thirds, dropping from around 220,000 acres to 61,000 acres. Dr. Thomas Weaver, chairman of the Department of Resource Economics at the University of Rhode Island in South Kingston, believes that the acreage of good cropland—land that is responsive to fertilizer, is well-laid, and the like—declined by a similar margin over the same period to less than 30,000 acres.[6] Figure 6.1 and Table 6.1 show the decline of farmland acreage in Rhode Island from 1880 to 1982.

Most of the farmland which has been converted to other uses was abandoned to woodland and not developed.[7] Indeed, Dr. Weaver believes that of the approximately 60,000 acres taken out of production between 1960 and 1974, only about 1,000 acres of prime farmland were converted to developed uses.[8] Thus, much of the state's resource in primary and secondary agricultural soils still lies in forest lands, not in subdivisions and office parks. This land could be returned to production as well as developed, although returning such land to an agricultural use is not generally economically feasible after lengthy periods of abandonment.

Despite the overall decline in acreage, some statistics indicate that certain types of farming have remained viable in Rhode Island. A 1982 release of the

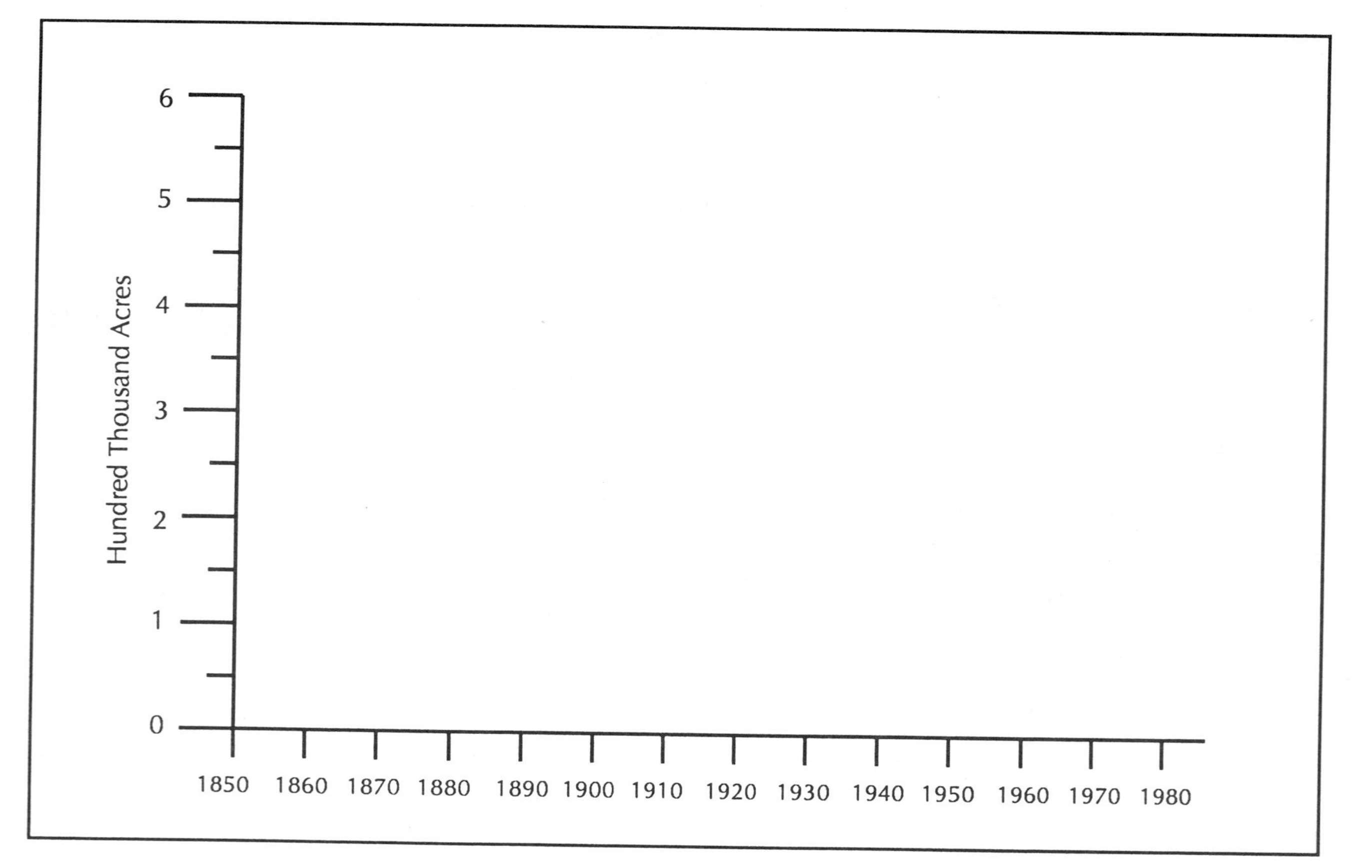

Figure 6.1. Land in farms—Rhode Island. *Source:* The Governor's Task Force on Agricultural Preservation, 1981.

Table 6.1
Statewide Change in Acreage of Farmland and Number of Farms in Rhode Island, 1880–1982

Year	*Acres*	*Number of Farms*	*Change*	*% Change*	*% Change from 1880*	*Average % Change per Year*	*Average Farm Size (acres)*
1880	514,813	6,216	—	—	—	—	82.8
1910	443,000	5,292	-71,813	-13.9	-13.9	-0.5	83.7
1940	221,913	3,014	-221,087	-49.9	-56.9	-1.7	73.6
1969	68,720	700	-153,193	-69.0	-86.7	-2.4	98.2
1978*	66,233	676	-2,487	-3.6	-87.1	-0.4	98.0
1982	62,466	728	-3,767	-5.7	-87.9	-1.4	85.8

*The 1978 Agricultural Statistics were revised in 1982 to compare more accurately to the 1982 and previous U.S. Agricultural statistics.

Source: U.S. Bureau of the Census, *Census of Agriculture.*

Rhode Island Department of Environmental Management indicated that much recently abandoned farmland was being cultivated by new farmers.[9] The 1978 Census of Agriculture reported that from 13,000 to 14,000 acres of Rhode Island soil were returned to agriculture between 1974 and 1978.[10] Although this change could be the result of the difference in the enumeration techniques employed by the Census for the two years,[11] the *Report of the Governor's Task Force on Agricultural Preservation* concluded that even if the techniques were kept constant, an 8 percent rise of land in farms is indicated.[12] Thus, it appears that the major declines in farmland acreage may have ended—which may be the result of farmland protection efforts.

In addition to the amount of farmland, the number of farms also appears to have increased during the 1970s. The 1978 Census determined that there were 866 farms in the state compared with 597 farms in 1974.[13] The Agricultural Task Force attributed this increase to a greater prevalence of part-time farmers who work smaller tracts.[14] Table 6.2 summarizes this trend.

Despite the increase of part-time farmers, about two-thirds of Rhode Island's farms are run on a full-time basis.[15] As discussed in Chapter 2, the average Rhode Island farm site is less than 100 acres,[16] making the state's farms the smallest, on the average, of all of the six New England states. It is quite possible that farming has continued because of the growth of part-time farming. Only about 1 percent of the labor force is "primarily engaged in" farming,[17] but agriculture as a second source of family income is becoming increasingly common.

Although the current acreage devoted to agricultural uses is very small, Rhode Island is noteworthy among the New England states for the diversity of its products. Dairy operations account for the greatest use of agricultural

Table 6.2
Number of Rhode Island Farms by Size, 1969–1978

Farm Size (Acres)	*1969*	*% Change 1969-1974*	*1974*	*% Change 1974–1978*	*1978*
1–49	195	(-10)	176	(+14)	201
50-499	256	(-11)	229	(+12)	257
500+	11	(+27)	14	(-29)	10
Total	462	(-9)	419	(+12)	468

Source: Report to the Governor's Task Force on Agricultural Preservation, January 1981, 29 (using Census of Agriculture statistics).

land, approximately 12,822 acres. The nursery business leads all other agricultural endeavors in cash receipts. Other important agricultural activities include: orchards, hog, and small animal production; vegetables; and potatoes.[18] Table 6.3 and Figure 6.2 illustrate the major cropland and traditional farm activities which remain in the state.

Agriculture makes an important contribution to the state's economy, largely because the industry is dominated by high-value, nontraditional farm commodities. Net annual sales total about $70 million,[19] but the 130 commercial nurseries of Rhode Island produce the greatest percentage of that sum, with an annual revenue of over $20 million.[20] Nurseries and turf farms both produce commodioties whose high value is the result of construction and landscaping demands caused by the urbanization of land which oftentimes was devoted to traditional agriculture.[21]

Despite the important role of agriculture in Rhode Island, the *Governor's Task Force Report* concluded that this activity is in jeopardy because of a land tenure pattern that is unique in the United States. A 1980 USDA report found that only 3.8 percent of Rhode Island's farmland was owned by the farmer-operators.[22] If accurate, absentee owners or retired farmers own nearly all of the state's farmland and may be waiting to sell their land. This situation concerns many people in the state because much of the farmland lies in the coastal regions where development pressures are greatest. The *Task Force Report* concludes, therefore, that "much of this land could be converted to other uses or simply withdrawn from farming virtually overnight." On the other hand, the same Report, which apparently relies on U.S. Census data, states that "the majority of farmers own the land they work... [and] the number of tenant farmers . . . accounts for only about 7% of all farmers."[23] Although these statistics appear irreconcilable, thereby rendering the land tenure issue an uncertain problem, Daniel Varin, chief of the Rhode Island Office of State Planning and the director of the Task Force, believes that the 3.8 percent figure was based on a nonrepresentative

Table 6.3
Rhode Island Agriculture: Farms, Acreage, and Value (1983)

	Number of Farms	*Acreage*	*Value of Crops ($millions)*
Dairy	74	12,822	6.6
Turf	14	1,803	8.7
Potato	14	2,670	5.5
Poultry and Egg	18	—	8.4
Orchards	29	702	1.6
Nursery	33	2,817	20–30 (approx.)
Fruits and Vegetables	94	2,731	4.2

Source: Rhode Island Department of Environmental Management, Division of Agriculture and Marketing.

sample, and in fact more than one-half of all active agricultural land is leased from a nonfarm landowner.[24] This degree of nonfarmer ownership helps make much Rhode Island farmland some of the most vulnerable to conversion in New England.

In sum, despite widely varying data, a few conclusions can be reached about the decline in Rhode Island's farmland acreage. First, the acreage devoted to farming in the state has declined tremendously since the Civil War, with the most dramatic decline occurring from 1940 to 1970. Second, much of this decline was because of the state's industrialization which made farming generally uneconomical, thereby causing the conversion of farmland largely into forest uses. Third, to meet economic realities, primarily as a result of competition from large Western farms, farming activities in Rhode Island have become increasingly nontraditional, specializing in turf, nursery, and other commodities needed for an expanding urban market. Finally, the decline in farmland acreage appears to have leveled off.

Concern over Spreading Urbanization

The decline of active agricultural acreage was viewed by many Rhode Island citizens as the direct result of changing land use patterns in which cities and suburbs were spreading haphazardly across once rural areas. Indeed, the initial and still most pervasive concern, leading to the adoption of the first agricultural protection programs in the state, was the desire to control urbanization, as many communities questioned the value of unmanaged growth.

The *Providence Sunday Journal* helped sound the alarm in 1966 with the statement: "The Rhode Island farm is disappearing. It is being replaced by golf courses, industrial plants, highway intersections, shopping centers,

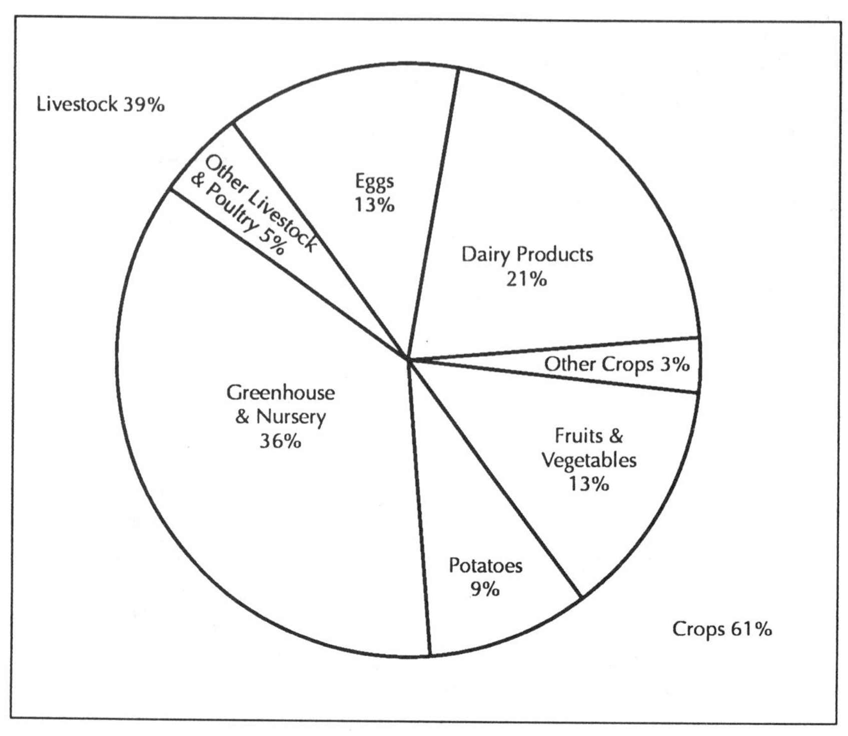

Figure 6.2. Distribution of cash receipts from Rhode Island farm marketings in 1983.

colleges, churches, and housing developments."[25] The basis for this fear of urban sprawl was clearly expressed in the 1975 report of the Rhode Island State Planning Council, which contains the state's long-range land use plan. This report, discussed more fully later in this chapter, states in part:

> Rhode Island's population density of over 900 persons per square mile makes it one of the most highly urbanized places in the world. Yet, throughout its history, rural land uses have overwhelmingly dominated the state's development pattern. In 1850, agricultural uses accounted for approximately 80 percent of the total land area. By1930, this figure dropped to 40 percent, but another 38 percent was still classified as "forest, sprout and scrub land." During the entire 325 years from 1636, the date of Rhode Island's first settlement to 1960, less than one-quarter of the total land area was developed for any form of urban use.
>
> Since 1960, the situation has suddenly changed as the pace of urbanization has gained momentum. It is projected that, at the present rates of urbanization, another one-quarter of the state's land area will be brought into urban use between 1960 and 1990. This rate is more than eight times as fast as the rate before 1960. It means that, if present trends continue, more than 110,000 acres

> of land in Rhode Island which were open space in 1960 will be in urban use in 1990. This amount is equal to the combined areas of Providence, Pawtucket, Central Falls, East Providence, North Providence, Cranston, Warwick, West Warwick, East Greenwich, and North Kingston. Much of the state's new population growth will occur in formerly rural areas, where its impact on the environment is most intensive.
>
> As this rapid transition takes place, Rhode Island lacks explicit policies at the state level as to how land should be developed and lacks any effective mechanisms for implementing such policies if they exist. In this policy vacuum, new urban development scatters over the countryside in virtually random fashion, without regard for suitability of land for various uses; the compatibility of neighboring uses; pollution of earth, air and water; conservation of natural and historic resources; spatial relationships between residences, employment, and services; or any of the other factors which determine the quality of urban life.[26]

Among the ways agricultural and other open space lands enhance Rhode Islanders' quality of life is by providing recreational opportunities. As the State Planning Council noted in 1975: "Passive as well as active recreation facilities are important for the enjoyment of citizens and for maintaining the natural resource base: rural open land, plants and wildlife, scenic beauty, and other natural features."[27] Agricultural lands were and are an essential part of the scenic beauty of the state that statistics cannot measure. Moreover, maintaining the various components of nonurbanized areas provides an important source of state revenue. The State Planning Council further noted that:

> Open space (in the form of agricultural and forest areas) also has an economic value, since it is the major factor in attracting the state's substantial tourist trade. In 1970, visitors to Rhode Island spent an estimated $80 million. It is figured conservatively that every million dollars spent by tourists turns over several times and produces $320,000 worth of state and local tax revenue.[28]

Thus, providing for recreational space, in the form of farmlands and other open space areas, was viewed as serving aesthetic, recreational, and economic ends.

Although farmland generally was not converted directly into urban uses, the abandonment, and sometimes the eventual development, of farmland was visible to Rhode Island citizens. This abandonment occurred at the same time that urban uses were growing nearby, rendering farming uneconomical because of considerations such as increased property taxation, discussed in Chapter 2. Thus, the link between urbanization and farmland decline became apparent to the public, and farmland protection policies and mechanisms were viewed as a means of controlling urbanization so that some green space areas would remain intact.

Concern over Food Production

In addition to spreading urbanization, other concerns fueling the desire to protect agricultural lands surfaced for legislative consideration in the 1970s in both the *Planning Council Report* and the *Governor's Task Force (Preliminary) Report on Agricultural Preservation.* Both reports, and especially the latter, stated that in addition to reducing urban spawl, the public viewed food production as a worthwhile goal of farmland protection efforts.

In the 1960s, improvements in farm technology kept pace with the growing food needs of the state. Statistics compiled by the Department of Food and Resource Economics at the University of Rhode Island's College of Agriculture revealed that the state's 1,115 farms of 1964 produced more and better food than its 6,200 farms of 1880.[29] In the 1970s, however, with an overall continuation of the conversion of agricultural lands, further shifts in the types of goods produced, and a steady growth in population, the food production capability of the state became questionable.

Rhode Islanders consider food production beneficial in many respects. Aside from providing some employment and income, locally grown foods are viewed by many consumers as more nutritious or healthier.[30] Also, Rhode Island citizens feared that food imports may not be as abundant in future years as they were in the early 1970s. For example, Dr. Thomas Weaver noted in 1980 that if Providence suffered from a transportation breakdown or sudden calamity, only a four- or five-day supply of food would have been on hand.[31] Moreover, in the wake of the international oil embargo of the mid-1970s, citizens became concerned that a local food supply was needed since western farmers experienced occasional water shortages and transportation costs were rising.[32]

Food production in Rhode Island also was considered valuable on a global scale. Although the state could not become self-sufficient, any production might allow previously imported goods to be exported abroad. There was a belief that this production would ease the United States' balance-of-payments deficit. More importantly, many citizens believed from a moral standpoint that any production would aid the problem of world starvation.[33] Thus concern over food production was another reason for citizens to desire to protect agricultural lands.

Concern over Environmental Quality

Another important objective of farmland protection efforts in Rhode Island, expressed in both the *Planning Council* and *Governor's Task Force Reports,* was the need of a healthy environment. Although environmental concerns in the1960s and 1970s for the protection of wetlands and coastal areas tended to encourage agricultural uses rather than development in those areas,[34] agricultural lands themselves were not viewed as a critical resource. As recognized in Connecticut, agricultural lands serve as ground water recharge areas[35] which purify and collect water needed for nonagricultural pursuits.

There were special-interest concerns as well. For example, forest owners wanted their tax burden reduced through agricultural-type preservation laws. Nonetheless, the general public sentiment was for adoption of effective protection measures to benefit citizens, tourists, and even victims of famine abroad.[36] The breadth of these concerns is described in the following finding of the Task Force:

> Certain agricultural land should be preserved because it is best suited to that use, and adequate land is available elsewhere for other activities. Similar objectives are to protect farmland in order to offer relief from urbanization, to shape development patterns, to maintain an ecological balance, to control pollution, to provide for recreation, to maintain valued lifestyles, and to create an attractive environment which may, in turn, attract other forms of economic activity. These objectives range from the more scientific environmental considerations to such intangible motivations as the feeling of wanting the next generation still to have first hand acquaintance with farms.[37]

In summary, the three predominant concerns leading to agricultural land protection in Rhode Island were to control urbanization and its effect on the quality of life, to enhance food production, and to provide a healthy environment. These concerns fueled the desire to protect agricultural land specifically. However, the genesis of many of the early efforts to protect agricultural lands arose from a more general concern to protect environmental quality. The following section discusses such early efforts and the evolution into focused agricultural land protection programs.

Conservation Programs and Early Efforts to Protect Agricultural Land

As in the other New England states, initial efforts to protect agricultural land in Rhode Island grew primarily from a broad concern over general environmental quality. In the 1960s, a number of private organizations, such as a "Save Rhode Island Farms" group, and other concerned citizens' groups worked to bring about legislative changes that had some positive impact on farmland retention. Among these changes were the creation of Conservation Commissions, the adoption of a Green Acres Act, and a differential taxation law.

Conservation Commission Act (1960)

Largely through the efforts of organizations such as the Rhode Island Audubon Society and the Rhode Island Wildlife Federation, the state passed Conservation Commission enabling legislation in 1960. Rhode Island thus became the first state to follow the lead of Massachusetts in allowing local communities to establish these bodies "to promote and develop the natural resources, to protect the watershed resources and to preserve natural aesthetic areas."[38] Several local commissions soon formed, and by 1984 thirty-three of the state's thirty-nine municipalities had active Conservation Commissions.[39]

Despite the assistance of the Department of Environmental Management (DEM), the effectiveness of Conservation Commissions in their conservation efforts has been limited by a lack of coordinated state administrative and financial support.[40] Additionally, some local governments encourage development to increase local tax revenues and, therefore, have not always welcomed an active commission. Nevertheless, these commissions' success has been marked on several occasions. In the late 1960s, for example, the Conservation Commission in Lincoln, a town eight miles north of Providence, convinced the town government to set aside about 40 percent of its land area as open space. Pursuant to the state's enabling act, the Commission received donations of land and, with private, state, and federal funds, purchased the development rights or the full title to much of the town's remaining conservation lands. In 1984, the land was still under Conservation Commission management, largely in the form of ponds, marshes and woodlands.[41]

Regarding farmland protection efforts, Conservation Commissions helped gain support for the Green Acres program of the 1960s and the Farm, Forest and Open Space Land Act of 1968. The Rhode Island Association of Conservation Commissions encourages its member commissions to work with local property tax assessors in implementing the 1968 Act, as the commissions have express statutory authority to give testimony in classification and assessment appeals under the Act. In general, however, Conservation Commissions play only an advisory role in the formation of local decisions which affect agricultural lands.

The Green Acres program, another legislative movement of the 1960s, was designed to protect land for recreational and conservation purposes, and specifically land for agriculture. Like the Conservation Commission enabling legislation, it did not focus primarily on agricultural land.

Green Acres Land Acquisition Act (1964)

In 1964, the Rhode Island General Assembly established a Green Acres program, allowing the state to acquire and improve lands for public use and also to make grants to localities for the same purpose.[42] The legislation provides that the lands involved must be devoted to recreational and conservation purposes. Among those purposes explicitly listed in the Act is "the use of lands for agriculture."

In 1966, the voters approved a $5 million bond issue to fund the program. Funds were soon expended largely on seventeen state acquisition projects, totaling approximately 3,100 acres, as well as seven recreational improvement projects. In addition, thirty-one different municipalites, including Lincoln, in conjunction with its Conservation Commission, acquired 1,600 acres in seventy-two projects, and sixty-nine local projects were authorized.[43] The vast majority of the lands involved were devoted to recreational uses. A few agricultural parcels, notably in Cranston, were purchased, but most of the tracts were not purchased for long-term farming purposes and later became part of the state's recreational facilities. In 1968, a second bond issue was narrowly defeated by the voters, probably because Rhode Island was expending its federal funds for land and conservation projects in the mid-1960s.[44]

The Green Acres Land Acquisition Act is still "on the books" and could play a role in farmland protection efforts in the future. However, Daniel Varin believes that with more than one-half of the state devoted to open space uses, the public has not discerned a real need to allocate state revenues into a general conservation and land acquisition program.[45] What the voters have been willing to allocate funds for, however, is the protection of agricultural lands. The Farmland Preservation Act, discussed later in this chapter, is a more recent, limited land acquisition program providing for the state purchase of farmland development rights, which is widely supported.

The Farm, Forest and Open Space Land Act (1968)

Conservation Commission enabling legislation and the Green Acres program were not focused primarily on agricultural lands. In 1968, however, the Rhode Island legislature passed a law providing for use-value taxation of farm, forest, and open space lands.[46] This legislation was largely the result of efforts by several environmentally concerned groups, including local Conservation Commissions. Modeled largely after similar legislation adopted in Connecticut in 1963, this tax law was the first state program in Rhode Island to have a significant impact on farmland retention.

The purpose of the Farm, Forest and Open Space Land Act is expressed in the following legislative declaration:

> a. that it is in the public interest to encourage the preservation of farm, forest and open space land in order to maintain a readily available source of food and farm products close to the metropolitan areas of the state, to conserve the state's natural resources and to provide for the welfare and happiness of the inhabitants of the state.

b. that it is in the public interest to prevent the forced conversion of farmland, forest and open space to more intensive uses as the result of economic pressures caused by the assessment thereof for purposes of property taxation at values incompatible with their preservation as such farm, forest and open space land.[47]

Like most differential assessment laws, the Act was designed to ease the farm landowners' property tax burden by assessing farmlands at their value in agriculture, rather than their fair-market value; thus, these owners could keep their land in agriculture. Such legislation was needed because of Rhode Island's high farm real estate taxes.

The original Act established an annual valuation of enrolled parcels by local assessors for both development and agricultural purposes. If the use of the land changed to development, then a "rollback" tax would become due. The rollback penalty was the difference between the tax assessed at use value and the tax assessed at development value for the year of development. If the parcel was enrolled for three consecutive years, then the two prior years would be included as well.

Various problems arose with the program once it was implemented. The State Planning Council reported some of these administrative problems and made recommendations to address them in its 1975 Report. More important to reform, however, were the preliminary recommendations of the Governor's Task Force on Agricultural Preservation, which proposed revisions in the Act. Acting upon these suggestions, the legislature redesigned the legislation to improve its effectiveness, for lands that came under the program after December 31, 1980.

The primary problem with the 1968 legislation was its administration. The 1968 Farm, Forest and Open Space Land Act failed to define clearly the categories of land falling under the law. Moreover, as with the Connecticut current-use assessment law, local assessors had wide discretion in determining whether a parcel of land qualified under the Act. As a result, assessments were not uniform,[48] and at times the need to raise tax revenues resulted in classification decisions that did not reflect local or state policies on land use and the protection of agricultural lands.[49] The natural tendency on the part of local assessors to resist the erosion of the tax base probably was reinforced by the increase of urban dwellers moving into the rural areas of the state. While some argued that taxation of farmland at a lower value was equitable because such land required less in the way of public services, others pointed out that recent years had brought a well-educated, articulate population to rural Rhode Island who were demanding the same services that had been available in the suburbs and cities.[50]

In addition, because the reduction of tax revenues fell directly upon municipalities, some towns simply avoided using the Act's enabling powers by not allowing use-value assessments.[51] The law proved inequitable, therefore, because farmers in different towns could be producing the same goods on similar parcels of land, and yet the difference in their tax bills could be substantial.

Aside from inequitable local administration, another problem with the 1968 Act was that it failed to provide for simple appeal procedures. A dissatisfied applicant was forced to bear the cost and delay of court proceedings.[52] This defect in the law compounded the frustration of a landowner who felt he was taxed improperly.

Finally, the original Act was criticized heavily because its penalty provision allowed for abuse of the program. "Speculators, posing as farmers, [took] advantage of the law in order to save on taxes."[53] Investors would allow the infrastructure needed for development to be built while paying low assessments. Later, they would convert the land to a developed use. Observing such activity has caused Daniel Varin to believe that the "predominant" use of the Act was "in urban fringe communities by well-known developers for purposes obviously not related to agriculture." This abuse of the program occurred primarily because the penalty provisions simply were not strict enough, and in fact the short two-year rollback period encouraged use of the law as a tax shelter, rather than as a device for protecting farmland. The problem was heightened by the many local tax assessors who frequently did not bother to make both development and agricultural assessments each year, rendering the rollback tax difficult, if not impossible, to apply fairly. Moreover, appraisals often were not kept current, some parcels in the state not having been reappraised for fifty years or more.[54] Without full valuation, the penalty tax could be quite small. These assessment practices also deterred many farmers from participating in the program because they were unwilling to risk paying a high rollback penalty if they were later forced to stop farming for economic reasons. As the Task Force on Agricultural Preservation commented in February 1980: "the provisions for collection of deferred taxes if the use of the land was changed were difficult to administer, unfair to the community and discouraging to the prospective applicant."[55]

To rectify these flaws in the use-value taxation program, the Task Force recommended to Governor Garrahy that the categories of land covered by the Act be more carefully defined and that the director of the Department of Environmental Management (DEM), rather than the local assessor, determine which lands would qualify as farmland under the law. Furthermore, the Task Force recommended that an administrative appeals procedure be established and a change-of-use tax penalty be substituted for the two-year rollback provision.[56] The amendments suggested by the Task Force were submitted to the legislature in 1980 and adopted in the same year.

Under the 1980 amendments, a landowner can apply to the director of the DEM seeking to have his land classified as farmland under the law.[57] The DEM has promulgated rules and regulations to determine what can be classified as farmland.[58] Generally, the parcel must be five acres or more and "actively" farmed. The land is "actively" farmed if $2,500 in gross income is produced in one of two prior years. There are exceptions to the income criterion as well as conditions on the nature of the agricultural activity.

When the director determines that a tract of land constitutes farmland, the local assessor is so informed. The landowner then applies to the local assessor to have the land taxed at its use value in agriculture rather than at its fair-market value. The Department of Community Affairs must publish and make available information on land values to local assessors, although the assessors are not bound to utilize this data in determining use value.

Provision is made for appeal in the event the local assessor either refuses the application or sets the use value at too high a level, or if the DEM finds that the land use has changed. The landowner is entitled to a hearing before the local board of assessment review or the city or town council and can bring an appeal to the state superior court. The court reviews all of the evidence when making its decision and may also order the local review process to be reopened if new evidence is discovered.

The penalty provision in the new legislation provides for a 10 percent land use change tax, replacing the rollback tax in the 1968 version of the Act. This tax works on a sliding scale. It equals 10 percent of the fair-market value of the land for the first six years of use-value taxation. The rate declines to 9 percent the seventh year, 8 percent the eighth year, and so forth, up to a point where no penalty is assessed in the sixteenth year. The first five years of this period can be voided if the land was farmed continuously for five years prior to classification under the Act. Fair-market value is determined by either the sale price, if a sale occurs, or by established appraisal techniques. A change of ownership starts the computation period running anew, except for certain types of inheritance or interfamily transfers.

The sanction's aim is to penalize an owner or purchaser who develops the land, thereby recovering back-tax revenue, and to deter land speculators from using the program to avoid paying their fair share of infrastructure costs. It was hoped that with DEM participation and the stiffer penalty for converting land uses, only bona fide farmers would benefit from current use taxation. The six-year period for the flat 10 percent tax was selected precisely because, according to Daniel Varin, "available information showed that speculators usually held land to 'ripen' for development for five years or less."[59] If the penalty proves ineffective, other means of deterrence may be adopted, such as coupling exclusive open space or agricultural zoning with use-value taxation. As a result, would-be developers would be forced to attempt the sometimes difficult task of getting their land rezoned.[60]

The DEM repeatedly has emphasized that the law does not create a tax exemption because the tax rate at use value can equal the current tax rate at market value on some parcels.[61] The Act has worked best, therefore, in urban fringe areas where the land is rezoned to a higher use or when local government revalues the property.

In the first year after the new amendments, the DEM received over 140 applications for the 1981 tax year and designated 5,054 acres as farmland in nineteen of Rhode Island's thirty-nine towns and cities. Most of these applications were from landowners whose property recently was revalued.[62] By the beginning of 1984, Stephen Morin, assistant to the director of the DEM, reported that only about 23 percent of all farms that qualified under the Act were enrolled in the program, although the number of participants had increased to over 225, covering about 11,000 acres.[63] At the same time, four towns had not revalued as they are required to do by the end of 1984, pursuant to the Property Tax and Fiscal Disclosure Act.[64] Participation is expected to rise because farmers in about fifteen towns were revalued in 1983. For example, one 100-acre farm in Cranston was revalued in 1979 for the first time since 1954. Partly because of industrial zoning, the property tax bill jumped from $1,100 to $72,000![65] Thus, farmland owners who wish to keep their land in agriculture may be induced to join the program, and if the penalty provisions prove effective, development will be delayed on such tracts for some time.

Initially, many people felt this law was effective in halting urban sprawl, but more recent opinion is to the contrary.[66] Although many farming operations may become economically viable under the program, many observers believe still stronger measures are needed. Because the tax program is completely voluntary, a farmer can choose not to participate and subsequently sell his land for a new use. Moreover, development pressures could become so great that even owners of farmland participating in the use-value tax assessment program, as amended, could yield to these pressures and still profit after payment of the land use change tax, especially if over time the land use change tax will be small. Addressing these problems, beginning in 1980, the Rhode Island legislature has enacted stronger farmland protection measures in direct response to the reports of the State Planning Council and the Governor's Task Force on Agricultural Preservation.

Recent Agricultural Land Protection Programs

Public concern over protecting agricultural lands in Rhode Island came into sharper focus in the 1970s. With the 1975 *State Planning Council Report*, agricultural lands were identified as an important component in the mix of land uses that make up the state's landscape. In 1979, the Governor's Task

Force Report on Agricultural Preservation made clear that agriculture was no longer considered simply one aspect of conservation legislation or general land use plans. Rather, as a result of the efforts of farmland protection advocates, a number of initiatives and programs specifically designed to protect farmlands emerged. The most significant of these programs are the 1981 Purchase of Development Rights law, a 1982 right-to-farm law, and various zoning and marketing efforts.

The State Land Use Policies and Plan, issued in 1975 by the State Planning Council, views agriculture as playing a limited but important role in the state's future landscape. Following is a description of its evolution and findings.

State Land Use Policies and Plan (1975)

In 1964, the Rhode Island State Planning Council was established by Governor John H. Chafee to plan for the physical, economic, and social development of the state, and in 1978 this program was codified.[67] The State Planning Council was responsible for the program and was composed of various local, state, and federal representatives. The Council began working on a long-range state land use plan, which included the role of farmlands. In 1969, a preliminary land use plan was issued. In 1975, the Council issued the "State Land Use Policies and Plan," as one element of the State Guide Plan. This report set forth the state's land use policy and plan for the years from 1970 to 1990. Rhode Island is the only state that was able to do this.

The report describes four general goals for land use:

1. Relate state land use policies to a population ceiling of 1.5 million.
2. Make efficient use of available land and water, producing a visually pleasing, coherent and workable environment.
3. Sustain economic growth at a rate adequate to support the state's population, in a manner consistent with the state's characteristics, capabilities, and environmental objectives.
4. Continuously improve the structure and operations of governments and their responsiveness to their citizens in the area of land use planning and management.[68]

The role of agriculture was considered in formulating these goals. The report stated that a fifty-fifty balance in open space and developed areas was appropriate and desirable for the 1.5 million population ceiling. Measuring supportable population by the sufficiency of native food products was not considered feasible because of the state's low and rapidly declining agricultural production. Yet among the environmental objectives was the goal of conserving and protecting "desirable existing residential, commercial, industrial, and agricultural areas." In addition, the Council articulated a specific policy goal: "Preserve and protect open space, including recreation and conservation areas, rural and open land, and selected agricultural forest areas, so as to enhance the total quality of the environment."[69]

The report views agriculture as playing a limited but important role in the state's future landscape. The prime farmland that was designated for protection was for "dairy and poultry farms, nurseries and greenhouses, field crops (chiefly potatoes), fruit (chiefly apples), and apiaries," and its value was first and foremost as an urban buffer. The general decline of agricultural activity, however, was viewed as inevitable. Three types of good agricultural lands were considered likely to remain in production only until 1985; the fastest declining agricultural lands were pressured by urbanizing forces such as "tax competition, spreading residential areas, and construction of new highways."[70] This somewhat bleak illustration of the decline in active agricultural lands was expected, however, because of the focus and scope of the Council's inquiry. Stephen Morin of the DEM explains that the study was concerned mainly with the lands near urban areas which were facing development pressure, not agricultural lands in more remote regions.[71] Moreover, the Plan did not account for legislative initiatives, subsequently adopted, which were designed to protect farmlands. Nevertheless, one aspect of the state's official land use policy had become clear by 1975: the best agricultural lands should be protected.

Governor's Task Force on Agricultural Preservation (1979)

The State Planning Council's State Land Use Policies and Plan discussed various aspects of the physical, economic, and social development of the state. One area about which the Council expressed concern was agricultural land protection. After the report was released, the public felt that the farmland issue deserved greater attention. Accordingly, in 1977 the legislature established the Department of Environmental Management (DEM), replacing the Department of Natural Resources, with several mandates, one of which was to study the issue of farmland conversion. Two years later, Governor J. Joseph Garrahy issued Executive Order No. 79-6, based largely on the recommendation of the DEM, which created the Task Force on Agricultural Preservation.

The Task Force members represented four different perspectives: active farmers, special-interest groups (including those in marketing, conservation, real estate, and the Rhode Island Farm Bureau), local government (including a tax assessor, a town council president in a rural community, and a Conservation Commission member), and state government (including representatives from the Department of Economic Development, the DEM, and the Office of State Planning). The governor charged the group with five major tasks:

1. To assess the need for a comprehensive Food and Agricultural Policy for the State.
2. To investigate the effect of Federal, State, and local taxes on farming, specifically estate and gift taxes; sales tax on farm equipment; property taxes; income tax; capital-gains tax; and the Farm, Forest, and Open Spaces Act.
3. To review the current marketing situation in Rhode Island.
4. To examine all Federal, State and local laws, ordinances and policies that impact on farming and farmland.
5. To review its findings and make recommendations to the Governor on the need for a balanced program of state, local, and private-sector initiatives to protect and promote farming and to encourage the preservation of farmland in Rhode Island.[72]

In response to this charge, the Task Force set forth to answer the twofold question of "What must be done to first, stabilize the present level of agricultural activity, and, second, create the conditions that would permit future growth?"[73] Numerous meetings and substantial research took place. The members of the Task Force decided that the first step toward resolving this question would be to establish a means of farmland protection.

The Task Force proposed numerous programs in its 1981 final report. The proposed revisions in the Farm, Forest and Open Space Land Act, discussed earlier in this chapter, were adopted in 1980 after issuance of preliminary recommendations on that program. The proposed adoption of the Farmland Preservation Act (a PDR program), a Right-to-Farm Law, and several marketing approaches also resulted in legislation. Other recommendations of the Task Force have had legislative consideration but have not been enacted.

A major proposal of the Task Force on Agricultural Preservation was an "agri-bond" program under which the state would provide long-term, low-interest financing for farmers interested in buying land and major pieces of equipment.[74] While this program may help farmers get started, or get a new start, it does not ensure that farmland operations will remain profitable. The state's Port Authority and Economic Development Corporation could operate the program under existing legislation.[75]

Many proposals, including educational programs, have been made by the Task Force and other groups which address the farmland-protection issue, but have not been adopted by the legislature. Most of these programs would play only a minor or coordinating role in retaining active agricultural land.

The Task Force also proposed to inform landowners who are eligible for conservation restrictions of the benefit of such action.[76] Under Rhode

Island law, owners of land can give the development rights to a public or private organization for tax benefits.[77] The Task Force also recommended that state government agencies be requied to file an agricultural impact statement when farmland is taken under a state eminent domain proceeding, which would establish that the land was to be taken for the public welfare and that no other prudent and feasible alternative exists.[78] This proposal is similar to the Massachusetts statute involving state land takings and, of course, would only protect those farmlands threatened by state governmental action.

The Task Force also addressed the growing problem of soil erosion. Extensive sedimentation of water bodies results each year from erosion in Rhode Island, most of which came from agricultural operations. The Task Force recommended a change in the state property tax law as an incentive for erosion and sedimentation control measures.[79] While this type of program may help protect agricultural lands as a resource base, it does little to make farming economically viable.

Because of a similar problem—soil nutrient depletion as a result of farmland conversion and the turf and nursery industries—the Task Force also recommended that composted sludge from municipal sewage treatment plans be used as a soil conditioner.[80] While this measure was not adopted, sludge is beginning to be used in some farming operations.

The 1982 legislature enacted the Soil Erosion and Sediment Control Act.[81] Essentially, this measure is an enabling act allowing municipalities to adopt an erosion control ordinance. The model ordinance requires local approval to develop land. Among the exceptions is when a "development project" disturbs less than one-half an acre of land in planting season, on land that has a slope of less than 10 percent and is not within one hundred feet of a watercourse; and further, the building official must render an opinion that no erosion will take place. For the farmer, an exemption is recommended for "accepted agricultural management practices such as seasonal tilling and harvest activities associated with property utilized for private and/or commercial agricultural or silvacultural purposes."[82] Thus, farmers are encouraged to use their land for agricultural pursuits, and local governments may be able to forestall or prevent the development of agricultural lands under such an ordinance. Several state agencies are using local workshops to aid municipalities in adopting local ordinances.

Another recommendation of the Task Force was to establish special programs to help the small farmer. This recommendation was viewed as desirable because part-time family farms as small as twenty acres are considered important to protect land in agriculture or prevent its conversion.[83] In a state such as Rhode Island where the average farm size is only ninety-eight acres,[84] the protection of so many small farms can have a significant impact. These operations are also important because they support the agricultural infrastructure (such as supplier, marketing, and processing facilities) which helps make up the critical mass necessary to encourage larger operations. Also, they serve to break up the urban nature of Rhode Island. Accordingly, the Task Force proposed that financing, training, and marketing assistance be afforded to such farmers,[85] but no legislation has yet been adopted for this purpose.

Without doubt, the Governor's Task Force Report on Agricultural Preservation has become the central document for agricultural land protection efforts in Rhode Island. Its publication marked the full recognition of the importance of agriculture in the state and the beginning of comprehensive efforts to protect agricultural lands. As a result, a number of initiatives and programs emerged specifically to protect farmlands. Following is a discussion of the most significant programs: the Farmland Preservation Act, the Right-to-Farm Act, and zoning and marketing efforts.

After reviewing the various means which have been employed throughout the United States for protecting agricultural land, the Task Force concluded, among other things, that Rhode Island should initiate a program for the purchase of development rights. According to the Task Force's report, such a program:

> is the only effective way of dealing with land under the immediate threat of conversion. This most frequently occurs on the fringe of urban communities but can occur anywhere. It also provides a means for the farmer to realize some of the appreciation in value of this land without requiring that he dispose of it for developmental or speculative purposes. Conversion of part of this value into cash can permit the farmer to improve his operations, expand his farm, or retire.[86]

The DEM and various agriculture industry groups agreed with the Task Force that a purchase-of-development-rights program was the only way to save Rhode Island farmland under acute threat of conversion.[87] However, the adoption of such a program was considered by some, such as Dr. Thomas Weaver, to be too costly. Much of the state's prime agricultural land lies in the coastal regions, and property in that area had rached values as high as $30,000 to $40,000 per acre by 1980.[88] With development rights in other New England states characteristically being valued at two-thirds or more of the total value of a parcel of land,[89] such a program clearly can become quite expensive. Concerned advisors, such as Dr. Weaver, felt that a more efficient means of protecting agriculture in the state could be utilized, such as allocating more state funds for research at the state's Agricultural Experiment Stations to develop energy crops which produce gasohol or to develop new irrigation technology.[90] But no alternative, like gasohol, has proven to be a viable, effective tool for farmland protection. The Task Force, therefore, did not agree with such alternative strategies. The state legislature ultimately followed the recommendation of the Task Force by adopting a statewide PDR program, known as the Farmland Preservation Act (FPA), in May 1981.

The express legislative purpose of the Act is to "maintain farming, productive open spaces, and ground water recharge areas." The FPA attempts to achieve these goals by "acquir[ing] the development rights to the bulk of the remaining [farm] land most endangered by development."[91] Thus, as in Massachusetts, jeopardy is the primary criterion. Specifically, the Act establishes an Agricultural Lands Preservation Commission made up of the directors of the DEM and the Department of Administration, a representative of the Rhode Island League of Cities and Towns, six appointees of the governor (one recommended by the speaker of the House and one by the president of the Senate), the dean of the College of Resource Development at the University of Rhode Island, and a state conservationist of the USDA's Soil Conservation Service. The eleven-member Commission is charged with the responsibility to:

1. develop the criteria necessary for defining agricultural land
2. make a reasonably accurate inventory of all land in the state which meets the definition of agricultural land
3. prepare and adopt rules for administration of the purchase of development rights and criteria for the selection of parcels for which the development rights may be purchased, and the conditions under which they will be purchased
4. draw up and publish the covenant and enumerate the specific development rights to be purchased by the state
5. inform the owners, public officials and other citizens and interested persons of these provisions[92]

In fulfilling these tasks, the Agricultural Lands Preservation Commission determined that a parcel of land must be five contiguous acres or more and suitable for agricultural use to be considered under the program. The inventory of agricultural land was completed in 1980. With respect to the program's administration, in June of 1983 the Commission revised its rules dealing with its operating procedures and evaluation criteria, based largely on the guidelines used in the Massachusetts and Connecticut programs.

Moreover, the Commission has broad powers to hire consultants, obtain records, and accept federal grants to fulfill its obligations. The development rights which can be purchased are defined under the Act, in part, as the right of a landowner "to develop, construct on, divide, sell, lease, or otherwise change the property in such a way as to render the land unsuitable for agriculture."[93] The price the Commission is authorized to pay cannot exceed the average of two independent appraisals of the value of the development rights. This value is defined as "the difference between the value of the property for its highest and best use and its value for agricultural purposes."[94] Thus, the Commission has specific, although broad, authority to fulfill the goals of the FPA.

The Commission's rules provide that after applications are received, the Commission will evaluate them in three stages. In the initial stage, scores are given for the "primary criteria" of soil type, development pressure, and economic importance. Soil quality is determined, in part, by reference to the USDA's Soil Conservation Service publication of July 1981 entitled "Soil Survey of Rhode Island." Included in this evaluation are crop yield and soil management techniques. Soils yielding specialty crops like fruits undergo an evaluation based primarily on past production. Development pressure is analyzed in terms of (1) the characteristics of the land, such as slope, drainage, prior subdivision, and scenic features; (2) location, such as community policy toward growth, infrastructure, and population; and (3) type of farm operator. Finally, the economic importance of farmland is determined by the amount of farming activity, the types of goods produced, and economic viability.

Applications which receive a prescribed minimum score after the initial evaluation stage are re-evaluated under "additional criteria." These criteria are the significance of soil management, protection of watersheds or ground water recharge areas, preservation of open space, and aesthetic values and cost. After the evaluation of the primary and additional criteria in the first two stages, the Commission combines the scores obtained. Voting to purchase parcels is based on the combined score of "geographic diversity, mix of farm types, and cost of development rights."[95]

By July 1985, a total of seventy-one purchase applications had been received under the FPA Program, and fourteen of these had made it to the second stage of evaluation. Purchase-and-sale agreement negotiations for the development rights to six farms were under way, and four purchases were expected to be completed by the end of the summer. Daniel Varin, a member of the Commission, hopes that the average cost per acre will be $2,000, in which case no more than seven farms are likely to be restricted with current fiscal resources.[96]

To fund the program, the legislature authorized an initial issuance of state bonds in a total amount not to exceed $2 million. This bond issue, along with eight other bond proposals, came before the voters in a November 2, 1982 referendum. Rhode Island voters approved the measure by an almost two-to-one margin, the highest acceptance of any of the referendum issues, including the bottle bill and a nuclear freeze. Every Rhode Island city and town voted to fund the Act.[97]

In the same November 2 referendum, the town of New Shoreham (Block Island) approved a measure to raise revenues for its own local farmland PDR program. The legislature had authorized $1 million in municipal indebtedness for such local purposes, and Governor Garrahy signed the bill approving a local sales tax, directed primarily at tourist activities, to purchase $500,000 in development rights and $500,000 in fee-simple land acquisitions for New Shoreham. The farmland on Block Island is to be protected largely for its scenic or aesthetic, as opposed to productive, value. The procedures for purchase are entirely local.

Although these state and local programs have enjoyed strong initial public approval, future funding is not guaranteed, and the experience of the other New England states indicates that funding will always be an issue. Yet Stephen Morin of the DEM believes that the majority of voters are not likely to change their mind suddenly because support for the program has been very strong in Rhode Island since the mid-1970s. Prior to the adoption of the FPA, a 1980 telephone survey of six hundred Rhode Islanders revealed that 87.9 percent of those questioned supported the idea of protecting farmland in the state.[98] When initially adopted in 1981, the FPA passed with only three dissenting votes in the House and unanimous support in the Senate. One year later, in March 1982, the legislature also voted unanimously to pass the Right to Farm Act. Then, in January 1984, Governor Garrahy proposed a $14 million "Heritage Bond Issue," $3 million of which was to be earmarked for the PDR program. The bond issue was approved by the legislature but voted down by the voters in a November 1984 referendum. Some analysts believe a possible reason for this defeat is that the Heritage Bond Issue was Question 13 on the ballot, and over 100,000 fewer total votes were cast that far down the ballot. Since the issue lost by 20,000 votes, such a lack of votes could have been significant. It is also widely felt that the issue of farmland protection should be separate from the issue of the Heritage Bond. Subsequently, the legislature approved a $2 million bond issue exclusively for the PDR program to go before and was passed by the voters in a November 1985 referendum.

Land restricted under the program probably will be protected for farming or other open space use for some time, although repurchase provisions do exist. The repurchase provisions of the Act require that the farmland owner petition the Commission after two-thirds of the city or town council has approved a proposed development. Seven of the Commission's eleven members then must be convinced "that there is an overriding necessity to relinquish control of the development rights" after a hearing is held in the town where the land lies.[99] The landowner must return to the Agricultural Preservation Fund the value of the development rights at the time of sale. As a practical matter, local and Commission approval is not likely to be granted easily because of public pressure to retain open space.

The Rhode Island statute also helps ensure the permanent retention of land that comes under the Act by requiring that state and local agencies limit the use of their eminent domain power. Land designated as agricultural land under the program and farmland which actually is restricted under the program cannot be condemned by the state government unless the state or local agency establishes "extreme need and lack of any viable alternatives."[100] The agency must file a report with the Commission establishing this necessity after holding public hearings and receiving the agreement of the governor. This provision is extremely broad because it applies to all farmland classified under the program, whether the development rights are purchased or not. The limitation of eminent domain powers is unique to Rhode Island. The only other state which has a similar provision is Massachusetts, and the Bay State's law is purely advisory.

The repurchase and eminent domain provisions of the Act underscore how strong the concern to protect agricultural land is in Rhode Island, whether the development rights be purchased with state or local funds. Both the legislature and the Commission were aware that nothing more than open space will be protected if the profitable use of the land is not maintained. Drawing upon the experience of the other New England states, the Commission's evaluation criteria focus on highly productive, jeopardized farmland. Environmental and aesthetic considerations are also involved in the evaluation process. These considerations accord well with the overall public concerns in Rhode Island to protect agricultural lands.

Right-to-Farm Act (1982)

The year after the FPA was adopted, the legislature sought to aid farmers laboring in a largely urbanized state by adopting a right-to-farm law. The Governor's Task Force on Agricultural Preservation had recommended adoption of such a law, believing it would ease the conflicts between agricultural and other uses which arise from the "haphazard scattered" development patterns in the state.[101] On March 11, 1982, Governor Garrahy signed into law the Rhode Island Right-to-Farm Act. This law is similar to those passed by other New England states since 1979 and seeks to block nuisance suits which result from conflicts between agricultural and urban land uses.

In adopting the Right-to-Farm Act, the legislature found the following:

a. that agricultural operations are valuable to the state's economy and the general welfare of the state's people;
b. that agricultural operations are adversely affected by the random encroachment of urban land uses throughout rural areas of the state;
c. that, as one result of this random encroachment, conflicts have arisen between traditional agricultural land uses and urban land uses; and

d. that conflicts between agricultural and urban land uses threaten to force the abandonment of agricultural operations and the conversion of agricultural resources to nonagricultural land uses, whereby these resources are permanently lost to the economy and the human and physical environments of the state.[102]

Many observers believe the statute will not work as well as in other New England states because of several shortcomings in its language. First, unlike Vermont's statute, the Rhode Island measure only immunizes certain types of agricultural externalities, namely: "(1) odor from livestock, manure, fertilizer or feed, occasioned by generally accepted farming procedures, (2) noise from livestock or farm equipment used in normal, generally accepted farming procedures, (3) dust created during plowing or cultivation operation, (4) use of pesticides, rodenticides, insecticides, herbicides or fungicides." Therefore, for example, noise from a poultry operation or irrigation overflow are not explicitly protected by the law. Second, the Rhode Island law does not protect agricultural operations which may be considered malicious or negligent, or those which violate certain state or federal laws. This contrasts with the Vermont law, which protects all agricultural activities except those that are proven to have "substantial adverse effects on the public health and safety," regardless of the operator's state of mind or conduct. Third, unlike the Vermont statute, the law does not create a rebuttable presumption that there is no nuisance if the farmer has adhered to good agricultural practices before the conflicting urban use(s) arrived. Thus, it is easier for a person in Rhode Island to sue a farmer for a nuisance because the complainant has less to prove. Finally, only nuisance suits brought under state law and local ordinances promulgated under state law are prohibited by the Right-to-Farm Act. Thus, other forms of nuisance action, such as common law nuisance suits, can be brought. In short, the focus of the law is on a few situations and on a farmer's conduct, not on the actual effect of that conduct.[103]

Despite the comparatively limited protection granted to farmers by the law, Rhode Island courts have construed it liberally to further the legislative purpose of helping farmers. In early 1983, for example, the Newport superior court found in favor of a Middletown dairy farmer because of the statute. A neighbor had sued the farmer for $100,000 because flies generated by the operation numbered thirty-five per square foot on the roof of the neighbor's house, making eating outside rather unpleasant. The farmer prevailed at trial by proving that he was engaged in normal farm practices, even though the statute does not refer specifically to flies. After the suit, however, the farmer moved his operation to Massachusetts because of community hostility.[104]

About the same time as the Middletown case went to court, a hog farmer in East Greenwich brought suit against the town when it passed an ordinance limiting the number of hogs a farmer could have. The Right-to-Farm law, therefore, was used as a sword to act offensively instead of as a shield to defend against a nuisance suit brought by a newcomer. The farmer was successful, and the ordinance was declared illegal.[105]

Another hog farm that has created a substantial public controversy is in Tivertown, a town that has become a suburb to Fall River, Massachusetts within the last decade. Odors generated by some 2,800 hogs when the farm's waste disposal system malfunctioned in late 1983 resulted in protests by nonfarm residents on both sides of the state border. The farm lies in a rural residential zone. Massachusetts Congressman Barney Frank declared, in a Tiverton town meeting, that he would introduce federal legislation in Congress to prevent cross-state farm nuisances because the Massachusetts state courts have no jurisdiction over Rhode Island land. Throughout the controversy, the farmer has asserted his rights under the Rhode Island Right-to-Farm law because he conducts a generally accepted farm practice. No suit had been filed against him by 1984.[106]

In sum, the language of the Rhode Island statute is limited when compared to the wording of the other New England states' laws. Because of the legislative intention behind the law, however, the courts appear likely to give the state's farmers as broad a protection as farmers in other states.

Zoning

Following a trend that is growing throughout New England, there is an increasing awareness in Rhode Island that conflicts between farming and other land uses can be reduced through careful local planning. Current state law, in the form of general zoning enabling legislation, does not expressly allow localities to zone solely for agriculture.[107] This original enabling legislation was adopted largely in 1921 and primarly is designed to regulate buildings, structures, and floodplain areas. As a result, as the State Planning Council reported in 1975:

> Valuable open space and ecological features are not protected, except in a few communities which have open space zones (usually for existing parks) or flood hazard zones . . . urban sprawl and "wasteful land development practices" are actually encouraged rather than prevented. Uniform lot restrictions applied rigidly over a large area result in a monotonous sprawl pattern, obliterating open space areas and disregarding natural site features.[108]

To address this lack of flexibility in the state enabling act, the State Planning Council proposed that residential cluster and planned unit development be allowed. The concept behind this approach, as in every other New England state, is that intensive uses of land should be concentrated in some areas to allow for open space in other areas. This program can be more difficult and costly to administer than traditional zoning, but a few communities, such as South Kingston, have made use of it without express statutory authority to protect agricultural parcels.[109] South Kingston awards developers a 25 percent density bonus for certain rural cluster developments.

Unlike incentive zones, exclusive agricultural zones have not been allowed in Rhode Island. One of the primary recommendations of the Task Force on Agricultural Preservation was that municipalities be allowed to establish such zones.[110] Despite the introduction of legislation in the 1970s that would have granted such authority, no such bill was adopted.

This legislative inaction may have been the result of the nature of the proposals rather than opposition to agricultural zoning. Between 1976 and 1981, most of these proposals were part of broad changes in the Zoning Enabling Act, often requiring the institution of new land management concepts such as conforming land use to the natural characteristics of land and allocating uses on the basis of planned services.[111] Most farmland owners, of course, oppose agricultural zoning. Because this type of legislation could lock land into agricultural use without the farmer's consent or could result in open space protection without farming if agriculture is not profitable, some members of the Task Force did not recommend such legislation.[112]

In 1982, legislation was introduced which provided that a zoning ordinance may be designed to "preserve agricultural land."[113] Additionally, the bill would have required consideration of the value of lands in farms whenever there is a petition for change in zoning that would affect such lands. Thus, under this proposal localities would have been forced to be aware of agricultural uses and could provide for protection of such lands. The bill was introduced too late in the January 1982 session of the legislature, however, to receive approval.

Revised and resubmitted in 1983, the new version declared that the state's land must be carefully managed in order to "discourage conversion of productive agricultural land . . . to other uses."[114] Like its 1982 counterpart, the bill provided that a zoning ordinance could be designed to "preserve agricultural land" but the 1983 version did not require a consideration of the value of farmlands when there is a petition for a zoning change. This omission came about because local officials had felt the 1982 version was too burdensome. The 1983 bill failed to pass in the Senate, largely because building and real estate lobbies (such as the Rhode Island Builders Association) opposed the specificity of development requirements. In addition, local Boards of Zoning Appeals, which have organized into a State Association, opposed the bill because it mandated that they established new record-keeping procedures.[115]

A new zoning bill drafted largely by the Rhode Island League of Cities and Towns and the Rhode Island Section of the American Planning Association, and with Governor Garrahy's support, also failed to be adopted. But efforts are continuing for the state to enact a flexible zoning law that can be used to protect agricultural lands.

If agricultural zones were created under new legislation, their effectiveness in protecting agricultural lands would depend, in part, upon whether they were exclusive and whether variances, which break up the zone, were allowed only in rare cases. If these and other safeguards were in place to assure that the critical mass for agriculture remained intact, then landowners within the zone would be forced to use their lands for agricultural activities or let their land lie undeveloped. The farmland resource would be protected and, unlike the PDR programs, there is no direct cost to the taxpayer. Complementary legislation, however, may still be required to maintain the economic viability of agriculture. Equity questions are raised. Therefore, combining zoning with economic assistance measures also is more likely to receive support from the farm industry.

If a new zoning enabling act is not adopted, localities continually will be faced with wrestling with the traditional zoning enabling statute to accomplish agricultural land protection objectives. One town that has made progress in this manner is South Kingston. In addition to its rural cluster development ordinance, the town in 1984 was considering downzoning about 10,000 acres of open space land from two- to five-acre rural residential lots.[116] This "rural low density zone" would apply to lands mapped as containing farms, coastal area open space, and aquifers. Most of the farmland in South Kingston is devoted to turf or potato production, but towns with other kinds of operations could follow this approach. Proposals such as the tabled 1984 amendments to the enabling act, however, would allow greater flexibility in initiatives of this character.

Marketing

Recognizing the importance of a healthy and viable agricultural industry to assure the protection of agricultural land, the Rhode Island legislature has adopted several additional programs which indirectly protect farmland by helping farmers continue farming. Among the most important of these indirect efforts are state-supported farm marketing programs.

The Governor's Task Force summarized the farm marketing structure in 1981 when it found that although "[m]any retailers are not in the habit of dealing with local producers," marketing arrangements "are generally adequate" for farm produce given the current level of production.[117] Nevertheless, because the Task Force and farmland protection advocates believed that effective marketing and the prosperity of farmers are complementary, certain measures were recommended to the legislature and state agencies to ensure that marketing arrangements were improved and designed to accommodate greater production.

As part of statewide farm marketing efforts, the DEM has begun working with various farm associations to promote domestically grown products in numerous ways. First, a state-owned logo, "Grown in the Biggest Little State in the Union—Rhode Island Just For You," was developed to help consumers identify and select local produce.[118]

In addition to the use of logo stickers, the DEM is engaged in an advertising campaign consisting of posters, public service announcements, and brochures. The Task Force found that because "[m]any farmers market at least a part of their production through farm stands," direct sales should be encouraged. Accordingly, the DEM, the Office of State Planning, and the Department of Economic Development prepared informational brochures for the public. One brochure describes various "apple country" scenic tours for the spring blossoming and the fall harvest. Another brochure maps out fifty-one produce stands scattered throughout the state. This latter brochure lists thirty different kinds of fruit and vegetables grown and sold in Rhode Island. Still another brochure states that no Rhode Islander is more than ten miles away from a farmer's market.[119]

According to Stephen Morin, the logo, brochures, and other features of Rhode Island's farm marketing program have generated an increased level of citizen support for the local farm industry.[120] Despite this genuine concern of the citizenry over food independence and the agricultural economy, it is unlikely that advertising alone will have much effect on the decline of active farmlands. Only certain agricultural products can be marketed freshly in Rhode Island, and there is no legal obligation on the part of state institutions or public agencies to purchase locally grown products, as in Vermont, Maine, and New Hampshire. Moreover, the increased revenue any individual farmer is likely to receive as a result of the advertising campaign is unlikely to make farming more profitable than selling the land for development. In short, the state's promotional effort, although important, does little by itself to ensure that the family farm will be protected. State policies have considered marketing to be but one element of a broader range of program protection strategy.

In an effort to increase the impact marketing has on the state's protection of farmlands, the Task Force on Agricultural Preservation proposed three additional measures. First, storage facilities for seasonal fruit crops were suggested. These facilities could be partially funded and designed by the state, like other development projects. Such facilities would ease dependence on out-of-state suppliers during the off-season by making certain local fruits available year-round. Second, following the lead of the three northern New England states, the Task Force recommended that state operations which serve food to institutional populations be required to buy local produce if it is cost-competitive. Again, this would provide a strong incentive to grow products that are currently only marginally profitable because of out-of-state competition. Finally, following the lead of Massachusetts, the Task Force recommended that local and regional farmers' markets be established and promoted through advertising to allow city residents to purchase local produce directly from producers.[121]

Even if all three of the Task Force's recommendations were adopted, however, there are other considerations affecting farming operations in Rhode Island that could reduce farm profitability and prevent expansion. As the Task Force recognized, "issues of access to markets, imperfect competition among buyers, inadequate information, and obstacles to efficient distribution" cannot be completely resolved by the limited role the state government can play in the marketing effort.[122] Analysts in other New England states, such as New Hampshire and Massachusetts, believe that marketing is a major cornerstone of farmland protection, but Rhode Islanders have taken the view that such a traditional form of state assistance is unlikely to reverse the trend of agricultural conversion by itself, particularly when urbanization is the cause of conversions. Therefore, the state legislature has adopted only the more direct farmland protection programs discussed previously and is considering others in recognition that a range of programs is needed.

Pending Agricultural Land Protection Programs

Recognizing that the programs already in place may not be enough to protect farmlands adequately, several agricultural protection bills continue to be introduced in the legislature. The most significant is a bill to create municipal trusts.[123] These trusts, which would be created by municipal ordinances, would be exempt from taxation and any special assessments. The property acquired by the trusts would be held solely for agricultural uses, open space, or in some cases public recreation. The bill also gives municipalities the authority to levy tax upon real estate transfers. The proceeds from this tax would be deposited into a municipal revolving fund, which can also receive various voluntary contributions, grants, or loans. The trust fund in turn can be used to purchase property and to provide for its upkeep or management.

The trust funds would give the municipalities greater autonomy about future growth and allow them to choose what land is important to protect. Supporters of the bill feel that it is politically feasible because of the strong statewide support for farmland protection and open space preservation of the PDR referenda in 1983 and 1985. The passage of this bill could give Rhode Island a significant boost in protecting its farmland resources.

Future of Agricultural Land Protection Programs in Rhode Island

Rhode Island's location within the northeast urban corridor has made, and will continue to make, farming very difficult. Unlike most of the New England states, Rhode Island has no major farming area, like Champlain Valley, Connecticut River Valley, or Aroostook County, which is profitable for farming and at least partly removed from urbanization pressures. A high degree of urbanization, high land costs, and a high percentage of absentee farmland owners make much of the state's farmland ripe for conversion.

In the 1960s, this urbanization problem was not addressed adequately in the piecemeal legislation which sought to protect open spaces generally. Indeed, the 1968 Farm, Forest and Open Space Land Act as originally adopted did not have a high farmland owner enrollment and actually encouraged farmland speculation because of its weak penalty provisions.

In the 1970s, land uses still were controlled by economic forces. As the more significant agricultural activities converted to operations which supported urbanization (such as the turf and nursery industries), the public became concerned about the loss of traditional agriculture. Many activities of state government, however, continued to address agricultural land protection as an economic issue. The Governor's Task Force was devoted to "Agricultural Preservation" rather than to "Agricultural Land Preservation."

In the 1980s, the public has come to recognize the importance of farm*land*, and with the adoption of the Farmland Preservation Act the most significant step in that direction has occurred. The eminent domain limitation in that Act and the Soil Erosion and Sediment Control Act of 1982 indicate a re-evaluation of the concerns at issue and the programs needed to address those concerns. These developments are somewhat startling in a state that was on the road to having no farms just a few years ago. Daniel Varin has commented:

> A highly urbanized, industrialized and densely developed state such as Rhode Island obviously has many other considerations that, in terms of their overall effort, outweigh preservation of agriculture. Nevertheless, the importance of maintaining nonurban activities and land uses and of strengthening these is receiving greater recognition and agriculture is specifically identified as one of the most important activities in this category. Gaining recognition of the current situation and its problems was a major function of the Task Force and one that appears to have been successfully accomplished. This recognition can lead to significant results if the follow-up activities are continuously and strongly supported, both politically and financially.[124]

Because Rhode Island is a "highly urbanized, industrialized, and densely developed" state, urban pressures will continue to be a major concern. Urban type responses, therefore, seem to be the key to successful farmland protection programs in the 1980s and 1990s.

As suburban development continues in the 1980s and 1990s, states such as Rhode Island will become more and more developed with a "uniform" landscape pattern. For this reason, agricultural land is also valuable for its "parkland" attributes and its food production capabilities. From this perspective, planning for agriculture can be an important component of a total urban plan.

Now that there is widespread recognition of a need to protect agriculture and agricultural land uses, proposals for farmland protection programs are receiving substantial support and will probably continue to receive such support through the 1980s.

Such a political mandate was given to the detailed state Land Use Policies and Plan in Rhode Island, and the success of this Plan continues to be dependent on community support. Rhode Island's Land Use Plan is essentially a comprehensive city plan because the state is so small. Rhode Island's size is precisely the reason why the Plan is able to be so effective, unlike Vermont's Act 250. This type of "urban" response to urbanization pressures could be used more effectively in the future.

The idea of urban green belts, supported by public funds and containing the critical mass necessary to support agriculture, could be implemented. PDR programs could also be very effective in these circumstances. Innovative zoning would contribute to a total plan, but amendments to the zoning enabling act are needed to provide regulatory support for the protection of farmlands.

Coordinated with the statewide Land Use Plan are the local community planning efforts. Rhode Island's example in communities such as South Kingston and Block Island show the success of local efforts when combined with state support. Passage of a bill to create municipal trusts and real estate transfer taxes as a mechanism to acquire agricultural land would represent a significant step forward in farmland protection efforts.

Furthermore, the agricultural industry itself has and should continue to tailor its commodities to the needs of its expanding urban population and for "urban" land uses (i.e., turf farms). The production of nontraditional crops in recent decades helped to assure the protection of agricultural land and the agricultural industry in Rhode Island. Although these activities will not completely control the outside forces that have brought about such a dramatic change in Rhode Island's land use pattern since World War II, they will mitigate those forces and help ensure that a healthy agricultural industry remains a critical part of the state's landscape.[125]

Notes

1. "Vanishing," *Providence Sunday Journal*, January 10, 1966, 4.
2. *Report of the Governor's Task Force on Agricultural Preservation*, Report No. 38 *(Task Force Report)* (Providence: Office of State Planning, January 1981): 19–23.
3. Ibid., 21–23.
4. Ibid., 23.
5. Using *Census of Agriculture* methodology, the Rhode Island Department of Environmental Management (DEM) reported that there were approximately 76,000 acres in farms in 1970 and 75,000 in 1981. *Rhode Island Agricultural Statistics* (Providence: Department of Environmental Management, 1981): 23. A USDA Soil Conservation Service study, however, using data compiled by the Rhode Island Conservation Districts, identified 38,165 acres of agricultural land in 1979, including 10,708 acres in tilled farmland; 18,976 acres in hay or pasture; 1,680 acres in nurseries; 633 acres in orchard or vineyard; and 6,168 acres in open land. *Soil Conservation Study* (USDA, March 1980), reprinted in *Task Force Report*, A-12. Yet a 1982 report released by the DEM states there are 74,585 acres in farms (approximately 11 percent of the state's land area), and 36,434 acres of this land are cropland. "Where Have All the Farmers Gone?" (Providence: Department of Environmental Management, 1982): 1. These statistical discrepancies illustrate that the importance of agriculture and the severity of the decline of farmland use varies according to the source of information. The Census and DEM statistics generally include the acreage of wood and wastelands surrounding farm tracts while the USDA study does not. Nevertheless, all authorities agree that most of what was once farmland in Rhode Island is now in nonagricultural uses.
6. Thomas F. Weaver, chairman, Department of Resource Economics, University of Rhode Island in South Kingston, comment, Lincoln Seminar 1. One report concludes that in just four years, between 1965 and 1969, farm acreage dropped by one-third. *State Land Use Policies and Plan, Report No. 22* (Providence: Office of State Planning, January 1975): 6.
7. *Task Force Report*, 39.
8. Thomas F. Weaver, comment, Lincoln Seminar 1.
9. "Where Have All the Farmers Gone?," 2.
10. *1978 Census of Agriculture*, 451.
11. Stephen G. Morin, assistant to the director, Rhode Island Department of Environmental Management, comment, Lincoln Seminar 2.
12. *Task Force Report*, 27.
13. *1978 Census of Agriculture*, 451. These figures could be misleading, aside from enumeration techniques, because inflation brings more farms into the category of selling $1,000 or more of agricultural products. Similarly derived DEM statistics indicate that the number of farms grew from about 820 in 1970 to 860 in 1981. *Rhode Island Agricultural Statistics*, 23. On the other hand, the DEM has recently released figures of both 1,000 farms ("Where Have All the Farmers Gone?," 2) and 608 farms (*Rhode Island Agricultural Facts* [Providence: Department of Environmental Management, January 1982]: 3), adding confusion to the real nature of farmland decline in Rhode Island.
14. *Task Force Report*, 30.
15. "Where Have All the Farmers Gone?," 2.
16. *1978 Census of Agriculture*, 451; *Farmland Preservation Survey* 3 (Bethesda, Md.: Farmland Preservation Institute, Inc., February 1982): 2; *Rhode Island Agricultural Statistics*, 23. The *Task Force Report* puts the figure at 104 acres. *Task Force Report*, 26.
17. *State Land Use Policies and Plan*, 128; Stephen G. Morin, comment, February 24, 1983.
18. *Rhode Island Agricultural Facts*, 5.
19. Ibid.. The *Task Force Report*, again drawing on U.S. Census data, states the figure in 1978 was "about 25 million," which may be due to underreporting. *Task Force Report*, 26.
20. *Rhode Island Agricultural Facts*, 6.
21. The DEM reports that nursery operations account for approximately 29 percent of annual commodity values while sawmills and fuelwood contribute 21 percent, turf farms 12 percent, egg and poultry 10 percent, dairy 10 percent, potatoes 7 percent, vegetables 5 percent, and other commodities 6 percent. *Rhode Island Agricultural Facts*, 3.

22. Nelson L. Bills and Arthur Daughtery, "Who Owns the Land?" *A Preliminary Report for the Northeast States, ESCS Staff Paper No. 80-8* (USDA, 1980): 11 (Table 4), reprinted in *Task Force Report*, 23, 39, 66.
23. *Task Force Report*, 29, 39.
24. Daniel W. Varin, comment, February 24, 1983.
25. "Vanishing," 4.
26. *State Land Use Policies and Plan*, 1 (footnotes omitted).
27. Ibid., 16.
28. Ibid., 17.
29. "Vanishing," 5.
30. Stephen G. Morin, comment, Lincoln Seminar 2.
31. Thomas F. Weaver, comment, Lincoln Seminar 1.
32. *Task Force Report*, 25.
33. Ibid., 37, 38, 40.
34. Ibid., 25.
35. Rhode Island Statutes (R.I.S.) 42-82-1(a).
36. Stephen G. Morin, comment, Lincoln Seminar 2.
37. *Task Force Report*, 41. The State Planning Council used similar language in its 1975 report: "Preserving agricultural land is important for both economic and land use reasons. The economic reasons are not as strong in Rhode Island as in other parts of the country, since agriculture contributes relatively less to this state's employment and income. If other states stop exporting food within the next decade, however, as some projections have indicated, then agricultural production will be a more serious concern. Maintaining some amount of local food supply also makes economic sense considering transportation and energy costs and the costs of developing new agricultural land to replace that which is lost to urban development.

"Farmland should be protected, more importantly, on the basis of land use objectives. Certain land should be retained in agriculture since it is well suited to that use and adequate land is available elsewhere for other activities. Farmland preservation is important for maintaining a valued lifestyle and an attractive environment which may in turn attract other economic development to the state. (If recent trends continue, farms would disappear in Rhode Island in the 1980s, thus eliminating part of the state's heritage.) Agricultural land should also be protected for some of the same purposes as other open space areas . . ." *State Land Use Policies and Plan* (June 1975 Addendum): 6.

38. R.I.S. §45-35-1.
39. Richard Pourier, president, Rhode Island Association of Conservation Commissions, comment, February 13, 1984.
40. Duddleson, *Supplementary Report* to Scheffey, *Conservation Commissions in Massachusetts*, 171–74. The DEM is required to cooperate with, advise, and guide local Conservation Commissions. R.I.S. §42-17.1-2(g).
41. Roth Tatrow, Lincoln Conservation Commission, comment, February 10, 1984.
42. R.I.S. §32-4-1 et seq.
43. *State Land Use Policies and Plan*, 178.
44. Daniel Varin, interview, February 24, 1983.
45. Ibid.
46. R.I.S. §44-27-1 et seq.
47. R.I.S. §44-27-1(a), (b). This declaration is nearly an exact duplication of the declaration adopted in Connecticut's differential taxation law. C.G.S. §12-107(a).
48. *State Land Use Policies and Plan*, 221.
49. *Task Force Report*, p. 10.
50. Thomas F. Weaver, comment, Lincoln Seminar 1.
51. *State Land Use Policies and Plan*, 187.
52. *Task Force Report*, 10.
53. *State Land Use Policies and Plan*, 219.
54. Daniel W. Varin, interview, February 24, 1983.
55. *Task Force Report*, 10.

56. Ibid., 10–11.

57. Under the Act, the term "farm land" means "any tract or tracts of land, including woodland and wasteland constituting a farm unit." R.I.S. §44-27-2(a)(1).

58. These rules are reprinted in Stephen Morin and Jacqueline McGrath, *A Citizen's Guide: Farm, Forest and Open Space Act* (Providence: Department of Environmental Management, n.d.).

59. Daniel W. Varin, comment, February 24, 1983.

60. *State Land Use Policies and Plan*, 221.

61. *A Citizen's Guide: Farm, Forest and Open Space Act*, 1; *Rhode Island Agricultural Facts*, 8.

62. *Rhode Island Agricultural Facts*, 8.

63. Stephen G. Morin, comment, January 23, 1984.

64. *Task Force Report*, 9–12.

65. Stephen G. Morin, comment, January 23, 1984.

66. *Task Force Report*, 9–12.

67. R.I.S. §45-11-10.

68. *State Land Use Policies and Plan*, 5–14.

69. Ibid., 6–8, 16, 17. Subpart (e) of this goal states: "Preserve selected land areas in agricultural and forest use in order to provide a limited agricultural base, to provide a long-term land reserve, and to protect rural areas." This land "provides a limited agricultural base in dairy farming, field crops, pomology and forestry; furnishes a land reserve for long-term future needs; and protects rural areas which offer a wildlife habitat or which give shape and order to urban growth."

70. Ibid., 49, 128.

71. Stephen G. Morin, comment, Lincoln Seminar 2.

72. Executive Order 79-6.

73. *Task Force Report*, 7.

74. *Task Force Report*, 50.

75. The problems which can arise from such low-interest state financing of unprofitable agricultural operations is illustrated by the case of the Bouchard farm in South Kingston, Rhode Island. The owner of the farm applied to the Farmers Home Administration (FmHA) of the USDA and obtained a loan secured by a second mortgage. Later, as farm economic conditions deteriorated, the farm became heavily in arrears on the loan payments. The property was foreclosed by FmHA, which must abide by USDA policy to preserve farmlands. The agency found itself in a position where the land had to be resold for development to cover the cost of the loan it had made and the price it paid at auction to the holder of the first mortgage. The FmHA's total indebtedness is approximately $800,000. Although some of this sum could be recovered by the sale of the farm's machinery, the agency views its role as a banker and seeks to sell the property at development value for a subdivision as the only way to recoup its investment. Although the fate of this parcel was not decided by the summer of 1983, this example illustrates how a loan program can run against a farmland protection policy unless it is administered carefully. An important consideration for all low-interest financing programs such as the proposed agri-bond program, therefore, is that farm loans be made on the basis of actual farm value. If foreclosure later becomes necessary, the state may recoup its fund while selling the land to a qualified farmer at a price which operations will support. "Foreclosing Farms and Foreclosing Future Options," *Farmland Preservation Survey* 3 (Bethesda, Md.: Farmland Preservation Institute, Inc., July 1982): 1–4.

76. *Task Force Report*, 44–45, 57, 58.

77. R.I.S. §34-39-1 et seq.

78. *Task Force Report*, 45–46.

79. Ibid., 58–59; R.I.S. §44-3-3.

80. *Task Force Report*, 56–57.

81. R.I.S. §45-46-1 et seq. The legislative finding is as follows: "The general assembly finds that excessive quantities of soil are eroding from certain areas of the state that are undergoing development for certain nonagricultural uses such as housing developments, industrial areas, recreational facilities, commercial facilities, and roads. Erosion occurring in these areas makes necessary costly repairs to gullies, washed out fills, roads and embankments. The

resulting sediment clogs storm sewers and road ditches, soils streams, and deposits silt in ponds and reservoirs. In some of the state's waters, silt resulting from erosion has become a major water pollutant and threatens water supply, recreational, aesthetic, and wildlife habitat values associated with these waters."

82. R.I.S. §45-46-5.
83. *Task Force Report*, 53.
84. Crop Reporting Board, Soil Conservation Service, USDA, December 1981.
85. *Task Force Report*, 14, 53.
86. Ibid., 42.
87. "Where Have All the Farmers Gone?," 2.
88. Thomas F. Weaver, comments, Lincoln Seminar 1, and interview, January 24, 1984.
89. *Task Force Report*, 42.
90. Thomas F. Weaver, comments, Lincoln Seminar 1, and interview, January 24, 1984.
91. R.I.S. §42-82-1 (a). The complete legislative purpose of the Farmland Preservation Act reads as follows:

> (a) The general assembly recognizes that land suitable for food production in the state has become an extremely scarce and valuable resource. The amount of good farmland has declined so dramatically that unless a comprehensive program is initiated by the state to preserve what remains it will be lost forever. It is in the best interest of the people that the state identify and acquire the development rights to the remaining land, most endangered by development so as to maintain farming, productive open space, and ground water recharge areas.
>
> (b) The general assembly finds that productive farmland is being converted to other uses because its development value at present far exceeds its value for agricultural purposes; that agriculture is an important part of the state's economy, environment, and quality of life; and that local food production will become increasingly important to the people of the state. It also finds that agricultural preservation will allow more orderly development and permit the cities and towns to plan for and provide services more adequately and at a lower cost. Therefore, the general assembly hereby establishes an agricultural land preservation commission to conduct inventory and acquisition of development rights to farmland in this state.

92. R.I.S. §42-76-5(a).
93. R.I.S. §42-82-2(d).
94. R.I.S. §43-83-5(b).
95. *Rule I, Operating Procedures* (Providence: Agricultural Lands Preservation Commission), p. 5.
96. Daniel Varin, comment, January 18, 1984.
97. The results of the referendum are reported in "State Updates," *Farmland Notes* (Washington, D.C.: NASDA Research Foundation Farmland Project, November 1982): 3; *Farmland Preservation Survey*, 2–3.
98. Stephen G. Morin, comments, January 23, 1984, and Lincoln Seminar 2.
99. R.I.S. §42-82-5(e).
100. R.I.S. §42-82-6.
101. *Task Force Report*, 56.
102. R.I.S. §2-23-2.
103. R.I.S. §2-23-5; V.S.A. 12 §5753; *Pucci* v. *Algiere*, 106 R.I. 411, 420 (1970).
104. Lee Gardner, University of Rhode Island Extension Agent, comment, Newport, February 10, 1984.
105. Stephen G. Morin, comment, January 23, 1984; see R.I.S. 23–19.2-1.
106. Comments by Daniel Varin, January 1984; Lee Gardner, February 10, 1984.
107. R.I.S. 45-24-1 et seq. In addition, the courts may not be disposed to help maintain the integrity of rural areas. The Supreme Court of Rhode Island ruled in 1977 that the town of Burrillville could not bar the proposed use of a thirty-five-acre tract as a summer campground in a farm and rural residential zone where the landowner seemed to have met the requirements for a special exemption and the town did not come forward with sufficient evidence of adverse effects on the zone such as traffic congestion. *Perron* v. *Zoning Board of the Town of*

Burrillville, 117. R.I. 571 (1977).

108. *State Land Use Policy and Plan*, 203.
109. Ibid.; Daniel Varin, comment, February 24, 1983.
110. *Task Force Report*, 43–44.
111. Daniel Varin, comment, January 18, 1984.
112. *Task Force Report*, 44.
113. R.I.S. §45-24-28(f) (proposed); 82-52530.
114. R.I.S. §45-24-26(a)(9) (proposed); 83-50627.
115. Daniel Varin, comment, January 28, 1984.
116. Glen Anderson, resource economist, University of Rhode Island in South Kingston, comment, February 9, 1984.
117. *Task Force Report*, 51–52.
118. This theme parallels a campaign to improve the image of the state which has met with questionable success. "Rhode Island's Economic Pickle," *Boston Sunday Globe*, April 4, 1982, 1.
119. "Where Have All the Farmers Gone?," 1–2; *Rhode Island: Apple Country* (Department of Economic Development and Department of Environmental Management, n.d.); *Rhode Island: Farm Produce Stands—Fruits and Vegetables* (Office of State Planning, Department of Economic Development, and Department of Environmental Management, n.d.).
120. Stephen G. Morin, comment, Lincoln Seminar 2.
121. *Task Force Report*, 52–53.
122. Ibid., 51.
123. "Rhode Island Proposed New Tools for Local Farmland Preservation Efforts," *Farmland Notes* (March 1986): 2.
124. Daniel Varin, comment, February 24, 1983.
125. Ibid.

7

New Hampshire Programs for the Protection of Agricultural Land

As in the other New England states, concern over the decline of New Hampshire's active agricultural lands began with a concern over the loss of open space and the deterioration of the natural environment. Historically, a multitude of groups has labored in New Hampshire to protect the countryside. The Society for the Protection of New Hampshire Forests played a leading role in these efforts, as well as the Association of Conservation Commissions, the Lake and Stream Association, New Hampshire Fish and Game Clubs, the New Hampshire Environmental Education Council, the Land Use Foundation, the New Hampshire Farm Bureau Federation, and many others. Perhaps no other state has a better organized citizen base to protect the environment than New Hampshire.

Despite this high degree of environmental awareness and concern since the mid-1960s, the Granite State has lagged behind its New England neighbors in enacting new legislation to protect the environment effectively against the accelerating growth which began to change the unique physical character of the New Hampshire landscape. This slow adoption of land use protection programs and policies, including those relating to agricultural land, is largely attributable to the conservative political nature of the state. As one political journalist wrote in 1976:

> The rise of Loebism, the parsimony of the state government, and stiff opposition by influential private interests seeing a threat to their profits and activities, have all played a part in the picture. In a broader sense, one could almost say that as soon as environmental protection became a really important political issue, the conservationist organizations discovered how thin their base of support had always been....The early decline of agriculture in New Hampshire accounts for much of this.[1]

The state's conservative response to agricultural land protection initiatives could be expected considering the political consensus that affects two aspects of state government. In New Hampshire, probably more than in any other New England state, there exists a genuine hostility toward government regulation. The state has long tried to adhere to strict principles of free enterprise, by which land is viewed as a commodity to be traded and sold freely. Individuals are seen as independent merchants not easily subject to state regulation or control in their commercial dealings. Thus, the state's policies reveal an apprehension about land use controls, which historically have been almost nonexistent to allow for a form of economic freedom.

Another conservative position evident in the state's government is the small size of the state budget relative to its population and compared to its New England neighbors. Few state programs are heavily funded. New Hampshire prides itself on being the only state in the continental United States with no sales or income tax. Government services, therefore, are funded largely by the property tax and "sin" taxes on gambling, liquor sales, and tobacco sales. This regressive tax structure has resulted in property taxes at an almost "confiscatory level"[2] to pay for local services. U.S. Department of Commerce statistics for 1980 reveal that New Hampshire had the eighth highest per capita property tax in the nation, at $405.63. (The national average was $302.42).[3] As a percentage of personal income, however, the tax puts New Hampshire fourth in the country. The limited tax base of the state limits the ability to fund costly new land use programs.

The state's revenue problem has been heightened by three recent circumstances. The most important of these is the Reagan administration reductions in federal money grants, which made up a high percentage of the state budget. Second, in the summer of 1980, Rockingham Park, the state-run horse-racing and betting facility in Salem, was destroyed by fire. The facility generated only 1 percent of state revenues, but the sum was important. No new revenue-generating facility had been built on the Park's land by mid-1983. Third, the state's bond rating dropped from AAA to AA in March 1982, resulting in higher interest payments. Ironically, the new rating is the result of the budget deficit and the lack of a broad-based tax.[4]

Because of these fiscal problems, the property tax burden has increased to a point such that on November 9, 1981, the voting delegates of the New Hampshire Farm Bureau Federation endorsed a statewide income tax plan that would return half of its revenue to local government. Other groups also have endorsed such a tax. Such initiatives led John Andrews, head of the New Hampshire Municipal Association, to report in January 1982: "Talk of income taxes or broad-based taxes used to be sort of like cancer; people didn't talk about it because they were afraid they would get it. But now that's not necessarily the case."[5] Nevertheless, the current governor, John

Sununu, is credited with winning the November 1982 election largely because he took "the pledge" not to institute new taxes; many of the state's fiscal problems continue.

In this politically conservative state with its small tax base, the desire to preserve agriculture for food production and economic independence were the primary forces for legislative action to protect farmland. Although New Hampshire recently has instituted some farmland protection programs, the hostility toward government regulation and the lean state budget have meant that the programs are adopted slowly, and in a "very realistic environment."[6] Land use programs in the other New England states have not been influenced the way New Hampshire's have by the fear of a big-brother government and high tax rates.

Public Concern to Protect Agriculture

New Hampshire has experienced the most substantial conversion in agricultural acreage of all the New England states, with much of the farmland being converted directly to development. While the concern over loss of open space and the change in community character has been most pronounced among the state's conservationist groups, the desire to maintain agriculture for food independence and economic well-being has been the major impetus for legislative action.

Decline in Farmland Acreage

The conversion of land devoted to active agriculture, particularly in the past twenty-five years, has been more dramatic in New Hampshire than in any of the other New England states. During this period, New Hampshire farmland acreage decreased over 60 percent.[7]

Although urbanization has been responsible for much of the recent decrease in acreage, in the past most of the converted agricultural land simply was abandoned and is now woodland. In addition, New Hampshire has developed a strong manufacturing industry which provided greater financial rewards than New England farming. Agriculture has declined more dramatically in New Hampshire than in Vermont and Maine, where the manufacturing industry was virtually absent.

The year 1860 is generally considered to be the high point of agricultural land use in New Hampshire. In that year, the total acreage of all land in farms was 3,744,625 acres, or 67 percent of the state's total land area of 5,564,000 acres. By 1880, there were 3,721,173 acres in farms. By 1950, the figure had dropped to 1,713,731 acres, a 54 percent decrease in farmland acreage in ninety years, or a .6 percent loss per year. Much of this decline, as in other areas of New England, was in the less productive hill farms and on land that was not ideally suited to farming. Such lands could not compete economically with the large new farms in the western United States. By the early 1950s, however, the decline of farmland use was stabilizing despite some localized decreases. In the *New Hampshire Idle Farm Land* study, published in 1953 by the Agricultural Experiment Station of the University of New Hampshire, the following findings were made:

> Most New Hampshire farming is not in the production of products for which nearness to markets is an advantage; much of New England farm produce is no longer competing directly with that from the more level and more fertile lands further West. Also, many of the more difficult farms have been abandoned; many of the farms that remain are at least fairly well adapted to modern farm machines. However, there is evidence that the process of abandonment and adjustment is still in progress.[8]

The *New Hampshire Idle Farm Land* report did not foresee how substantial the abandonment of farmland acreage would be over the next twenty-five years. Since 1953, when the report was issued, the process of "abandonment and adjustment" has continued at record rates. By 1982, the Census of Agriculture reported that total land in farms in New Hampshire declined to 545,000 acres on 3,400 farms, an average decrease of 33,000 acres or 2.1 percent per year since 1953. This represents a decrease of over 65 percent since 1953. Perhaps more strikingly, in the thirty-two years between 1950 and 1982, an amount of land equal to the total New Hampshire farmland in production in 1980 was abandoned or converted. Table 7.1 summarizes the decline in productive agricultural land from 1860 to 1982.

Although the 1.35 million acre decline in total farmland for the twenty-five year period 1900 to 1925 was the greatest in New Hampshire history, the more recent decline generated more concern throughout the state because it has occurred in areas of highly productive farms that contain high-quality soils. Prior to 1950, the conversion of less productive farmlands was of less concern and was in fact considered to be in the best interest of the farm industry and the general economy. The 1953 report on *Idle Farm Land* commented:

> ...in the past, much land abandonment has taken place and that, in the long run, is as it should be if people are to seek higher standards of living. In this study of idle farm land we have tried to exclude land which cannot be farmed so as to provide the farmer with a living is good as he can obtain on other land or at other occupations. We were interested in idle land which could be profitably farmed with modern methods and without having to be reclaimed from the woods.[10]

Table 7.1
Statewide Change in Acreage of Farmland and Number of Farms in New Hampshire, 1880–1982

Year	*Acres*	*Number of Farms*	*Change*	*% Change*	*% Change from 1880*	*Average % Change per Year*	*Average Farm Size (acres)*
1880	3,721,173	32,181	—	—	—	—	115.6
1900	3,609,864	29,324	–111,309	–3	–3	–0.15	123.1
1925	2,262,064	21,065	–1,347,800	–37	–39	–1.5	107.4
1940	1,809,314	16,554	–452,750	–20	–51	–1.3	109.3
1950	1,713,731	13,391	–95,583	–5	–54	–0.5	128.0
1959	1,124,312	6,542	–589,419	–34	–70	–3.8	171.9
1969	612,750	2,902	–511,562	–54	–84	–5.4	211.1
1978*	484,631	2,508	–128,119	–21	–87	–2.3	193.2
1982	469,582	2,757	–15,049	–3	–87	–0.8	170.3

*The 1978 Agricultural Statistics were revised in 1982 to compare more accurately to the 1982 and previous U.S. Agricultural statistics.

Source: U.S. Bureau of the Census, Census of Agriculture.

The *Idle Farm Land* report reveals that by 1953, the abandonment and conversion of high-qualty farmland in New Hampshire had already been recognized as an important land use issue. As the report pointed out, previous depletion of less productive lands was not the central concern. In fact, the purpose of the *Idle Farm Land's* study was to determine the quantity of high-quality farmland lying idle across the state and to understand the reasons for that land going out of production. The study found that within twenty representative towns examined, there was an average of 631 tillable acres per town of idle, high-quality, potentially productive farmland. The study also found that "largely residential uses have in a sense outbid agriculture for this land, although the process has been as much one of default by agriculture."[11] Urbanization thus was recognized to be one important cause for the declining agricultural land base in New Hampshire. Since 1953, urbanization has continued to be an important cause of the conversion in agricultural acreage, although general abandonment has accounted for the larger share. During the twenty-three year period from 1952 to 1975, about 60 percent of the farmland converted to other uses in the state was the result of abandonment, and approximately 40 percent of farmland taken out of production went into development.[12]

The pattern of conversion in the the state is also significant. In southern New Hampshire adjacent to the Massachusetts border, most of the abandoned agricultural land ultimately went into development. The urbanization in the southern half of the state has been the result of the substantial population growth that has occurred there, largely because of the expansion of high technology industries along Massachusetts Route 128, which allows interstate commuters to live in New Hampshire with no state income or sales tax while enjoying an easy commute to work in Massachusetts. According to the 1980 census, 89 percent of New Hampshire's population growth between 1970 and 1980 was in the southern half of the state. Only 11 percent of the population growth was in the three northern counties, causing one observer to comment that the census map of New Hampshire looks like a "bottle whose contents have all settled to the bottom like a thick residue."[13] Following this pattern of growth, most abandoned agricultural land in the northern half of the state went into forest uses rather than into urbanization.

The impact of these demographic trends on agricultural land use patterns was further analyzed in 1978 in a comprehensive statewide mapping of major land uses for the years 1952 and 1975. A community-by-community summary of this mapping program was published in a technical report entitled "Agriculture, Forest, and Related Land Use in New Hampshire, 1952 to 1975," prepared by the New Hampshire Agricultural Experiment Station, University of New Hampshire, in cooperation with the USDA Soil Conservation Service. The data in the survey summarized the extent of land use change taking place statewide, the nature of such change, how the rate of change differed in various parts of the state, and the amount and quality of agricultural land still available for farming.

The study showed that although roughly 27 percent of the 1950 land in agricultural use had been changed to other uses by 1975, only 8 percent of the 1980 farmland acreage went into development. In line with population distribution trends, the southern counties of Hillsborough and Rockingham lost 16 percent and 19 percent, respectively, of their Class I-III agricultural lands to development, whereas the northern counties of Cheshire, Coos, Grafton, and Sullivan lost less than 7 percent of their Class I-III agricultural land to development.[14]

In 1979, the New Hampshire Office of State Planning made a more in-depth analysis of this data as part of the New Hampshire Rural Development Planning Program. This analysis was designed to determine what was happening to agricultural lands in the state. The twenty-year land change for every community in the state was analyzed. Areas where agricultural lands were being converted to developed uses were identified. More importantly, however, two additional areas were identified: land not leaving agricultural use, and land leaving agricultural use but not converting to development use. This analysis was undertaken to understand the critical mass of agriculture in the state.[15] State planners hoped to locate areas of New Hampshire with the greatest potential to develop the agricultural industry in order to target supportive state services to assure a critical mass of agriculture in certain areas.

On the basis of this computer analysis, three categories of agricultural conditions were identified, depending on local and regional variations in land use conditions. The categories are:

> *Viable agriculture.* Areas where agriculture is currently producing successfully and competitively in the local, regional, and state economy.
>
> *Agriculture under pressure.* Areas of agriculture with land use trends over the time frame of the study (1952 to 1975) evidencing significant pressure for conversion to development use.
>
> *Agricultural potential.* Areas of unutilized agricultural potential in which good agricultural lands exist (Class 1, 2, 3) but are not currently used for agricultural production.[16]

A comparison of these conditions provides some insight into the nature of the decline in farmland acreage in New Hampshire and the growing concern for the protection of agricultural land since the mid-1960s.

There are essentially three viable agricultural areas in the state. The largest is the Merrimack River watershed; the greatest mass of agricultural lands lie in the central and north central portion of the river basin. The second area is a relatively contiguous but narrowly defined band along the entire length of the Connecticut River from the Massachusetts to the Canadian border. The third viable agricultural area of any significance is the lowlands in the southeastern corner of the state.

The areas where agricultural land is under the greatest pressure for conversion to nonfarm uses are in the south central and southeastern portion of the state, areas which also have viable agricultural lands. During the 1950s and 1960s, the conversion of farmland to development was the most pronounced in this region. This pressure for conversion of agricultural lands continued unabated for most of the 1970s, despite legislation to provide tax incentives for farmland owners. These conversion pressures may be expected to continue through the 1980s. A comparison of population trends revealed in the 1980 census and patterns of urban pressure on agricultural lands indicates that they are coincident.

Finally, the state has several areas of unutilized agricultural potential. Of interest is the rather large area of agricultural potential in the coastal lowlands in the southeastern corner of the state, adjacent to existing viable agricultural lands. Questions arise as to the likelihood of these lands becoming productive in agricultural use, considering the pressures for conversion that exist. As the development of the Boston surburban fringe along Routes 128 and 495 continues, the intensity of conversion pressures is likely to grow. It is, therefore, unlikely that these areas of unutilized agricultural potential will become productive unless measures are taken to encourage such uses.

The kinds of farm products produced in New Hampshire are similar to those of the other New England states. Table 7.2 and Figure 7.1 describe New Hampshire's crop and livestock cash receipts from 1981 to 1983.

Concern over Loss of Open Space and Community Character

Although much of New Hampshire's recent effort to protect agricultural land has been focused on a desire to increase food production, initial public concern was focused generally on the loss of open space and the deterioration of environmental quality. The decline of farmland acreage was viewed as an incidental part of a general environmental deterioration that by 1965 had caused the influential Spaulding-Potter Charitable Trust to warn: "Beautiful, rural New Hampshire faces a calamity. The qualty of life is threatened by the fastest growing population in the nation—by spreading urbanization."[17]

Indeed, since World War II, rapid urbanization was changing the face of New Hampshire dramatically. New interstate highways were constructed through the valleys, condominium developments sprang up in the White Mountains, and shopping centers and roadside developments began to spread throughout the rural countryside. Rapid population growth and urbanization continued unabated through the 1960s and 1970s. The 1980 census revealed that New Hampshire's population increased 24.8 percent statewide over the previous decade, the fastest rate of any New England state and nearly six times the rate for the region as a whole.[18]

In 1970, a citizens' Task Force which reviewed the management of the state's government reported that the environmental degradation caused by this urbanization was hurting the state's economy:

> Unspoiled surroundings are essential not only to our peace of mind and our health, but also to our economy. Unless our incomparable environment—our fields, forests, lakes, streams, mountains, and the air—is protected against further damage by uncontrolled encroachment of man-made ugliness and pollution, our recreation and tourist industry will deteriorate with the environment.[19]

In an effort to slow the depletion of open space and encourage the preservation of an attractive and healthful outdoor environment, a collection of environmental groups and the New Hampshire Farm Bureau worked to establish a differential tax assessment law for farm and open space lands. The program sought to:

> encourage the preservation of open space in the state by providing a healthful and attractive outdoor environment for work and recreation of the state's citizens, by maintaining the character of the state's landscape, and by conserving the land, water, forest and wildlife resources.[20]

Table 7.2
Cash Receipts from Farm Marketings in New Hampshire, 1981–1983 (thousands of dollars)

Commodity	*1981*	*1982*	*1983*
Crops			
Hay	1,767	1,952	2,223
Misc. Vegetables	4,309	4,606	4,650
Apples	8,089	8,739	7,544
Berries	887	1,350	1,330
Misc. Fruits	98	223	203
Maple	2,107	1,774	1,778
Forest Products	2,676	2,770	3,940
Greenhouse/Nursery	11,300	11,300	11,950
Misc. Crops	1,962	2,082	2,332
Total Crops	33,195	34,796	35,950
Livestock			
Cattle/Calves	5,060	9,266	8,085
Hogs and Pigs	1,268	1,010	682
Sheep and Lambs	186	208	291
Dairy Products	49,785	52,395	55,005
Chickens	454	217	318
Eggs	10,933	10,098	10,340
Misc. Poultry	2,856	2,076	1,818
Misc. Livestock	1,431	1,329	1,576
Total Livestock	71,973	76,599	78,115
Total Commodities	105,168	111,395	114,065

Source: New England Crop and Livestock Reporting Service.

Despite these public concerns for open space protection, land use control for largely aesthetic ends (with indirect economic benefits) never has been welcomed in New Hampshire, which has limited political and statutory alternatives available to open space advocates. Therefore, open space and farmland protection activity has been focused on food and economic independence, which supplies an effective political foundation.

Concern over Decline of Food Production Potential

In the 1970s, the focus of public concern over loss of open space and community character shifted to a concern over the declining local food-producing ability of the state brought about by the conversion of active agricultural land. New Hampshire's agricultural productivity had fallen from producing 38 percent of the state's food needs in 1930, to producing only 14 percent in 1975.[21] This declining food-producing capability became a concern of both farm industry specialists and the public at large. As Gerald Howe, Community Resource Development Specialist of the Cooperative Extension Service of the University of New Hampshire, explained:

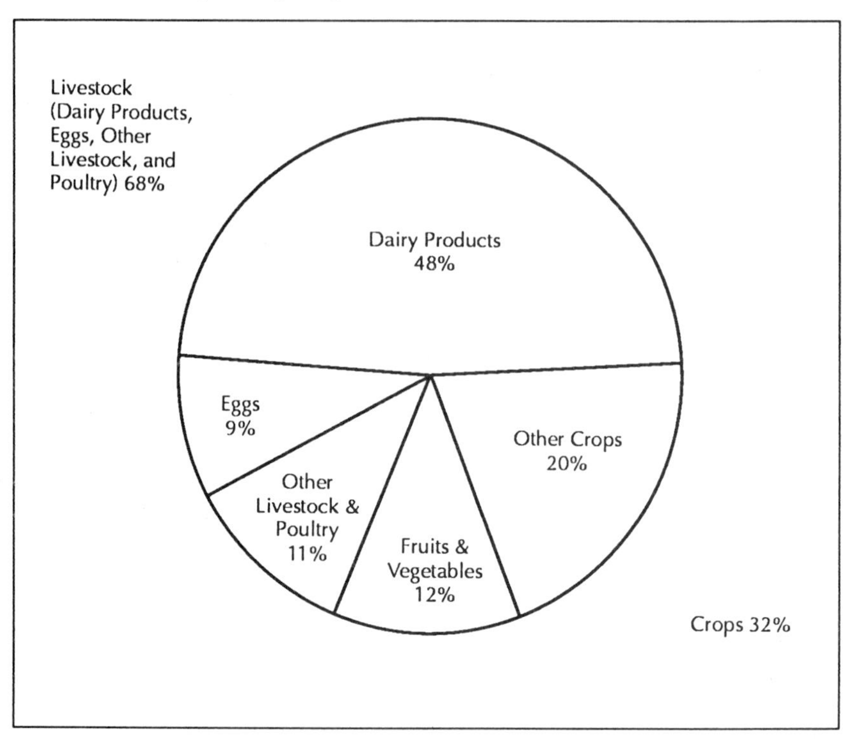

Figure 7.1. Distribution of cash receipts from New Hampshire farm marketings in 1983. *Source:* New England Crop and Livestock Reporting Service.

> From a realistic standpoint, we know we will never be self-sufficient in food and we really don't want to be. We simply would like to increase our self-reliance, and we think we can do that by our policy decisions on food and land use.[22]

Undoubtedly, the OPEC oil embargo brought to light the possible problems of food dependency. In 1979, the Cooperative Extension Service published "Recommendations for a New Hampshire Food Policy." The report, known as the Food Policy Report, stressed economic independence through greater food production in New Hampshire. The problem of food dependency is highlighted in the first paragraph: "New England is at the end of the food and transportation supply line. The state and the New England region are without stored food surpluses, other than supermarket stocks, which would be needed in the event of a crippling storm or other natural disaster."[23]

The possible loss of the agricultural industry also was viewed as an economic injury to the state. Moreover, farmers are considered by many as an archetype of honest, hard-working citizens—pillars of the community whose loss of occupation and way of life are vital to New Hampshire's tradition. These specific reasons for the concern over declining food independence were expressed in House Concurrent Resolution 13, also discussed later in this chapter. That Resolution states, in part:

> [A]griculture represents an important industry for New Hampshire, as well as a vital heritage for the state; and . . . the decline of agriculture in New Hampshire represents not only a threat to our citizens' well-being but also a continuing loss of an occupation and a way of life that is central to New Hampshire's tradition of independence and self-reliance.[24]

Because land use controls often produce negative reactions, farmland advocates seized upon the food dependency issue to muster broad public support for land protection. As Howe explained:

> We decided, as advocates of the public interest, we would look at food—not land, but food—because we felt it was the least common denominator. Not everyone is a landowner, but everyone does eat. We felt that we would not focus primarily on the preservation of agricultural land because if you preserve agricultural land and you don't preserve agriculture, you're simply preserving open space. We felt that all agricultural land is open space, but not all open space is agricultural land. Therefore, let's go for the smallest category and concentrate all our efforts on that. And we did that through food and food policy.[25]

Food independence, therefore, was linked with economic well-being and has become the primary basis for the adoption of farmland protection programs in New Hampshire. In a politically conservative state, this approach has proved highly effective.

Conservation Programs and Early Efforts to Protect Agricultural Land

Although programs to increase food self-sufficiency have recently become the most important basis of farmland protection, initial efforts to protect agricultural land in New Hampshire were not directed specifically at farmlands but were part of broad environmental efforts. Since farmland conversion was of little concern until recently, either to the agricultural industry or the general public, actions which resulted in the protection of agricultural land were usually the byproduct of plans or programs which protected open space, wildlife, or other environmental resources. Many of these programs and institutions have expanded their roles to include farmland protection, either directly or indirectly, in recent years as awareness of the decline of agricultural activity has grown. Indeed, some of the early conservation programs were instrumental in the implementation of recent farmland protection programs.

Private Trusts

Many of the early statewide land conservation and environmental protection programs grew out of the conservation efforts of various private charitable trusts. The Spaulding-Potter Charitable Trusts (1957–1972), established in 1957 from the bequests of wealthy New Hampshire and Boston families, were instrumental in local land acquisition programs during their fifteen-year lifespan. During this time, the Trusts expended about 60 percent of their $16.8 million of available funds for environmental projects in New Hampshire. Among its many projects, the Trust made money available on a matching basis to New Hampshire towns for permanent land acquisition funds. In 1961 the New Hampshire Charitable Fund was established. This fund enabled various family bequests to be applied to educational, cultural, health, and environmental projects in the state. Notably, these two Trusts provided funds to Conservation Commissions for land acquisition programs where state assistance was lacking.

Although not established for agricultural protection purposes, the Spaulding-Potter Charitable Trusts and the New Hampshire Charitable Fund were instrumental in promoting an awareness of the natural environment of the state. An important activity of the former trust, undertaken in 1969, was a project known as "New Hampshire Tomorrow," a "multi-faceted program to alert New Hampshire people to the threats to their environment and encourage them to confront those threats."[26] Through "New Hampshire Tomorrow," over fifty projects were undertaken, including studies of environmental threats to the coastline, mobile homes, access to public waters, the siting of power plants, and the eutrophication of lakes. The results of these demonstration projects were discussed at a final conference in October 1972. The demonstration projects and the final conference underlined the sense of need for united action to save the state's natural heritage. This probably was the first time in New Hampshire's history that conservation groups shared a feeling of cooperation.

Shortly after the conference the Environmental Council was created (also known as SPACE—Statewide Program of Action to Conserve our Environment). Since its formation, SPACE, with participation by the leading conservation groups in the state, regularly prepares a list of legislative priorities and fights for land use bills in a coordinated manner. This type of approach has been a major factor in the adoption of the state's farmland protection programs.

Conservation Commission Enabling Act (1963)

As in Massachusetts, possibly the most important early legislative program to evolve from the environmental concern of the early 1960s in New Hampshire was enabling legislation allowing the establishment of local Conservation Commissions.[27] Initially, local Conservation Commissions were primarily established to conserve and protect natural resources, such as wetlands, water bodies, and critical wildlife areas. In recent years, however, as awareness of the decline of agricultural acreage has grown, local Conservation Commissions have expanded their roles to become important vehicles for local agricultural protection efforts. Conservation Commissions now play an important part in implementing state programs such as the New Hampshire Agricultural Land Preservation Act of 1979, which allows the state to purchase farmland development rights.

The movement to adopt Conservation Commission enabling legislation in New Hampshire began at the New England Conference on Conservation held at the Harvard Business School in Cambridge, Massachusetts in 1960. The conference was held primarily to promote the recently enacted Conservation Commission legislation in Massachusetts. However, a group of New Hampshire citizens who attended began working through various state and local groups to press for Conservation Commission enabling legislation. In 1963, the New Hampshire legislature adopted such legislation, modeled after the Massachusetts statute.

As in Massachusetts, the growth of the Conservation Commission movement initially was slow but has steadily gained momentum. By 1969, more than 85 of the state's 235 cities and towns had Conservation Commissions. The Commissions were widely scattered through the state and concentrated in rapidly urbanizing areas and the lakes region.[28] By 1981, 184 New Hampshire cities and towns had Conservation Commissions.

Unlike Massachusetts, the Conservation Commission program in New Hampshire has never had significant financial assistance from the state. Although the enabling legislation authorized the State Department of Resources and Economic Development (DRED) to assist communities which established Commissions, funds were never appropriated to assist these bodies in land acquisition and administrative activities the way the Self-Help Program did in Massachusetts. Consequently, New Hampshire's Conservation Commissions have made less significant headway in acquiring land, since they rely more heavily on gifts and grants from local sources or private trusts.[29]

Over the years, Conservation Commissions have expanded their role in protecting open space. Initially, open space protection was accomplished by adopting local subdivision regulations which require certain set-asides for open space within new subdivisions. Unfortunately, these set-asides were not always ideal and did little to protect agricultural land. Local Conservation Commissions were often donated wet or muck lands. These could not be developed, and they provided a significant tax benefit to land subdividers or developers and little benefit to the town or the Commission. Eventually, the public began to question this type of open space preservation.

As concern over the conversion of farmland grew, open space advocates began to use local Conservation Commissions as a vehicle of farmland protection. As one observer noted:

> People concerned about open space through the Conservation Commission movement all of a sudden said "let's get agriculture and let's do something to protect agriculture," realizing that farmland is open space too, just a certain desirable kind of open space.[30]

This sentiment breathed new life into the farmland protection movement, and many local Conservation Commissions have begun to prepare regulations to restrict the amount of development that may occur on agricultural land. Moreover, the Conservation Commissions have been identified as participants in the Agricultural Lands Preservation Act, which was adopted in 1979. Thus, the Conservation Commission program, which was not initially established to protect New Hampshire farmland, has grown to become a significant element of farmland protection efforts.

Citizens Task Force

The increased action for open space protection and the general environmental protection is further illustrated by actions taken in 1969. In that year, as a result of a 1968 campaign promise, then-Governor Peterson persuaded the legislature to authorize a thorough management review of New Hampshire's state government.[31] A Citizens Task Force was appointed to oversee the study. By the beginning of the 1970 legislative session, the Task Force identified a plethora of statewide concerns and proposals for reform, ranging from executive reorganization and legislative reform to improved welfare programs and tax reform. On the subject of the environment, the Task Force recognized the importance of farmland, forests, and the natural environment for recreation and tourism.

Many citizens of New Hampshire felt that the Task Force recommendations represented the minimum reforms necessary to give New Hampshire an adequate state government. Few of the recommendations, however, resulted in legislation. Among those proposals enacted by the legislature was a 1970 act which transferred the Office of State Planning from DRED to the governor's office, thereby increasing the visibility and importance of the planning office. This action symbolized the concern for planning and the overall economic, social, and environmental character of the state that was becoming widespread. The Office of State Planning has since proved instrumental in many of the farmland protection programs discussed here.

Critical Lands Bill (1973)

Realizing that comprehensive land use regulations such as Vermont's Act 250 or Maine's Site Location Development Act were not politically possible, a coalition of concerned environmental and agricultural groups proposed a Critical Lands Bill in 1973. This bill would have controlled construction on prime agricultural lands, flood plains, high elevations, and unique natural areas—roughly 6 percent of New Hampshire's land area. Despite its limited scope, the bill was defeated by a margin of almost two to one in the legislature.[32]

The defeat of this bill highlights the conservative home-rule nature of New Hampshire politics. In most legislative sessions since the early 1970s, environmentally concerned groups have sought comprehensive land use legislation but repeatedly have seen it defeated by groups fearful of losing control over the land.

Historically, New Hampshire farmers have strongly opposed comprehensive land use regulations. Most farmers have invested the majority of their human and financial resources in their land. Despite this heavy personal investment, average farm income is low. As a result, a farmer's hope for the future, for retirement or in the event of illness, is the market or development value of his land. Since regulatory measures sometimes can restrict the ability of the farmer to sell his land when needed, such measures have lacked the support of New Hampshire farmland owners. The distrust of regulatory measures is not unique to New Hampshire farmers, but the state's conservative atmosphere increases the impact of such sentiments.[33]

Gradually, however, some of the strongest opponents to land use controls have begun to reconsider. Farmers sometimes have found that the critical mass of land and support has vanished, with newcomers disliking their way of life. Also, many development interests in the state have begun to realize that lacking state controls, many of the most desirable towns were adopting zoning bylaws which exclude practically any new development. As such exclusionary laws were adopted, there was no way for developers to force their way into a town.[34] Consequently, since the defeat of the Critical Lands Bill, there has been a waning of opposition to some form of statewide land use controls. Although by 1985, comprehensive land use legislation had not been adopted, the citizens and legislature have indicated their willingness to accept and pay for some open space and farmland protection programs in recent actions that have been taken.

Recent Agricultural Land Protection Programs

The work of private groups, land trusts, and the Conservation Commissions did not prevent the conversion of active agricultural land during the population boom years of the 1960s and 1970s. The citizenry of New Hampshire, therefore, looked at the efforts to protect agricultural land taken by neighboring New England states. The voters subsequently amended the state constitution in 1968, allowing for passage of a current-use taxation law in 1974. This program helped encourage the retention of various kinds of open space, but firmer measures were needed for farmlands. The Food Policy Report, which stresses economic independence and food production, provided the impetus for the establishment of a purchase of development rights program and of an Agricultural Task Force. The Task Force, in turn, successfully urged the use of local zoning and better marketing programs and supported passage of a right-to-farm law. Thus, a pattern of strong local planning and implementation with state leadership and support is emerging in New Hampshire, like the other New England states. In light of New Hampshire's fiscal problems, however, most recent efforts focus on farm economics and require little or no direct expenditures of state funds. The result today is that there exists a variety of direct and indirect farmland protection programs designed to foster the food and economic independence of the state.

Current-Use Tax Act (1973)

The property tax rate in New Hampshire, with no sales or income tax, is among the highest in the country. The population boom in southern New Hampshire also brought about a significant rise in farm and other land values, which further increased property taxes. This population increase also caused the tax rate to increase to provide better public services to the growing population, many of whom were from urban areas and had grown accustomed to a high level of public service. Farm income, however, was not rising enough to make many farm operations profitable.

Realizing that owners of certain undeveloped land needed property tax relief if the rural character of the state was to be protected, citizens worked toward altering these landowners' tax bills. A coalition of environmental groups and the New Hampshire Farm Bureau Federation worked to place a referendum on the 1968 ballot amending the state's constitution to allow for use-value property taxation. On November 5, 1968 this referendum was passed.[35] In 1973, the legislature passed a statute recommended by the open space committee of SPACE. This statute, which went into effect on April 1,

1974, relieves various open space owners of some of the pressure to sell their land. The legislative declaration of purpose makes this aim clear:

> It is hereby declared to be in the public interest to...prevent the conversion of open space to more intensive use by the pressure of property taxation at values incompatible with open space usage, with a minimum disturbance of the concept of ad valorem taxation.[36]

The Current-Use Tax Act establishes a special taxation scheme allowing owners of farm, forest, recreational, and other types of undeveloped land to receive a lower property tax assessment in return for keeping their land in "open space." The program is broader than those in some states because it is not limited to specified land uses such as agriculture. Like current-use tax assessment programs in other states, however, the Act is designed to lower the tax burden of land based on its value in certain specified uses. Land which meets certain use, acreage, and productivity requirements of the statute is assessed at its value in its *current use* rather than at its full use value in its "highest and best use." Thus, the Current-Use Tax Act provides an economic incentive (or removes a disincentive) to encourage the protection of open space by tying property value to the value of the land in its current use. (The Act also allows a flat 20 percent assessment reduction if the landowner allows public trespass for recreational purposes.)[37]

The statute lists farmland, flood plain, forest land, open space land, recreation land, wetland, and wildland as qualifying land uses. The Current-Use Advisory Board, created by the statute, has developed specific qualification guidelines. Classifications involving agriculture are:

> *Farmland.* Any tract of land with at least ten acres actively devoted to agriculture or horticultural use or which has an annual gross income totaling at least $2,500 regardless of acreage.
>
> *Inactive farmland.* Land of at least ten contiguous acres which is being kept open by generally accepted methods, but not cropped, if the intent is to preserve scenic qualities, improve wildlife habitat, and maintain a agricultural land reserve, or if recommended to be maintained as open space land by the municipal Conservation Commission or other body designated by the Board of Selectmen. It must have been or could be used as farmland and must be devoid of woody growth.
>
> *Productive wild land.* A tract of unimproved land of at least ten contiguous acres which is capable of producing commercial agricultural or forest crops and which is now and has been left in its natural state for five years without interruption of natural ecological processes.[38]

To qualify for current-use tax assessment, in most cases the landowner must file an application before April 15 of the year for which tax benefits are to apply. The determination of whether a given parcel of land qualifies for appraisal at current use value is made annually by the local Board of Assessors or Selectmen in accordance with the above guidelines. A right of appeal to the Board of Taxation and the courts exists if the landowner feels that the denial of classification was improper.

Assessments are based on criteria and land use valuations which are determined each year by the Current-Use Advisory Board. This schedule assigns a range of values for the various classifications of land use. In the case of farmland, a variety of categories exists, including pasture, cropland, or horticulture. Each of these categories of agriculture is assigned a range of values from which the local assessor may assign a value to the particular parcel in question. The actual commodity being produced in a particular category makes no difference. For instance, in the agricultural category of forage, whether an acre of land is used for the production of hay or silage corn is unimportant. A range of values, such as $150 to $350 per acre for forage, is established annually by the Current-Use Advisory Board, and the local assessor then establishes a value based on his interpretation of the value of the commodity or commodities being produced. Thus, the local assessor might conclude that corn has a higher value to a dairy farmer and assign a value of $300 to $350 per acre to land producing corn for silage.

This valuation and assessment procedure, which bases land values on general categories of land use, differs from similar programs in some states such as Massachusetts, which establish statewide farmland value assessments based on the value of the actual farm commodity produced. The New Hampshire program allows a unique degree of local discretion. In conservative New Hampshire, this was considered a political necessity. At times, it is also a source of local conflict between farmers and town officials over the appropriate value to be assigned to farmland.[39]

There are two other procedures a landowner can use to qualify for the program. First, if a parcel participates in the purchase-of-development-rights program, which was adopted in 1979, it automatically receives use-value assessment.[40] Second, if the land does not fall into a current-use category, then the landowner can apply to the municipality where the land is located and grant the muncipality a discretionary easement.[41] Under this procedure, the town must determine if the land's use is consistent with open space objectives. If so, the landowner can grant an easement not to subdivide, develop, or otherwise change the use of the land for at least ten years in exchange for use-value assessment. An early release is allowed for in cases of "extreme personal hardship," if the landowner pays a penalty tax.

The penalty provisions of the New Hampshire Current-Use Act are the key to its impact on protecting farmland and open space. A 10 percent land use change tax is levied when land under the program is changed from open space to a nonqualifying use. The tax is based on a reassessment of the land at its full market value at the time of the use change. The new assessment applies to that tax year, regardless of an earlier assessment. There is no liability for the land use change tax whenever the use is changed to another qualifying use. In such cases, however, a new appraisal of the land will be made when the annual application is made to the board of assessors.

The type and timing of a change in use is critical because when land value changes, so does the amount of the tax. The statute provides that a change occurs when there is actual construction, excavation or grading, any other act consistent with construction (except construction of roads or buildings for agriculture), or extraction of topsoil, gravel, or minerals. The New Hampshire Supreme Court has ruled that the change occurs at the time of a physical change, not when a subdivision is approved. Also, if the landowner clears lots to get subdivision approval, then the change has occurred.[42] Thus, a landowner wishing to minimize his penalty tax liability may alter the open space use so as to make it appear unprofitable, pay the penalty tax, and then build a subdivision.

The penalty has often been criticized as being too permissive for several reasons. Because the penalty is triggered only when a change actually occurs, speculators can buy or sell land with impunity if they do not alter the use. (This criticism may not be valid since the real objective of the law has been met—no change in land use). Unlike the Massachusetts program, the statute does not grant municipalities a right of first refusal if the land is sold for development. Also, some land use changes have not been detected and reported, and there is no additional penalty for the delay of tax payment aside from the cost of interest. The penalty has also been criticized as being too stringent because, unlike in some states, the rate of the land use change tax does not decrease over time.

A recent study of selected municipalities indicates that the penalty has not been very effective in reducing the conversion of agricultural land.[43] In all current-use categories, less than 1 percent of the participating lands came out of current-use status and into other uses. In areas of high conversion pressure, however, the rate was higher. Merrimack, a quickly growing town, led the municipalities studied with a 10.6 percent rate of conversion from open space to another land use. Moreover, about one-fourth of all conversion was on farmlands. Finally, the earlier the entry of the land into the program (and therefore the greater the tax savings), the greater the rate of conversion. Given that the study covered the years from 1974 to 1980, it appears that the rate of conversion may increase. Thus, the penalty is not sufficient to keep participating lands in agriculture in many cases.

When first adopted in 1974, 80 percent of the current-use applications resulted in farmland valuations which were higher than the value previously established by local assessors.[44] Many of the communities in the state are or were predominantly agricultural, and so there was a tendency by local officials to assess well below market value. The Current-Use Tax Act, however, set values based on 100 percent valuation of land in its current use, which was often close to full value. Consequently, the program initially received few applications.

Recently, however, New Hampshire communities have moved to have property-tax assessments based on 100 percent valuation, which has resulted in more landowners applying for assessments under the Current-Use program. Greater information about the program also has been considered responsible for increased participation.

In 1974, 112 of the state's 235 cities and towns participated in granting use-value assessments to 254,503 acres.[45] By 1980 some 209 towns and 1,604,840 acres of open space lands were enrolled in the program. Average parcel size was 106 acres, and the sum total represents about 30 percent of all private, undeveloped land in the state. Only 130,000 acres of farmland were enrolled, although some farmland may be classified under productive wildland (333,800 acres). The official statistics of the Department of Revenue and Taxation for 1982 reveal that 145,833.6 acres of farmland were enrolled, representing about 8 percent of the total acreage under the program (1,930,312 acres and 17,596 landowners), and less than one-third of all farmland in the state.

The reason for the low enrollment of farmland is twofold. First, because local governments rely on the property tax for most of their revenue, local officials tend to set farmland values at the top end of the range established by the Current-Use Advisory Board without looking at the actual productivity of a particular parcel or other considerations affecting its value. As a result, many farmland owners would do no better under the program than with full-value assessment, especially in more remote areas. As one state analyst commented:

> I don't think [local] officials are looking at long-range plans. They are looking at short-run cost minimization. If the range established by the Current Use Advisory Board is 150 to 300 dollars per acre, they're going to say it's 350 dollars per acre because at least it keeps the rest of their taxes down. If it's set at 150 dollars per acre, and they have to raise a certain amount of money to support town services, then that additional $200 value is going to be pushed off on some other property owner just to get the assessment up to keep the tax rate down. So they tend to set assessed value at the highest dollar value within the range.[46]

Thus, local assessors have helped create an inability to bring land under the program.

A second reason farmers do not enroll in the program is that some prefer to speculate on the increased value which development of their land may bring. Especially in southern New Hampshire, a farmer must be highly dedicated to his labor to turn away offers to sell his land. The Act works best, therefore, when the value of the land is too speculative to forego a tax benefit with restrictions.

The New Hampshire Current-Use Tax Act has been somewhat unpopular among the general population. Many wealthy landowners who do not need a tax benefit and do not provide direct public benefits with their classified lands are enrolled in the program. And, as in other differential assessment programs, people often perceive an inequitable shift in tax burden from one landowner to another. Although the program's supporters argue that an increase in taxes would result to all landowners if open space lands were developed (from the cost of services), this argument has less appeal in New Hampshire than in other parts of the country. As noted earlier, the property tax is at nearly confiscatory levels and the state's financial crisis of recent years has caused many to feel that full assessment of open space land would be an easy way to yield revenues. It is no surprise, therefore, that a bill is introduced annually in the state legislature to repeal the Act.

Despite the opposition, most analysts agree the Current-Use Act will remain on the books. The program has sometimes made the difference between profitable and unprofitable farm operations. For example, one New Hampshire farm recently was reassessed under the program from $99,000 to $12,700.[47] The smaller tax liability significantly altered the farmer's profit-and-loss sheet.

The only amendments to the Act have been technical. If a further change in the law is to be made, it may be to broaden the tax. Farm buildings are exempt from the Act, and local assessors tax them heavily.[48] As a result, ground-storage facilities are increasingly replacing traditional grain silos. If the citizens want to preserve traditional New England, a change in the law is necessary.

Whether farmland owners will increase their participation under the Act probably will depend on the fiscal status of the state and future development pressures. Participation will be desirable only if farmland owners receive a lower assessment under the Act than with full-value assessment. Moreover, farmland owners must be willing to resist the financial pressures to develop their land. In any event, the program to date certainly has helped protect some agricultural land.

Food Policy Report (1979)

The Cooperative Extension Service's 1979 "Recommendations for a New Hampshire Food Policy" became the foundation for efforts aimed at economic independence through food production and agricultural land protection. The Food Policy Report sets forth a state policy by evaluating consumer needs, land use concerns, agricultural production, food processing and storage in the state, transportation of food products, and the marketing process. The report touches on all aspects of food policy and stresses the nutritional and economic concerns of consumers as well as the similar concerns of businesses in the food chain. Economic independence through greater food production and distributional autonomy is the message of the document. The Food Policy Report sets forth short- and long-range goals in areas related to agriculture.

The Food Policy Report recommended that the legislature work toward certain objectives in addition to agricultural goals. In the area of land use, the Report recommends that the state:

1. Encourage and assist private development and all levels of government to utilize soils poorly suited to agriculture.
2. Encourage publicly owned lands suitable for production to be made available for commercial, small, and part-time farming and for community gardens.
3. Compensate landowners who voluntarily choose to protect agricultural lands permanently.
4. Encourage the revision of zoning ordinances to allow agricultural uses in residential areas.
5. Encourage good conservation practices on all agricultural land to insure long-term productivity.[49]

Objective 3 has proved to be the most important, for the Report recommended that to achieve this objective the legislature should pass the Agricultural Encouragement Act, a program to encourage local agriculture, and the Agricultural Land Preservation Act, a PDR program, as well as continue the Current-Use Assessment Act. The legislature agreed with the Report only on the Current-Use Assessment Act and the Agricultural Land Preservation programs.

Food Policy Resolution

With slight modification, House Concurrent Resolution 13 incorporated the goals of the Food Policy Report into a clear statement of legislative goals and objectives. This resolution was approved by the New Hampshire House and Senate, setting forth the state's agricultural production and food policy. The nine goals read as follows:

1. Assure an adequate, wholesome food supply, emphasizing locally grown foods at reasonable cost;
2. Develop a level of public knowledge and understanding of foods and nutrition that will promote optimal nutrition, health, and well-being;
3. Preserve as much land as possible that is currently in agricultural use.

4. Expand agriculture to land capable of agricultural production;
5. Recognize agriculture as an important component in the balanced growth of New Hampshire;
6. Maximize New Hampshire's commercial food production and encourage small and part-time farming and home production to enhance family and community self-sufficiency;
7. Establish processing and storage facilities and services that will adequately serve the needs of New Hampshire's farmers and to have processing storage facilities that will insure an adequate supply of food for New Hampshire throughout the year;
8. Provide efficient transportation of food and feed products from producer to consumer in a timely manner;
9. Enhance the food marketing systems in New Hampshire by facilitating the access of consumers to products produced in New Hampshire and New England region to benefit the producer, the consumer and the general economic welfare of the state.[50]

Resolution 13 states that all state actions should relate to at least one of these goals, and it further encourages the citizens of New Hampshire to determine responsible present and future food policies from an educated viewpoint. Shortly after the adoption of Resolution 13, two important proposals which would stimulate farmland protection in New Hampshire were introduced—the Agricultural Encouragement Act and the Agricultural Land Preservation Act.

Agricultural Encouragement Act

The Agricultural Encouragement Act was designed to promote native agriculture by:

1. Advertising and marketing (i.e., of pick-your-own locations).
2. Requiring the purchase of New Hampshire food products by state programs in certain cases (state universities, subsidized school lunch programs, state mental health divisions, and similar programs).
3. Instituting a "Grown in New Hampshire" labeling program.
4. Limiting the liability of landowners who have pick-your-own operations.[51]

Many supporters of the two bills felt that politically the time was not right for the passage of an expensive farmland protection bill such as the Agricultural Land Preservation Act. There was a feeling that the $3 million PDR program would fail, but the Agricultural Encouragement Act, funded at $96,000 over two years, would have a better chance of passage. The reverse actually occurred. Gerry Howe commented:

> To be a weatherman in New Hampshire is difficult. To be a political forecaster in New Hampshire is even more difficult. The purchase of development rights passed with $3 million; the Agricultural Encouragement Act...did not. That was for a political reason, not that the program wasn't supported. The money did turn up, however, and we have most of the things going that were in the Agricultural Encouragement Act, without the passage of it.[52]

Despite the failure of the Agricultural Encouragement Act, marketing tools instituted through the Department of Agriculture achieved most of the goals of the Agricultural Encouragement Act; they are discussed in a later section of this chapter.

Other recommendations of the Food Policy Report also have been initiated through federal, state, and local agencies. Much of this activity has taken the form of policy and administrative changes.

In the conservative political climate of New Hampshire, the comprehensive view of agriculture presented in the 1979 Food Policy Report synthesized many of the concerns of consumers, farmers, the food industry, and others over the decline of agricultural land use in a single, coordinated document. Resolution 13 thus represents a firm declaration that agriculture in New Hampshire is an industry worth protecting. Many agricultural land protection programs, such as the Agricultural Land Preservation Act, which have been implemented since 1979 reveal that New Hampshire's Food Policy has been an effective foundation for legislative, executive, and administrative decision making in a politically conservative state.

Agricultural Land Preservation Act

As a result of the Food Policy Report and the Food Policy Resolution, public sentiment grew for the adoption of programs specifically designed to protect agriculture land. In 1979 the legislature took such action by establishing a PDR program and an agricultural task force;[53] the latter is discussed more fully in a separate section.

Support for a direct farmland protection program grew from the realization that such land was rapidly being converted to other uses. The Task Force noted in its January 1, 1981 Report:

> According to the 1980 census, New Hampshire has the largest growth in population of any state east of the Mississippi (excepting Florida). Population here expanded by 24.6%! Thus the need for more food and fiber from our farmers . . . at the same time this increased population is gobbling up farmland for development. Also simultaneously, costs of transporting food into the state rose dramatically due to hikes in gasoline prices. These events put tremendous financial pressures on farmland owners. A great many say they want to keep farming, but find it hard to resist the deals offered by speculators and developers. This is exactly the situation the Agricultural Land Preservation Program was enacted to combat.[54]

In other words, the Current-Use Tax Act was viewed as too indirect a means for retaining productive farmland because the owner could sell to a developer at a profit even after the land use change tax was paid. The Agricultural Land Preservation Act, therefore, focuses on the retention of the land. The Act's purpose is clearly stated in the preamble:

> to recognize the importance of preserving the limited land suitable for agricutural production, to safeguard the public health and welfare by encouraging the maximum use of food and fiber producing capabilities of the state's agriculturally suitable land and to ensure the protection of agricultural land facing conversion to non-agricultural uses.[55]

The Agricultural Land Preservation Act is a comprehensive statute. Although its main effect is to allow state purchases of development rights, it also provides that "development rights of agricultural lands may be acquired by any governmental body or charitable corporation or trust which has authority to acquire interests in land."[56] Because New Hampshire could draw upon the experience of states such as Massachusetts and Connecticut, the Granite State's PDR program is uniquely providing for many kinds of acquisitions in a single statute. Only the state acquisition program is discussed here.

To manage the program, the Act established an Agricultural Lands Preservation Committee composed of seven members: the Commissioner of the Department of Agriculture (chairman); the Commissioner of the Department of Resource and Economic Development, or his designate; the Director of the Office of State Planning, or his designate; the Secretary of the Agricultural Advisory Board; and three appointees of the Governor, two of whom must be farmers. In addition, the Dean of the College of Life Sciences and Agriculture at the University of New Hampshire and the New Hampshire State Conservationist of the USDA Soil Conservation Service sit on the committee in an advisory capacity as nonvoting members.[57]

The Act provides for certain general procedures to be followed in the application process and for certain broad considerations to be made in land selection decisions. More specific guidelines have evolved in the form of administrative rules and regulations.

The Act required the Commissioner of Agriculture, after obtaining the advice and consent of the Committee, to establish rules which municipalities must follow in evaluating farmland suitability and feasibility for the program; criteria to define and classify agricultural lands; procedures for purchase; and procedures for release of restricted parcels. Within a year, rules and regulations were promulgated under the Act after numerous public hearings, consultations with planners at the Office of State Planning and PDR officials from Massachusetts and Connecticut, and legal and technical assistance from the New Hampshire Attorney General's Office.

Under the statute and rules, the farm landowner who wishes to sell his development rights to the state must submit an application to the Commissioner of Agriculture and the municipality where the land is located. Much detailed information relevant to the need for a purchase is contained in this application. The local governing body, or an agency it selects, such as a Conservation Commission, is required to forward a statement within sixty days to the Committee and the applicant stating whether the proposal is recommended by the municipality and the extent to which the municipality may desire to participate financially in the purchase of the farmland development rights. In making its recommendation, the local governing body must include comments on the compatibility of the proposal with the local master plan, zoning bylaws, planned public works, and approved local ordinances. The failure of a municipal governing body to make a recommendation on the proposal or a negative recommendation by the local governing body, however, does not prevent the Committee from considering a proposal.

When the Commissioner receives the application, he may inspect the lands, negotiate with the applicant and the relevant municipality, and submit the application to the Committee. Within seventy-five days of receipt by the Committee, the land owner must be notified of provisional approval or disapproval.

The Agricultural Lands Preservation Act establishes three general guidelines for the Committee to follow in evaluating the merits of any proposed development rights purchase:

1. The degree to which acquisition would protect the agricultural potential of the state.
2. The suitability of the land for agricultural use, based on soil and other criteria.
3. Cost to the state.[58]

The rules established by the Commissioner of Agriculture identify six evaluation criteria and one bonus criterion. A point system was established to evaluate each proposal. A farmland unit must receive a minimum number of points before it can be considered for a purchase. These criteria and points are:

1. Important agricultural soils based on classification and soils productivity as derived from Soil Conservation Service data. (Maximum 30 points.)
2. Threat of conversion to a nonagricultural use. (Maximum 30 points, based on: written evidence of proposed sale, 10 points; conversion of other agricultural land within one mile, 10; and personal difficulties such as age, health, taxes, etc., 10.)

3. Cost to the state of New Hampshire to purchase the development rights. The more expensive the development rights, in terms of fee simple, the lower the point score. (Maximum 15 points.)
4. Economic viability of the farm including availabilty of wholesale and retail markets, availability of supplies and services and the concentration of adjacent farm and agricultural activity. (Maximum 15 points.)
5. Inactive or active farmland: If the land under consideration has been used to produce a crop over $5,000 gross income in one of the last three years. (Maximum 5 points.)
6. Owner-operated farm: The person or family corporation that has fee simple title to the land operates the farm. (5 points.)

   Bonus. Financial participation by city or town: If a municipality participates in the purchase of farmland development rights, the applicant is awarded one-half point for every 1 percent of the total cost assumed by the city or town. (Maximum 20 points.)[59]

Thus, the highest priority is assigned to protect the greatest amount of good agricultural land that is in the greatest danger of conversion.

Full appraisal can occur after provisional approval by the Committee. Initally, appraisals were based on the difference between fair-market value and restricted value. To clarify the law, it was amended to require appraisals be no greater than the difference between the land's market value and the top-value range for horticultural crops established by the Current-Use Advisory Board under the Current-Use Tax Act. This difference is the value of the development rights which the state purchases. The landowner must pay for the appraiser that he and the Committee jointly select.

Within forty-five days after a provisional approval, the Commissioner must resubmit the application for final approval. If given final Committee approval, the funds for purchase are reserved. The Commissioner must then give final approval. A party that is dissatisfied with the Committee's approval decision has a right to a hearing before the Committee.

The terms and conditions placed on the deeds of participating lands prohibit, or limit, among other things:

1. construction or placement of buildings except those used for agricultural purposes or for dwellings used for family living by the landowner, family members, or employees.
2. excavation; dredging; or removal of loam, sod, and other mineral substances in such a manner that would adversely affect the land's future agricultural potential.
3. other acts or uses detrimental to retention of the land for agricultural use.[60]

After the covenants, easements, or other necessary legal instruments are executed, the Commission pays the landowner the agreed price. A municipal official must survey the site when the deed is recorded; the landowner pays for the survey.

Once established, the restriction is held in perpetuity in the name of the State of New Hampshire but may be released by the Committee in certain cases. In general, the landowner must prove that the site is "no longer suitable for agricultural purposes."[61]

To obtain a release, the landowner must apply to the governing body of the municipality where the land is located and to the Commissioner of Agriculture. The municipality must inform the Commissioner within sixty days whether it approves the request. This recommendation can be based on information provided by Conservation Commissions and certain municipal officials. The Agricultural Lands Preservation Committee must hold a public hearing in the relevant municipality within 150 days after it receives the application from the Commissioner. A majority vote of the Committee will release the land.

In returning to the state the value previously received, the landowner must pay a sum not less than the difference between the fair-market value of the land at release and the fair-market value restricted to agricultural purposes when the development rights were acquired. If a municipality contributed to the initial purchase, it will receive a percentage of the release price proportionate to its contribution. This latter provision gives financially strained towns incentive to recommend release. The cost to the landowner of a development rights release can be significant. Figure 7.2 illustrates the projected future cost to the landowner of a development rights release.

Results of the Program. After the rules and regulations to govern the administration of the law were adopted in June 1980, the Committee established the period from July 1 through August 29, 1980 to accept the first round of applications for the program. During that period, twenty-eight applications from seven counties were received. Because the program was so new and the initial acceptance period was so brief, no application was received from a community wishing to participate financially.[62]

The process of land selection began in November 1980. It was obvious that development rights offered for sale by farmland owners far exceeded available funds. The twenty-eight applications received involved 1,868 acres of farmland with a total asking price of $4,787,350, or an average of $3,097 per acre. This included all acreage offered for sale on the farm property including woodland, wetlands, wasteland, and land under buildings.

Each application was "scored" according to the point rating system. Following preliminary evaluation, interviews, and on-site visits, the Committee eliminated sixteen applications. One of the rejected landowners exercised his right to a Committee hearing but no change resulted. Of the twelve applications that remained, the Committee instructed the Commissioner of Agriculture to proceed with negotiations to purchase preservation restrictions on eight of the farmland parcels. The Committee also authorized the Commissioner, at his discretion, to engage an appraiser to prepare an estimate of the range within which the market value would fall if a complete appraisal were prepared on one of the eight tracts of land. The Attorney General's Office later advised that under the law an appraisal must be made before any purchase of development rights occurs to ensure that the purchase price does not exceed the limits established by the law. Professional appraisals, therefore, were made on all applications prior to final closing. The Department of Public Works and Highways assisted with the title searches.

The first development rights purchase under the state program was completed on March 5, 1981. The land involved consisted of 68.72 acres of prime farmland located in Milford. By 1985, development rights had been acquired on fifteen separate farmland parcels, including two donated sites. In addition to these donated parcels, the state of New Hampshire had thus acquired the development rights to approximately 982 acres of agricultural land at an average cost of $2,937 per acre or a total cost of $2,885,523. Including the cost of appraisals and title searches, only $88,500 of the original $3 million authorized by the legislature in 1979 remained available for future purchases.[63]

According to the Annual Reports of the Agricultural Lands Preservation Committee, these acquisitions represented the best and most important agricultural lands under consideration. Although certain problems were anticipated, such as property boundary disputes and difficulties in determining the value of development rights, the program has run smoothly, partially as a result of the experience and assistance given by Connecticut and Massachusetts officials. A 1981 court decision upheld the discretion of the Committee to deny purchase when a parcel did not score well under the prescribed criteria.[64]

Problems common to the PDR programs in other New England states have also plagued the New Hampshire program. The most important of these is cost. In most cases the purchase of farmland development rights in New Hampshire has accounted for 70 percent or more of the total land value, making the program expensive to fund and raising the question of program feasibility. At what point does the cost per acre outweigh the actual value returned to the public by protecting a particular parcel of farmland? In a period of fiscal constraint and generally poor farm economic conditions, what is the threshold of allowable expenditure, given the benefits to be received? Questions also arise concerning the value of these expendi-

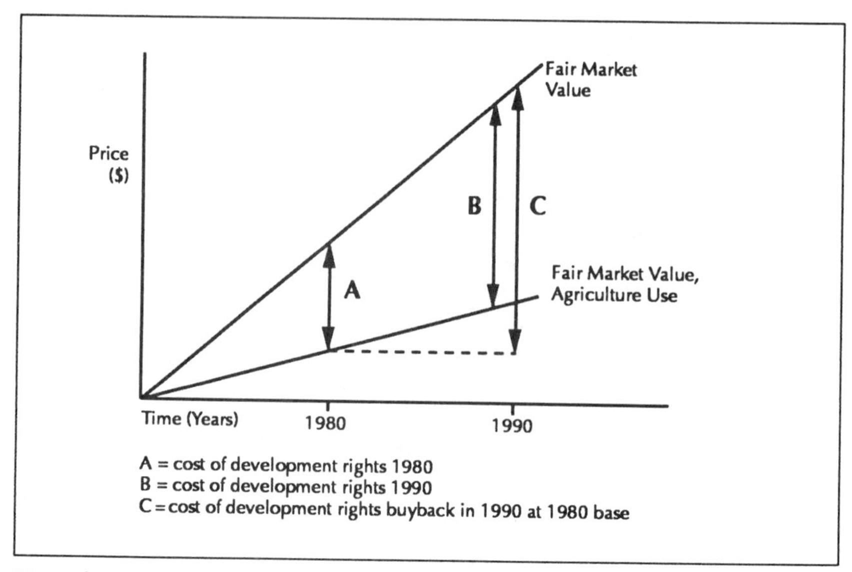

Figure 7.2. Cost of development rights.

tures to protect agricultural land since much of the problems facing New England agriculture are based on economic conditions and market imperfections beyond the boundaries of the state.

In New Hampshire, the response to such questions is mixed. Advocates of the PDR program claim that the rapidly growing population in the state requires increased food production from farmers and the public must permanently protect farmland to assure its availability for the future. The Agricultural Task Force took this position in its first report to the New Hampshire General Court of January 1, 1981. Support for the program, however, is not consistent throughout the state. Many people feel that expenditures to purchase farmland development rights are inappropriate at a time when severe economic constraints plague the state. Moreover, it must be questioned whether supporting primarily dairy operations improves the overall food independence of the state.

Since the adoption of the program, various legislative modifications have been proposed and/or adopted. One of the most important of these was passed in 1981 by the New Hampshire House—an ongoing funding measure in the form of a real estate transfer tax which was expected to yield $1 million a year. The Senate increased the tax rate but allocated the monies raised into the general revenues, not the PDR program. In 1983, legislation to provide a new bond measure of $3 million for the program failed to pass. However, in 1985 the New Hampshire legislature did pass a $2 million bonding measure for the PDR program, and the Agricultural Lands Preservation Committee accepted new applications for enrollment in that year. The Committee also is considering some application and evaluation changes for the program. These changes include initiating a matching grants feature which would encourage municipalities and private groups to join with the state in purchasing development rights.

The future of the program, therefore, may be limited because of funding problems. Moreover, because New Hampshire now is so concerned about farm economics, the prospect of protecting land that may prove unprofitable is hurting the program's popularity. If the state wants food independence, it must have the means to produce the necessary commodities. Purchasing development rights, however, may prove to be the best response in areas of rapid development, to save the most valuable and threatened land from conversion to other uses.

Agricultural Task Force (1979)

The Task Force, composed of farmers, legislators, and various state officials, was established in 1979 under the Agricultural Land Preservation Act. The Act required the Commissioner of the Department of Agriculture to "appoint a task force to consider mechanisms for the protection and promotion of agriculture in the state." Among these mechanisms were "the designation of agricultural districts, creation of land trusts and extension of the agricultural land preservation program to include other lands and alternative financing programs."[65] The Task Force was required to submit its findings and recommendations to the legislature by January 1, 1981.

Predictably, the Task Force is a leading advocate of farmland protection in New Hampshire. The 1981 Task Force Report made the following five recommendations:

1. Continued funding for the purchase of development rights to agricultural land is justified.
2. A designated source of continuing funding is preferable to bonding.
3. A permanent trust fund able to receive private donations as well as public funds should be considered for this program.
4. To protect its investment in agricultural land, the state needs to take certain complementary actions. (Among these are promotion of agriculture in the state, adoption of the Agricultural Encouragement Act, and support for farmers markets.)
5. The Department of Agriculture as presently constituted may require updating or possibly restructuring.[66]

The Task Force proposed additional actions, such as the consideration of agriculture in local planning and zoning and the passage of a right-to-farm law. Both of these recommendations have resulted in new programs.

The Task Force's report has served to clarify interesting agricultural land protection programs in the wake of the Food Policy Resolution and the PDR program. Unlike the other New England states, New Hampshire's Task Force was not subject to the directives or political pressures of an appointing governor; therefore, its evaluations proved objective and valuable.

Town Master Planning and Zoning (1981)

Throughout the 1960s and 1970s, the traditional opposition to land use controls resulted in widespread conversion to tract development in the southern part of the state. The critical mass once found in miles of contiguous acres was lost as farm operations became isolated in newly built suburban towns.[67] The newcomers sometimes made life difficult for the remaining farmers, bringing nuisance suits or actually causing destruction of crops. Those towns that did adopt zoning ordinances sometimes used the law to restrict agriculture and regulate farmers out of business.[68] As with property taxes, many farmers found themselves breaking from the ideology of conservative politics. It was becoming clear that municipal zoning was needed to protect agriculture land.

Both the Agricultural Task Force and the Food Policy Report advocated that municipal planning and zoning powers be used increasingly to protect agricultural activities throughout the state. The enactment of these suggestions proved significant by encouraging local decision makers to view agriculture as a vital component of the town plan. Some towns, such as Litchfield, have acted independently to pass zoning ordinances to protect agricultural lands. Litchfield's zoning ordinance limits single-family home building permits to fifty per year, both to save farmland and to control the cost of public services.[69] The state has also provided municipalities with substantial assistance in revising their planning and zoning bylaws.

To improve local agricultural planning further, the Agricultural Task Force has begun a program with the Office of State Planning and the Cooperative Extension Service to encourage municipalities to assure agricultural representation in town master plans. The objective of the program is to get local decision makers, including planning boards, selectmen, assessors, and Conservation Commissions, to view agriculture as a vital component of the total community plan.

The program began in 1981 with a pilot project of in-depth farmland inventories in selected communities. These inventories relied on the computer analysis of agricultural land uses that the New Hampshire Rural

Development Planning Program of the Office of State Planning had conducted. The results of these analyses were made available to the local bodies examining community growth and development. Seven pilot communities—Concord, Dumbarton, Hopkinton, Henniker, Stratham, Lee, and Strafford—were chosen to participate in the program. Of the seven pilot communities, it was hoped that at least two would develop creative ways of addressing agriculture within their master plan.

Although the program is young, results are encouraging. A proposal in Strafford utilizes a point system which would assign negative points for development on agricultural land.[70] Points are important to a developer because they determine the amount of density that will be allowed. If development on agricultural land is inevitable, points are awarded for development in a cluster-type arrangement or alternative methods which protect all or a portion of the farmland.

Innovative new proposals have emerged from other communities as well, such as proposals to revise subdivision regulations in such a way that for any new subdivision on agricultural land to be approved, it must set aside enough land to produce 25 percent of the nutritive requirements of its residents.[71]

Proposals such as these ultimately raise substantial questions of equity. Landowners who own flat agricultural land must set aside portions of their land free of development, whereas hillside landowners may subdivide their land without constraint. Precedents for this type of requirement may be found in regulations from new subdivisions to provide for parks, school sites, or other land uses required by the increased population. However, the exaction must be reasonably related to the impact created by the people who move into the new subdivision. Local government therefore has the burden of proving that exactions are based on measurable standards and that the standards are applied uniformly to landowners who propose to subdivide land. Meeting the burden of proof could be very difficult if the local government has to prove exactly what 25 percent of the nutritive requirements of a new subdivision would be, not to mention the difficult questions of whether it would be possible to produce that quantity or produce it economically. Despite the legal questions which such local proposals raise, the creative proposals of some communities may ultimately provide the best basis for a broad new range of farmland proposals. As Gerald Howe commented,

> The program to encourage agricultural representation in the town master plan seems to be the best way to go. It's easier to work in the smaller group and you get more immediate and more practical results if you can get the citizens of the community to realize the master plan can help them, and that people are affected by agricultural land more than just by its rural aesthetic value.[72]

Indeed, throughout the New England region, the belief is growing that many farmland protection objectives can be best accomplished at the local level through effective planning. In times of economic difficulty, both for farmers and government, effective citizen action may be the most positive means of maintaining viable agricultural land. At the same time, citizens can be more responsive to local needs than to a broad state program.

State government, of course, can play an important role in meeting farmland protection objectives. This is evidenced by the planning assistance which the New Hampshire Rural Development Planning Program is providing. With the broader resources that the state has at its disposal, it is possible to conduct detailed computerized analyses of agriculture that no municipality could afford and probably few could use effectively. It is, therefore, quite likely that less direct and less costly interventions in farmland protection such as the New Hampshire local planning will find greater favor and success in the future.

Coordination of FmHA Lending Programs

A pilot program which is closely related to the local planning and zoning programs discussed has been undertaken by the Office of State Planning under an agreement with the federal Farmers Home Administration (FmHA). Subsequent to the computerized map analysis, efforts to reconcile FmHA lending programs with agricultural protection efforts have begun since such lending programs have often encouraged the conversion of agricultural land to other uses. Although agricultural land protection is an important component of the program, the FmHA agreement covers areas other than agriculture, such as housing, infrastructure, and economic development. The purpose of the program is to identify the viable agricultural areas of sufficient importance and then to say, "this area is developable from an agricultural industry standpoint: let's keep Farmers Home Administration money for housing and non-agricultural purposes out of there and concentrate FmHA *agricultural* loans in that area."[73] The program would not only attempt to concentrate FmHA agricultural loans in agricultural areas but would also direct other specific types of FmHA loans to specific areas.

Critics of past FmHA loan practices point out that the agency has often stimulated the conversion of agricultural land to development, but hopes are high in New Hampshire that the agency will become a friendly partner in efforts to promote rural economic development *and* protect agriculture through coordinated local planning.

The computer analysis of statewide land use maps has been important to this program. In 1981, the implementation phase of the program was refunded by the FmHA, and currently the project to target FmHA funds is under way. By identifying areas of viable agriculture as well as identifying areas for other specific uses, the Office of State Planning is able to guide the FmHA to help assure that local objectives are more closely met. It also helps ensure that limited FmHA and other federal funds will be allocated most efficiently throughout the state.

As with other federal programs, New Hampshire has lost much financial assistance from FmHA since the Reagan administration took office in 1980. Thus, the value of this program has diminished, although it remains important.

Nuisance Liability Law (1981)

Another aspect of local planning for agriculture has to do with nuisance. Because zoning regulations often have allowed housing developments to be located on the edge of established farms, friction between the newcomer and the farmer often arises.[74] The new resident sometimes brings a nuisance suit to restrict normal farm practices such as farm machinery operation, the storage and spreading of manure, and the herding and movement of farm animals. To reduce this problem for New Hampshire farmers, the 1981 legislature passed a Nuisance Liability Law, commonly called a right-to-farm law, which limits when nuisance suits can be instituted against farmers.[75]

The need for consideration of a right-to-farm law was widely publicized in 1979 when the Food Policy Report recommended the revision of zoning ordinances to allow agricultural uses in residential areas. It also recommended that the Office of State Planning and regional planning agencies develop active programs to encourage agricultural production in residential, industrial, and business areas. Also in 1979, voting delegates at the New Hampshire Farm Bureau Federation State Convention instructed the organization's legislative committee to draft a right-to-farm bill. This bill was introduced into the legislature by Representative Marilyn Campbell, a dairy farmer from Salem. Two years later, the Agricultural Task Force, on which Campbell served, reiterated the need for statewide antinuisance legislation. Their January 1, 1981 report called for the protection of farms surrounded by developments which have instituted nuisance actions against them, making it difficult or impossible to farm. The support of the antinuisance legislation by the Task Force and other groups resulted in legislative action in March 1981.

New Hampshire's Agricultural Nuisance Law does not provide protection for new agricultural uses in existing residential areas. Rather, it protects agricultural operations that have been in existence for one year or more and were not a nuisance at the time the farm operation began. In addition, the law provides for exceptions where the agricultural operation is "injurious to public health or safety" where the nuisance results from "negligent or improper operation of an agricultural operation" or where actions to prohibit a nuisance are within the "duties and authority conferred upon the water supply and pollution control commission . . . or the commissioner of agriculture."[76]

As in the other New England states, the statute is a fairly broad prohibition on nonfarm newcomers' access to the courts for purposes of restricting existing agricultural operations. New Hampshire's version has a curious and unique provision, however. It states that to be protected under the right-to-farm law, the agricultural operation in question must not have been a nuisance when the farm operation began. Proof of a nuisance in a family farm of one hundred years or more would thus be very difficult. Nevertheless, a newcomer may be able to prove that the machinery used in the old days was even noisier than today or that manure was kept closer to what is known to be a water supply. Whether this anomalous provision will ever become important in litigation remains to be seen.

Governor's Advisory Council on Growth

The effect of growth and development on the conversion of agricultural lands to urban uses was reaffirmed by the Governor's Advisory Council on Growth. In October 1979, then-Governor Hugh Gallen established the Advisory Council to prepare for the governor's consideration and public discussion a set of "goals, policies, and recommendations which would increase the effectiveness and ability of New Hampshire state and local governments to respond to the challenges and problems of rapid growth."[77]

The Council was divided into five committees which concentrated on the subjects of housing, economic development and land use, natural areas, intergovernmental relations, and data needs. Although the recommendations of the five committees were prepared independently, certain conclusions were common to all the committees. The most important of these was the need to reorganize the various functions of state government. The Council also recognized a need to improve the relationship between state and local governments and to expand the executive branch's capacity to formulate policy and monitor programs. Finally, the Council expressed a "desire to protect and enhance those characteristics that make New Hampshire unique" such as "fragile environmental and agricultural areas."

While the natural-areas committee recommended the increased involvement of Conservation Commissions in natural-area protection, and the intergovernmental relations committee recommended supporting legislation that protects agricultural lands, the economic development and land use committee was primarily concerned with farmlands. This latter committee stressed the economic importance of agriculture to the state, a recurrent theme in New Hampshire.

Several objectives were identified, including increasing employment opportunities and attracting industry and tourists. To increase employment opportunities it was recommended that:

1. The Department of Agriculture expand promotion and marketing activities.
2. Financial assistance be given to farmers.
3. The state, counties, and municipalities attempt to purchase local food products for their institutional needs.

To attract industry and tourists to the state the Report recommended:

1. Continuation and improvement of the current-use assessment program.
2. Development of a comprehensive program to promote agriculture and forestry on those lands most suited to these uses.
3. Providing state and encouraging federal aid for land use development on important farmlands only if it is nonintensive.
4. Continuation and expansion of state funding for PDR on agricultural lands.
5. Adoption of state enabling legislation for the establishment of agricultural districts.
6. Adjustment of tax laws to reduce speculation in rural land.

The Report also recommends Office of State Planning guidance in local planning and zoning efforts, and a soil inventory of agricultural lands to help municipalities expand industry.[78]

Urban growth and development has played a large role in the conversion of agricultural land to other uses. The Advisory Council has recognized this concern and made recommendations for economic development while protecting valuable agricultural land. As discussed previously, some of these recommendations have already been enacted. Others are in the process of development. One such recommendation is to expand promotion and marketing activities for local food products.

Marketing

Much of the concern over farmland protection in New Hampshire is tied in with the profitability of farming and the state}s desire to increase its food production. As the Agricultural Task Force observed in its 1981 report:

> Saving farmland is essential to meet present and future food requirements of New Hampshire's citizens. But it is not enough. Farmers must be able to make a living on the land—and that requires active local markets, which only the state today has the ability to initiate and stimulate.[79]

The Food Policy Report highlighted the need for government assistance through a panoply of reforms in the food distribution, not just the production, process.

The Agricultural Encouragement Act was a proposal designed to deal with marketing issues. The Task Force believed it would "(a) increase marketing support for farm produce to be sold directly to consumers, (b) eliminate purchasing provisions or policies which make it difficult or impossible for public institutions to buy from local farmers, (c) protect use of the new 'Grown in N.H.' label—among several possibilities."[80] The bill failed to pass in the legislature in both 1979 and 1981, but the changes it proposed are nonetheless in effect or under way.

The most significant of these has been the creation of a marketing division within the New Hampshire Department of Agriculture, known as the Agricultural Promotions Division. This provides such services as promoting the agricultural industry to the public, working with the agricultural commodity groups on marketing and promoting their product, and supplying important marketing information to the individual growers. Much of this is accomplished at an agricultural trade and educational exposition put on each February. The exposition provides an excellent forum for panel discussions, comprised of people from all levels of the agricultural industry, concerning the direction and future of promoting and marketing New Hampshire's agriculture.

Many aspects of the Agricultural Encouragement Act are being administered by the New Hampshire Department of Agriculture and other state agencies. For example, the Department has developed a "Nature-ly New Hampshire" logo. Moreover, farmers' markets and roadside stands are being encouraged. The program urges farm groups who run the farmers' markets and roadside stands to capitalize on the "edible recreation" potential of direct farmer-to-consumer marketing. One agricultural official noted that "it's surprising how often you get people at the farmers' market from a recreational standpoint. Suddenly, they realize that this recreation is really food and it came from New Hampshire."[81]

Another marketing program that recently has taken shape evolved from the Connecticut River Valley Agricultural Project. Partially financed by the FmHA, the Vermont Department of Agriculture and the New Hampshire Office of State Planning jointly studied marketing problems in the Connecticut River Valley and released a report in February 1982.[81] The report notes that because of Valley population growth, demand for farm products is increasing and locally grown fruit and vegetables are becoming competitive with nonnative produce. The report stresses the need for aggressive marketing techniques with independent retail groceries, restaurants, and consumers. Recommendations are made to both growers and public agencies to ensure marketing success.

Aside from retail programs, the state is also seeking efficiencies higher up in the food distribution chain. The Apple Marketing Order[82] was passed in the 1983 legislative session. This enabling legislation allows apple commodity groups to tax their own product on the initiative of the apple growers. Currently, five cents per bushel of apples is assessed and is used to promote the product, establish quality controls, and conduct research. The goal of the law is to enhance producers' profits, although consumers should receive some benefits. The concern for the bill reveals a desire to provide assistance to an entire native industry rather than merely to particular retail operations.

Farm economics also involves cutting production costs and raising production efficiency. These goals traditionally have been furthered by the Department of Agriculture and the Extension Service at the University of New Hampshire. For example, encouraging part-time farming, diversification, or specialization may turn the balance sheet of a farm around.

In a conservative state such as New Hampshire, marketing through farm groups with technical assistance from the state is the most popular method of increasing farm profitability. Marketing has proved to be too indirect a measure of preserving agriculture and agricultural land, and more direct regulatory programs and incentives are increasingly seen as important components of New Hampshire's agricultural land protection program.

Future of Agricultural Land Protection Programs

New Hampshire long has been cherished for its natural beauty, especially in the White Mountain area. The state is also unique because its governors have taken "the pledge" not to institute new taxes. Both for aesthetic and eocnomic reasons, newcomers flocked to the Granite State during the 1960s and 1970s in record numbers. During this population boom, the lack of land use regulation permitted the conversion of thousands of acres of good farmland to developed uses. Farm economics also forced many farmers to abandon their earthly labor.

With the change in land use came a change in landscape and the ability to produce food. In the wake of the OPEC oil embargo, numerous private groups convinced the public and the legislature that food production and distributional autonomy were essential for the economic well-being of the state's citizens. Thus, food production and economic development became a major catalyst for farmland protection programs. Numerous state and local programs were instituted, with divergent but generally positive effects.

As the legislature in the 1980s continually tightens its budgets, programs that protect agricultural land, as opposed to enhancing farm profits, may come into question. It will be doubted why zoning or tax incentives should enable a marginal farm operation to survive when development arguably could bring more jobs and tax revenue. It will also be questioned why $3 million was spent to purchase the development rights of largely dairy-support lands. Low-cost marketing programs and local initiatives will catch the legislator's eye, much as they did before the last decade.

The answer to questions about the merit of recent land use initiatives in New Hampshire must relate to economics if these comparatively liberal approaches to farmland protection are to be successful and continued. Thus, the Governor's Advisory Council on Growth made clear that the farmland issue is not simply a matter of direct economic return. A rustic lifestyle is responsible for much of the industry that has relocated in New Hampshire and for the tourist trade. If such beliefs are correct, then the critical question becomes whether that allure will remain.

The programs and alternatives for action currently under consideration and in use in New Hampshire are not novel. The state has moved more slowly than its southern neighbors in government-assisted farmland protection efforts. This is largely because of New Hampshire's conservative political attitude on government regulation and monetary issues.

One recommendation that was stressed in both the Agricultural Task Force Report and the Governor's Advisory Council on Growth Report is the creation of agricultural districts. Such districts would provide the critical mass that helps make farming possible. Today, however, there are few contiguous areas of active agricultural lands. Moreover, the problems of public infrastructure are not as great in New Hampshire as in those states that have these districts. Aside from the question of need is the larger problem of political feasibility. Such a proposal could be welcomed only where the municipalities agree to restrict growth.

The fate of an agricultural districting proposal, like all others in New Hampshire, must overcome the problems associated with passage in a conservative legislature. Marketing or other problems that stress the economic viability of agriculture will continue to be the programs most seriously considered. And, as elsewhere in New England, local planning and implementation programs combined with strong planning leadershp and support at the state level have been most effective in farmland protection.

Notes

1. Peirce, *The New England States*, 358–359. "Loebism," as Peirce calls it, refers to the powerful conservative trend largely promoted by Manchester publisher William Loeb. As Peirce writes (p. 259): "In 1946, William Loeb, the president of the American China Policy Association, the strong right arm of the [anticommunist] China Lobby, had suddenly become the most powerful molder of public opinion in New Hampshire through his purchase of the *Manchester Union Leader*. Loeb quickly transformed the paper, New Hampshire's leading and only morning daily, into a propaganda organ for the China Lobby and his assorted other pet

peeves and causes . . . Loeb's purchase of the Union Leader was the paramount event in setting the tone and direction of New Hampshire public life since World War II."

2. Ibid., 286.
3. "New Taxes: No Longer a Dream," *Boston Sunday Globe*, January 31, 1982, 42.
4. "Lower Bond Rating Means New Headaches for N.H.," *Boston Sunday Globe*, March 7, 1980, 80.
5. "New Taxes," 41–42.
6. Gerald W. Howe, University of New Hampshire, comment, Lincoln Seminar 1.
7. Mark B. Lapping, *The Land Base for Agriculture in New England* (Burlington: University of Vermont, October 1979): 23.
8. W.K. Burkett, *New Hampshire Idle Farm Land* (Durham: University of New Hampshire, New Hampshire Agricultural Experiment Station, September 1953): 6. Statistical data from the Census of Agriculture 1850–1950.
9. *1982 Census of Agriculture.*
10. Burkett, *New Hampshire Idle Farm Land*, 8.
11. Ibid., 43.
12. Gerald W. Howe, comment, Lincoln Seminar 1.
13. "New Hampshire: Boom in South and Bust in North," *Boston Sunday Globe*, March 7, 1982, 50.
14. G.G. Coppelman, S.A.L. Pilgrim, and D.M. Peschel, *Agriculture, Forests, and Related Land Use in New Hampshire, 1952 to 1975* (Durham: University of New Hampshire, New Hampshire Agricultural Experiment Station, April 1978): 7.
15. For a discussion of the issues of "critical mass" in farming see Mark B. Lapping, "Agricultural Land Retention Strategies: Some Underpinnings," *Journal of Soil and Water Conservation* 34 (May–June 1979).
16. *First Progress Report: New Hampshire Rural Development Planning Program* (Concord: Office of State Planning, 1980): 15, 17.
17. Peirce, *The New England States*, 357.
18. "New Hampshire: Boom and Bust," 41.
19. Peirce, *The New England States*, 330.
20. New Hampshire Statute 79-A.1.
21. "And When We Went There the Cupboard was Bare!" (Durham: University of New Hampshire, Cooperative Extension Service, n.d.): 5.
22. Gerald W. Howe, comment, Lincoln Seminar 1.
23. *Recommendations for a New Hampshire Food Policy* (Durham: University of New Hampshire, Cooperative Extension Service, 1979): 1.
24. New Hampshire General Court House Concurrent Resolution 13 (May 16, 1979).
25. Gerald W. Howe, comment, Lincoln Seminar One.
26. Peirce, *The New England States*, 358.
27. N.H.S. 36-A: 1 et seq.
28. Duddleson, *Supplementary Report* to Scheffey, *Conservation Commissions in Massachusetts*, 185–186.
29. For example, until 1972 the private Spaulding-Potter Charitable Trusts offered grants up to $2,000 on a matching basis, $1 for each $2 allocated by a community to its conservation fund. Ibid., 186. In addition, the Trusts helped provide some technical assistance to the Conservation Commissions through a grant which helped establish a conservation specialist position in the Cooperative Extension Service at the University of New Hampshire. These assistance programs were instrumental in getting the Conservation Commission acquisition programs started which, in some cases, eventually resulted in the preservation of farmland.
30. Gerald W. Howe, comment, Lincoln Seminar 1.
31. Peirce, *The New England States*, 325.
32. Ibid., 360.
33. Kenneth Marshall, comment, October 1982.
34. Peirce, *The New England States*, 360.
35. N.H. Const. Pt. 2, art. 5-B makes an exception to Pt. 2, art. 5.
36. N.H.S. 79-A.1.
37. This provision has proven rather unpopular. Many landowners are hesitant to open

their land up for recreational purposes because of liability problems. Legislative efforts are under way to overcome this problem. Gerald W. Howe, comment, Lincoln Seminar 2.

38. "What You Need to Know About Current Use" (Concord: Society for the Protection of New Hampshire Forests, n.d.): 1.
39. Gerald W. Howe, comment, Lincoln Seminar 1.
40. N.H.S. 36-d:6.
41. N.H.S. 79-A:15-21.
42. *Frost v. Town of Candia*, 118 N.H. 923, 924 (1978). Appeal of the Town of Peterborough, 120 N.H. 325, 328 (1980).
43. "Current-Use Assessment: Does It Work?" (SPACE, n.d.).
44. Gerald W. Howe, comment, Lincoln Seminar 1.
45. "Current-Use Assessment," 4.
46. Gerald W. Howe, comment, Lincoln Seminar 1.
47. Kenneth Marshall, comment, October 1982.
48. James E. Hicks, Stafford Regional Planning Commission, comment, May 23, 1983.
49. *Recommendations for a New Hampshire Food Policy*, 5–6.
50. New Hampshire General Court House Concurrent Resolution 13 (1979 session).
51. House Bill No. 477 (1979).
52. Gerald W. Howe, comment, Lincoln Seminar 1.
53. N.H.S. 36-0: 1 et seq.
54. *First Report of the Agricultural Task Force* (January 1, 1981): 3.
55. Chapter 301:1 of the Acts of 1979.
56. N.H.S. 36-D:8.
57. James E. Hicks, correspondence, December 16, 1981.
58. N.H.S. 36-D:3 I.
59. N.H. Code of Administrative Rules, Part Agr. 705, 707.02 (effective July 1, 1980).
60. Ibid., Part Agr. 707.02.
61. N.H.S. 36-D:7 II.
62. *First Annual Report of the Agricultural Lands Preservation Committee* (March 12, 1982): 3.
63. *Third Annual Report of the Agricultural Lands Preservation Committee* (April 28, 1983): 4.
64. *Second Annual Report of the Agricultural Lands Preservation Committee* (March 12, 1982): 3.
65. Chapter 301:3 of the Acts of 1979.
66. *First Report of the Agricultural Task Force*, 2–7.
67. James E. Hicks, comment, May 23, 1983.
68. "Suburbia Collides with N.H. Farms," *Boston Sunday Globe*, March 1, 1981, 21.
69. "In New Hampshire, a 10-Year Boom Has Lost Its Zip," *Boston Sunday Globe*, December 12, 1982, 30.
70. Gerald W. Howe, comment, Lincoln Seminar 1.
71. Gerald W. Howe, comment, Lincoln Seminar 2.
72. Ibid.
73. Ibid.
74. "Suburbia Collides with N.H. Farms," 21.
75. N.H.S. 430-C:1 et seq.
76. N.H.S. 430-C:2-4.
77. Final Report of the Governor's Advisory Council on Growth (Concord: Office of State Planning, January 1981): i, 25, 27–29, 43, 51.
78. *First Report of the Agricultural Task Force*, 6.
79. Ibid.
80. Gerald W. Howe, comment, Lincoln Seminar 1.
81. *Selling Locally Grown Produce: Markets for Connecticut River Valley Farmers* (Connecticut River Valley Agricultural Project, Vt. and N.H. Departments of Agriculture and Vt. and N.H. Offices of State Planning, February 1982).
82. N.H.S. 341-A:15-22.

8

Maine Programs for the Protection of Agricultural Land

In many ways Maine is unique among the New England states. Its location in the extreme northeast corner of the country, combined with its historical, topographic, economic, and cultural background, sets it apart from its New England neighbors. The most obvious distinction is Maine's size. With a land area of 30,995 square miles, the state is only slightly smaller than the other five New England states combined. The sheer size of the state has resulted in a diverse development pattern that is unparalleled in New England.

The unique diversity of Maine is attributable partially to its location, which some view as a cul-de-sac in the eastern United States. Maine is currently experiencing the heavy direct pressures of urbanization only in York County, the southernmost coastal county and entry point to Maine from the populated urban corridor of the south. Like the coastal counties of all the New England states, York County has experienced significant urban growth since World War II. This phenomenon led one writer to observe:

> The natural sea and farm and wooded setting [of York County] is being invaded by ticky-tacky subdivisions and condominiums populated by people who work in the nearby coastal cities of Massachusetts' [Route] 128 orbit. This is the northernmost extension of the East Coast megalopolis, a county that rapid rail commuter service could put within 45 minutes of Boston.[1]

In addition, Maine is experiencing smaller scale development along the whole I-95 corridor.

By contrast, Aroostook County, at the extreme northern tip of the state bordered by the St. John River, is a vast area of forest, farmland, and wilderness. This county, which includes much of the Allagash Wilderness Waterway, is larger than the states of Connecticut and Rhode Island combined, accounting for 22 percent of Maine's total land area.[2] By 1870, when southern Maine and the rest of New England were well-developed regions of farms and towns, Aroostook County was still largely unsettled. Today, agriculture, specifically potatoes, is the foundation of the economy of Aroostook County, even though only one-tenth of the county is devoted to active agriculture. The glacial soils and the intense, cold winters of Aroostook County provide an excellent environment free of parasites for growing high-quality potatoes. Indeed, in 1969 and 1974 Aroostook County ranked number one nationwide in both the acreage and the hundredweight of Irish potatoes harvested.[3]

Aroostook County further illustrates the large size and diversity of Maine's landscape, in which farming plays an important, though secondary, role. Today 90 percent of the state's land surface is covered by forests and slightly over one-half of Maine consists of "wild lands," largely unsettled areas which are not within the jurisdiction of any city or town. These Unorganized Territories, as they are officially known, consist primarily of great tracts of forest identified only by their "township" and "range" designation.

The land use pattern of Maine thus stands in marked contrast to the neat and tailored look of the states of southern New England such as Massachusetts or Connecticut where the entire land area comes under the jurisdiction of one town government or another. Maine's unique physical chracter has had a profound effect, different in many ways from the other New England states, on the environmental concerns of its residents. In recent years, however, following the lead of its southern neighbors, Maine has begun modest efforts to protect agricultural land.

The earliest steps in this direction were taken in the 1970s, when Maine enacted the Site Selection of Development Act and the Land Use Regulation Commission Act. The comprehensive manner in which these laws tackled problems of land use and environmental quality are perhaps as significant to Maine as Act 250 is to Vermont. However, although these programs have had some importance in recent efforts to protect agricultural land in Maine, they were not originally intended as farmland protection measures. In fact, to date Maine has been slower than most of its New England neighbors in establishing specific programs to protect its agricultural lands directly. This laxity is despite a recent decline in farmland acreage which is comparable to, though perhaps not as severe as, the declines in the other New England states.

There are several reasons why the state has been relatively less active in establishing agricultural land protection programs. First, even though there generally has been some development activity on prime farmland in Maine, it has occurred on small parcels and the public is not yet fully aware of the cumulative impact. Thus, because large-scale urbanization is not perceived as a major cause of farmland conversion, programs to save Maine farmland from development have not been urgently sought.[4]

A second cause for Maine's legislative inactivity in farmland protection is the state's distinctive historical pattern of land use. Maine has never had as large a percentage of land area devoted to agriculture as the other states in the region. During the peak period of New England agriculture, around 1880, when agricultural land uses in Massachusetts, Connecticut, and Vermont represented 65 percent, 80 percent, and 84 percent of the total land area of each state, respectively, agricultural land uses in Maine accounted for only 34 percent of the total land area of the state.[5] Statewide the economy of Maine has never been as totally dependent on agriculture as were the other New England states. Forests and forest industries such as timber production, pulp, and paper have long been the largest users of land and the backbone of the statewide economy.

The third related reason for the comparative lack of farmland protection efforts in Maine is its large land area and small population. According to the 1980 Census, Maine has approximately 1,125,000 residents and a land area of 30,920 square miles; it is among the fifteen least densely populated states in the nation, with only 36.3 persons per square mile. Most of this population lives in the state's southern coastal regions. With thousands of squrae miles of forest lands and uninhabited wilderness, the pressure for stewardship of limited land resources is not as great as it is in the more densely populated southern New England states—not that the typical Maine resident does not cherish the woodlands, lakes, streams, and rural qualities of the state. In recent years efforts to preserve and protect this environment have grown immensely, along with efforts in the rest of the nation. But, to a large extent, in Maine people were slow to perceive the need for state intervention to protect this environment—partly because of the low population and vast areas of open space. Moreover, Maine residents have not become completely aware that certain types of land are in limited supply and are found only in particular areas and that there are competing uses and demands for the same land.

Finally, farmland protection and other land use protection programs have had less support in Maine because of the historically poor condition of its economy. Maine's average annual income is among the lowest in the nation. A 1982 report of the U.S. Bureau of Labor Statistics reports that average annual income for nonagricultural workers in Maine was $12,563, compared with $15,691 nationwide and $14,920 for the New England region as a whole. Only the state of South Dakota, at $12,099, was lower than Maine nationwide.[6]

The economic problems of Washington and Waldo Counties illustrate the economic problems which face the state as a whole. Washington County is a large county on the extreme north coast of Maine, adjacent to Aroostook County. Larger than the state of Delaware, with a population of less than 35,000, Washington County is one of the poorest counties in the state. In 1981, approximately 19.2 percent of the families in Washington County had incomes below the poverty level.[7] Earlier in this century, however, the Washington County town of Eastport boasted a healthy economy based on sardine fishing. As catches have become smaller through the years, the sardine industry has declined substantially and the market for sardines has dropped off. In 1981 the economy in Eastport drastically worsened when a local fish-processing plant closed. Of a total workforce of approximately 700 people, 300 lost their jobs. Unemployment in Washington County was among the highest in the nation.

Waldo County has been plagued by similar economic problems. Since 1980, the county has had the lowest per capita income of counties in Maine, and its income annually has grown at a slower and slower pace compared with the state per capita income; there actually has been a net decline in earnings. This began with a decline in the poultry industry in 1980. Belfast, the county seat, had been the "broiler capital." With the industry's decline, contract growers went bankrupt and thousands of people lost their jobs. This was the only large employer in the county, and its typically low wages had already made the county the second poorest. The county has not regained jobs for its residents.[8]

The economic problems facing Waldo and Washington Counties' residents are common through the state. Economic development programs of almost any kind are high on the priority list of Maine citizens and business organizations. Statewide land use programs, which are often perceived to inhibit the climate for economic development and cost the state precious tax revenues, thus generate considerable controversy in the state.

Despite the many reasons that have caused Maine to be relatively slow in adopting new farmland use protection laws, the state has taken modest steps in recent years to protect its dwindling acreages of active farmland. These efforts have been fueled by growing environmental concerns, particularly among newcomers to the state, and recent statewide food policies which are the result of growing concerns over food self-sufficiency. The following section discusses the growing concern for farmland protection in the state.

Public Concern to Protect Agricultural Land

As in all the New England states, public and private efforts to protect farmland in Maine began as public awareness that farmland acreage in the state was declining significantly. Also providing an impetus for action was the growing concern over environmental quality and the impact of man-made forces on the natural environment. The giant paper-manufacturing companies in the state, as well as the growth of heavy industry in coastal sections, raised questions about the need for government environmental protection regulations that would, at least indirectly, protect farmland. The most important concern that prompted efforts to protect agricultural land, however, was the realization that the state was no longer self-sufficient in food. Sitting at the end of a long, transnational food supply line, Maine citizens felt vulnerable to the unknown—they also felt that the changing economy threatened the family farm and traditional rural lifestyle of much of the population.

Decline in Farmland Acreage

As in other states in New England, during the late 1960s and the early 1970s, when environmental concerns were growing nationwide, the issue of declining farmland acreage became important in Maine. As early as 1961, the *Bangor Daily News* reported:

> Thousands of acres of good Maine cropland are falling before an invader armed with a power shovel and a gravel truck at a time when soil experts declare that the state is about to face a demand for land without precedent in the history of civilization....
>
> Statewide, Maine appears to be slated to sustain a loss by 1975 of 137,400 acres of land now being used for agricultural purposes, including forestry....
>
> Contrary to often expressed opinion, Maine does not have unlimited resources of first-quality land. The soil inventory conducted by the U.S. Soil Conservation Service credited the state with a cropland acreage of about 1-1/4 million in 1958. But only 19,500 acres of this is considered first-class agricultural land. All the rest has erosion problems, water problems, or is rocky....It may well be that the power shovels and the gravel trucks will have to extend their scope of activity to less valuable land. Otherwise, the business of feeding the U.S. in 1975 may be complicated, even more than it is now.[9]

These predictions of the future conversion of farmland acreage in Maine proved to be correct and, if anything, conservative, although urbanization has not been the primary cause. The conversion of active farmland to other uses in Maine exceeded 40 percent between 1964 and 1979.[10] From a larger perspective, this decline is part of a trend which has left Maine with less than 25 percent of the farmland acreage that existed in the state during the 1880s, even though Maine has retained more active farmland than any of its New England neighbors except Vermont.

The year 1880 is considered to be the statistical high point of farming in New England. At that time, Maine had over 6.5 million acres of land actively devoted to agricultural use in over 60,000 farms. By 1940, this figure had dropped to approximately 4.2 million acres on 39,000 farms, a decrease of over 35 percent in sixty years. This decline is modest when compared to the other New England states, considering that New Hampshire and Rhode Island both witnessed a conversion of over 50 percent of their active agricultural lands over a similar period. Considering Maine's large size, however, the decline in actual acreage was tremendous, more than any other New England state in the same period.

During the period from 1940 to 1970 Maine farmers continued to cultivate less and less agricultural land. By 1970, less than 1.8 million acres of active farmland in 8,000 farms remained, representing a 58 percent reduction in farmland acreage over the thirty-year period.

Since 1970, the total land area in farms has continued to decline in Maine, but at a lesser rate than in the previous thirty years. By 1982 the U.S. Census of Agriculture reported that approximately 1.5 million acres of land remained in farms. In conformance with national trends over the last hundred years, however, farms have decreased in number and grown in size.[11] Despite this general increase in farm size, in 1982 the average Maine farm was only 210 acres, well below the national average of 415 acres. Table 8.1 summarizes the decline in farmland acreage during the period from 1880 to 1982.

Unlike some of the more urbanized states in New England, most of the recent conversion of agricultural acreage in Maine has been caused by abandonment, as marginal and subsistence farmers have given up farming and migrated to jobs in the expanding industrial centers. In 1976, a study conducted by the Southern Kennebec Valley Regional Planning Commission analyzed shifting land use pressures in the greater Augusta growth area. Using air-photo interpretation, the study noted changes in developed lands and farmlands during the period from 1966 to 1974. In a twenty-town area, a shift from both active and abandoned farmlands to developed land was clearly evident. Agriculture and recently abandoned fields accounted for 20 percent of the land area in 1966 but accounted for only 14.5 percent by 1974. Developed areas increased from 5 percent to 9.5 percent in the same period. Other land uses, primarily forests and wetlands, remained constant. The report noted that agricultural lands were generally abandoned and subsequently developed.[12]

Table 8.1
Statewide Change in Acreage of Farmland and Number of Farms in Maine, 1880–1982

Year	*Acres*	*Number of Farms*	*Change*	*% Change*	*% Change from 1880*	*Average % Change per Year*	*Average Farm Size (acres)*
1880	6,552,578	64,309	—	—	—	—	102
1910	6,297,000	60,016	–255,578	–4	–4	–0.1	105
1940	4,223,297	38,980	–2,073,703	–33	–36	–1.1	108
1950	4,181,613	30,358	–41,684	–1	–36	–0.1	138
1959	3,081,987	17,360	–1,099,626	–26	–53	–2.9	178
1970	1,759,700	7,971	–1,322,287	–43	–73	–3.9	221
1978*	1,500,390	6,775	–259,310	–15	–77	–1.9	221
1982	1,468,674	7,003	–31,716	–2	–78	–0.5	210

*The 1978 Agricultural Statistics were revised in 1982 to compare more accurately to the 1982 and previous U.S. Agricultural statistics.

Source: U.S. Bureau of the Census, Census of Agriculture.

A 1979 report by the Maine State Planning Office also indicated that abandonment was the principal reason for the declining acreage of agricultural land. In that report, analysis of farmland trends in growth versus nongrowth counties revealed that declining farmland acreage has been greater in nongrowth areas than in growth and development areas. Further studies also show abandonment as the most important reason statewide for the decline in farmland acreage.[13]

In the 1950s federal agricultural policies had led to surpluses in potato production nationwide. As today, adjustments in farm production were necessary, and Aroostook County farmers were encouraged to participate in "soil-bank" programs. Policies such as the Conservation Reserve Programs were initiated to take land resources out of agriculture by encouraging farmers to plant more trees and change their land to forest. The statistics on the decline in Maine's farmland acreage reflect the impact of government policies which purposely sought to restrict the amount of active cropland in the state. According to Edward Micka, extension economist at the University of Maine in Orono,

> The fruits of some of those activities are starting to be more apparent today. It wasn't that we were going through a normal process of losing agricultural land. We were actually encouraging land to go out of agriculture.[14]

Related to this growing awareness of the decline in farmland acreage in Maine was the increased concern over environmental quality and the impact of manmade forces on the natural environment.

Concern over Quality of the Environment

The visible conversion of farmland to other uses, both from urbanization and from abandonment, was not in itself a major source of concern in Maine. The national concerns of the 1960s and 1970s were strongly felt in Maine, and controversies over land use control and environmental quality soon gained momentum. Much of this land use controversy centered on the vast forested lands of the state and the practices of the large timber products corporations and woodlot owners who controlled these lands. Approximately 8.5 million acres of the state's 16.8 million acres of forest lands are owned or managed by paper-manufacturing companies and other woodlot management groups. In fact, in 1981, seven paper-manufacturing corporations controlled about 6.5 million acres of Maine's timberland, roughly one-third of the total land area of the state. Another 2.5 million acres were owned by woodland owners with holdings of 5,000 acres or more. The remaining 6,000,000 acres of forest land are controlled by small woodlot owners.[15]

During the early 1970s, the corporate ownership pattern of Maine forest land led several environmentally concerned groups to take issue not only with the power which these companies exerted over Maine politics, but also with the environmental degradation which often resulted from their forest practices. The environmental controversy reached a national audience when a Ralph Nader study expressed the opinion that:

> Maine is a paper plantation—a land of seven giant paper and pulp companies imposing a one-crop economy, with one-crop politics, which exploits the water, air, soil, and people of a beautiful state.
>
> The goal of the paper industry is the maximum profitable extraction of pulpwood. And it dominates the state as it pursues this goal, with a tunnel vision unique even for large absentee corporations.[16]

The leading role of forest industries in the Maine economy and the minor role of farmland as part of the total Maine landscape are partial reasons for the absence of a strong mandate for farmland protection programs in Maine when compared to the other New England states. Land use and open space issues have often focused on forest practices and forest open space protection rather than farmland protection. As concern over forest land use practices and related environmental problems has grown, however, concern over other land use issues, such as the visible decline of farmland acreage, has grown as well.

The role of the timber companies has not been the only major land use or environmental issue to confront the people of Maine in recent years. In response to the region's chronic economic and energy problems, several major industrial proposals have been the focus of controversy. For example, efforts have been made to locate new deepwater oil port facilities with companion petrochemical plants along Maine's rugged coast. Bays and inlets along the Maine coast are the only areas on the entire eastern U.S. coastline capable of handling the supertankers currently used for the transport of oil. Over the last two decades, proposals for oil ports to be located at Machiasport, Eastport, and other locations have generated powerful statewide controversies. To date, none of these proposals has been implemented, but the controversy still rages.

Oil ports, timber practices, conversion of farmlands and pressures for second-home developments in the southern portion of the state have raised the issue of economic development versus environmental protection to a high level in recent years. Because Maine is such a poor state, economic development is also a very real concern. Thus, by the early 1970s, the debate over economic development versus environmental protection had reached a high pitch. On one side of the debate were the major corporate and business interests, labor unions, and traditional economic development officials who felt that the solution to Maine's economic problems was the attraction of large amounts of outside capital to generate new heavy industry and jobs. The opposing view, which was expresed by newcomers to the state, back-to-the-earthers, young people, academics, and newspapers such as the *Maine Times*, was that oil ports and other heavy industry would cause environmental damage that would outweigh any benefits to the economy.

Such a debate took place in northern Washington County following the closure of the fish-processing plant in 1981. Shortly after the closing, the *Boston Globe* reported that a major debate had erupted over future economic development in the county. Harold Keezer, director of the town of Eastport's small port was quoted, "What this county needs is not more welfare and food stamps but more manufacturing shops." These feelings were echoed by State Representative Harry Vose, a native of Eastport, who commented: "This town, this district, is in sad need of an industrial tax base, in sad need of jobs."

The alternative view was expressed by many of the newcomers to Washington County, 6,000 of whom arrived in 1980 alone. Some of those 6,000 are natives of Maine who had moved to Washington County to retire. Others are people who have never lived in Maine, referred to as people "from away" by the locals. They came for the natural beauty and because land is inexpensive and the cost of living is low. One of these newcomers, a numerical analyst who arrived from the University of Michigan eleven years prior, is representative of these people. He says the newcomers:

> ...have a new reality. They know what it's like to live in places where fish are poison to eat and the air poison to breathe....People here don't understand what they have...the energy of the county is in the sun that grows the forest and in the tidal estuaries that grow the fish...If I said, "Go grow a school of herring,"

> it would be impossible, but it's being done in the estuaries. If you're ignorant of this you say, "Let's build an oil refinery," and after one generation not only do you not have a refinery, you don't have the fish either.

The newcomers' protection efforts have been very effective in Washington County and other parts of the state and have often created bitter controversies. Eastport director Harold Keezer was quoted in *The Boston Sunday Globe:*

> Somewhere down the line the environmental people have to decide what is more worth having, that golden wing plover or chicken on the table....It is easy to be critical [of development] if you've got a check coming in every month....They're not in the real world. They left their real world and are coming here as an adventure, like going to Africa.[17]

The land use controversy in Washington County is a good example of the dilemma which has faced the entire state of Maine since the late 1960s. One-half of the state was still largely unprotected and unregulated in the late 1960s. Furthermore, at that time, only about 20 percent of Maine's cities and towns had enacted zoning ordinances and as recently as 1973 only 26 percent had enacted subdivision regulations.[18] Although plans for oil port facilities have never materialized, the discussions that occurred concerning these and other large-scale land use problems led to a general agreement among Maine citizens for the need for major state-level land use legislation. Amidst the growing concern over statewide land use and environmental issues, concern over the conversion of land in farms has also grown. Until the late 1970s much of the concern over the conversion of farmland was based on a desire to preserve existing environmental character and open space. Since the mid-1970s, however, a new set of concerns has also arisen.

Concern over Food Self-Reliance

As in most of the New England states, the self-reliance of the farm and food production economy recently has stimulated growing public concern for farmland protection in Maine. Since the mid-1970s, several organizations have stressed that the state is too dependent on nonlocal food producers. One of these groups, the Maine Consortium for Food Self-Reliance, has emerged as a leader in this effort and expressed this concern in March 1979:

> We do not equate self-reliance with total self-sufficiency....What self-reliance means to us is putting interdependence in the place of dependence. It means changing institutions, processes, and relationships that tend to leave us subordinate and vulnerable; not in control of the food production and distribution system on which we rely for one of life's necessities.[19]

Maine meets the needs of its consumers in certain commodities. The farms in the state yield a variety of products. Agricultural cash receipts are dominated by three major commodities: potatoes, eggs, and dairy products. These three commodities not only meet the needs of Maine consumers, but yield a substantial surplus for export. In addition, enough apples, wild blueberries, peas, buckwheat, and winter squash are produced to meet Maine's consumer needs. Maine is, therefore, unlike most of the other New England states since it produces a sizeable proportion of its own food needs in several commodities. Table 8.2 and Figure 8.1 summarize commodities produced in Maine.

Because of the state's large size, specific commodities are focused in different regions. Their location is largely a function of climate, soil conditions, and the accessibility of suppliers and processors. The decline in accessibility of these middlemen has contributed to Maine's economic decline. As some farming operations have terminated over the past few decades, so too have the agricultural support industries, necessary supplies, and a market for farm produce. As a result, most of Maine's best cropland, in terms of both soil quality and economic viability, is now located in a few areas where conditions have been right to maintain the critical mass necessary for faming. Most of these lands lie in eastern Aroostook County, where approximately 82 percent of Maine's remaining tilled cropland exists.[20] Other important agricultural areas include the Fryeburg floodplains area, southern Franklin County, northeastern sections of Oxford County, western Penobscot County, and the Bowdoinham area in Sagadahoc County.

Although Maine is an exporter of dairy, eggs, potatoes, and blueberries, over 70 percent of all food consumed in the state must be imported. Furthermore, over 90 percent of the feed grain used to produce two of these products comes from out of state. As discussed in Chapter 2, many of these production and consumption patterns are the result of national farm economic trends and the growth of large multinational agriculture industries. Recent corporate acquisitions of older "Down East" businesses such as Decoster Egg Farms, Burnham and Morrill Company (B&M Baked Beans), and Snow's Food Products (clam chowder) have helped to alert the people of Maine, however, to the decline of their food self-sufficiency and the loss of local control over food supplies. This concern led the Cornucopia Project to observe in 1981, "Our own supply of food is quite vulnerable. An energy cutback, a transportation strike, a freeze, drought, or labor dispute in California—any of these could mean a food crisis."[21] The fear of a food calamity has led many concerned people in Maine to feel that programs to encourage a greater diversity of food products, coupled with local marketing control, are necessary. However, to date little has been done to meet this

Table 8.2
Maine Crops: Acreage, Yield, Production, and Value (1982)

Crops	*Acres Harvested*	*Yield per Acre*	*Total Production (1,000)*	*Value of Production* (Thousands of $)*
Corn, all	32,000	16.5	528 tons	13,940
Oats	38,000	62.0	2,356 bushels	3,416
Hay, all	230,000	1.85	425 tons	27,200
Potatoes	94,000	235	22,090 Cwt.	130,331
Maple Syrup	—	—	53.5 gallons	1,193
Apples, commercial**	—	—	2,000 42-lb crates	13,008
Blueberries	25,000	1,790	44,653 lbs.	16,522
Veg. Proc.	11,278	—	— tons	5,800
State Total	430,278			211,410

Notes:
*Relates to marketing season or crop year.
**Production is quantity sold or utilized.
Source: New England Crop and Livestock Reporting Service, USDA, Concord, N.H, 1983.

objective, and overall it appears that the people of Maine consider the possibility of a food crisis remote. Even the Maine Consortium for Food Self-Reliance saw reasons why such a food supply calamity was unlikely:

> While numerous crises have erupted in parts of the U.S. food system in recent years, one cannot draw the conclusion that they imply a fundamental crisis of the system as a whole. One reason is that these factors (a winter frost, a drought, a truckers' strike, a farmworkers' action) tend to be transitory. Even when these factors bunch up together, as in California in recent years, human responses tend to blunt their effects. In particular, there are powerful vested interests in the status quo. Agribusiness corporations, farmers' organizations, and politicians will use their influence to create political, commercial and technological responses to crisis conditions, minimizing the negative effects on their interests.[22]

An additional, and perhaps more sensible, basis for the concern over food self-reliance in Maine is economic. Since marketing accounts for 60 percent of Maine food costs,[23] there has been growing opinion that direct marketing and farm cooperatives can keep native producers "in the black." This concern for a thriving rural, "Down East" economy is what brought many newcomers to Maine and has helped farmers in the legislature.

The desire for food self-reliance and its impact on the Maine economy was expressed by the legislature in 1979 when legislation to change the name of

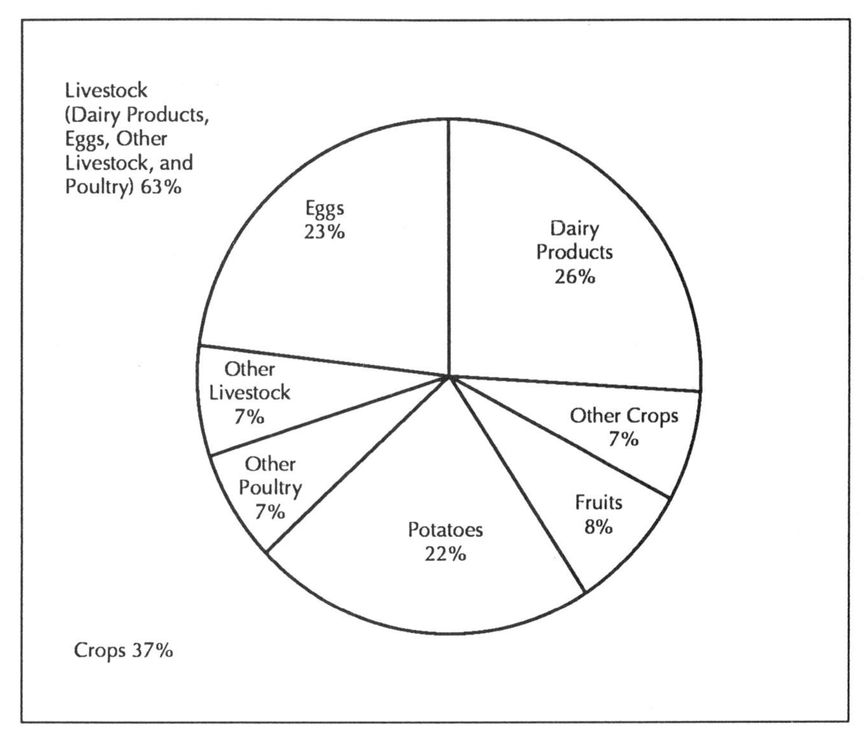

Figure 8.1. Distribution of cash receipts from Maine farm marketings in 1983. *Source:* New England Crop Reporting Service.

the Department of Agriculture to the Department of Agriculture, Food, and Rural Resources was adopted. At that time, the following declaration was made:

> The Legislature finds agriculture to be a major industry in the State, contributing substantially to the State's overall economy, essential to the maintenance and strengthening of rural life and values necessary to the preservation of the health, safety and welfare of all of the people of this State.
>
> The survival of the family farm is of special concern to the people of the State, and the ability of the family farm to prosper, while producing an abundance of high quality food and fiber, deserves a place of high priority in the determination of public policy. For this purpose there is established the Department of Agriculture, Food and Rural Resources.[24]

In accordance with this legislative change, the duties of the Commissioner of the Department were amended to require the promotion of "farm financing and rural development proposals; conservation and preservation of agricultural lands." Moreover, in 1984 the Maine legislature established a broad Maine Food Policy.[25] Thus, the concern over food self-reliance was directly tied to the protection of Maine's farmlands.

In summary, in recent years statewide concern to protect agricultural land in Maine has grown for similar reasons as it has in neighboring New England states—that is, a growing concern for issues of environmental quality, particularly open space and landscape character issues, and concern over food self-reliance.

Many people in Maine, including analysts within the agricultural industry, do not feel that the great concern over farmland conversion is justified. According to Dr. Edward Micka, extension economist at the University of Maine at Orono, most of the support for recent farmland protection programs has grown out of a concern for open space protection, rather than a genuine concern for agriculture:

> When we requested the constitutional amendment to get the farm and open space tax legislation, the Maine people voted for constitutional change, that they would support it if it was necessary. This is characteristic of the more populated areas. They may be saying that they would like to have a green belt.

In Maine, such a mandate from the public for open space preservation is not necessary, Micka asserts. With the vast areas of open space in the state, "people could drive through the rural areas, see the lobster boats, and so on. But they still supported the Farm and Open Space Tax Law."

Another facet of this concern for the preservation of open space in Maine has been the loss of active farmland to development. As the studies conducted by the Southern Kennebec Valley Planning Commission and the Maine State Planning Office indicate, however, most farmland has not been converted to development but rather to fallow fields. In addition, most of the land which was abandoned has been the less productive marginal lands incapable of competing in markets dominated by economic forces outside the state. Micka commented:

> I think we are not losing a lot of our really good farmland. We are, however, losing a lot of marginal farmland. For example, we have an area in Gorham, in the southwest corner of Maine where some of our greatest development pressures are, that has excellent cropland. We have not lost a lot of that good land.[26]

However, a study of nineteen towns undertaken by the State Planning office indicates the opposite: *good* land is being lost to both development and abandonment. Joyce Benson, senior planner at the State Planning Office, feels that "the attitude that only the poor land is lost has been used as a ploy by those who are against protection."[27]

Maine has a genuine problem of farmland loss: recently, analysts such as Micka and the members of the special 1979 Food and Farmland Study Commission are concerned about erosion, which is widely considered to be the most severe farmland depletion problem currently confronting the Maine agricultural industry. It is estimated that Maine's 303,000 acres of tilled cropland is losing an average of six tons of soil per acre per year, twice the USDA accepted rate. The most extensive erosion areas are in Aroostook County, while the highest rate of erosion is in the Knox-Lincoln County area.[28] Much of this soil erosion is the result of poor farming practices, and efforts have been undertaken to reduce this source of farmland depletion.

The public approach to land use in Maine has thus gone from an active policy of discouraging farming after World War II to one of benign neglect, to a concern among different segments of the population over general environmental quality, to open space protection, to protection of farmland for food production potential, and most recently to the concern over loss of agricultural land to erosion. Shadowing these concerns and their corresponding legislative response is the genuine need for economic development in a state that ranks among the poorest in the nation.

It is clear that the setting for farmland protection in Maine is unique in New England. Legislative concern to protect farmland has been less pronounced than elsewhere in New England for the following reasons:

1. Although thousands of acres of Maine farmland have been converted to other uses, the decline in farmland acreage has been less dramatic than in other New England states.
2. Low population density and open space have made residents feel there is no need to protect farmland even though there are competing uses for prime land.
3. Although some farmland areas are located adjacent to major urban areas, the development there has been on a smaller scale and there is a general lack of awareness of the cumulative effect.
4. Maine is primarily a forested state, having less farmland as a percentage of land area. Pressures for environmental and land use programs have thus focused more heavily on forest lands.
5. Historically, farming has played a less important role in the economy than other industries, notably forest industries.
6. Maine produces commodities the other New England states do not, such as potatoes and blueberries, for which it still maintains a unique market niche. Food self-reliance controversies have therefore been less intense than in other New England states.

7. Because of the poor economy of the state, there is more pressure for industrial development and less pressure for farmland and other land use programs which may serve to inhibit the climate for economic development. Moreover, farmers are hesitant to participate in any protection program which would restrict the use of their land or decrease its value.

Although the concern to protect agricultural land has been less intense in Maine than in the other New England states, several important land use and conservation programs have been adopted since the early 1970s, many of which have provisions for agricultural land. Very few programs, however, have been adopted that are solely for the protection of agricultural land. Furthermore, those farmland protection programs the state has adopted have not been unique. The state generally has followed Massachusetts, Connecticut, and Vermont by adopting a few similar policies and programs. Because of the special characteristics of local "Down East" politics, however, Maine has recreated these programs in its own way. The remainder of this chapter outlines more specifically the most important early conservation programs which had some impact on farmland protection and more recent programs specifically designed to protect agricultural lands.

Conservation Programs and Early Efforts to Protect Agricultural Land

Specific programs to protect agricultural land in Maine have been a relatively recent occurrence. As in most of the New England states, by the mid-1960s dwindling agricultural resources were not addressed by policymakers. Although the *Bangor Daily News* had sounded a warning in 1962, the citizens and the legislature took little action—aside from the efforts of small agricultural groups in a few municipalities. During this time most of the state activity concerning land use and environmental issues was confined to problems of water pollution and port development in coastal communities.

As in other New England states, the first land use programs after the increased attention to land use protection in the 1960s were not designed solely for farmland protection. They aimed instead to control adverse land uses and assure the protection and proper management of Maine's abundant natural resources. As community concern for open space and environmental character grew throughout the 1970s and into the 1980s, direct farmland protection plans followed.

During the 1960s, the general issue of statewide land use control was raised initially by the Natural Resources Council of Maine, a major statewide federation of fifty conservation groups. The Council played a leading role in generating interest in land use laws and other environmental protection measures, as a response to what was happening to Maine's environment. As a result, in 1965 Robert Patterson and other leaders of the Natural Resources Council proposed legislation authorizing municipalities to form Park and Conservation Commissions. The initiative to establish these Commissions grew out of the success of the Conservation Commission movement which originated in Massachusetts in 1957. Maine was the last of the New England states to adopt such a program.[29] Although the program was not originally intended to protect agricultural land, it was one of the first steps toward comprehensive protection of open space and community character in Maine. Later programs, such as the Farm and Open Space Tax Law, moved closer toward achieving the broad aim of programs such as the Park and Conservation Act of 1965.

Maine Park and Conservation Act (1965)

The Maine Park and Conservation Commission Act is unique among those in New England. Because of the conservative political climate in Maine, the statute was initially designed only to amend an existing law which allowed towns to establish local Park Commissions rather than attempt to create an entirely new commission. The result was to authorize hybrid park and conservation commissions, requiring that commissioners be elected, rather than appointed as in Massachusetts. In the initial years following the adoption of the new Act, there was very little support for the new Commissions. Clinton Townsend, a Skowhegan lawyer who drafted the original legislation, believes that the election requirement was an unwise aspect of the legislation. He said it was a hindrance to the establishment of Park and Conservation Commissions because people felt it only added cost and red tape to local government.[30] In 1969, the legislature amended the 1965 law to allow local boards to appoint Commission members. In recent years the movement has gained momentum, particularly as controversies over coastal and forest land uses have grown. Today, over two hundred of Maine's cities and towns have such Park and Conservation Commissions.

Although the original and amended acts did not specifically mention the protection of agricultural lands as an environmental objective, several local Conservation Commissions have become involved, to varying degrees, in efforts to promote and protect agricultural lands. Generally, however, any benefit to agriculture has been coincidental with other objectives of the program, such as the preservation of open space or the protection of natural resoruces. The general consensus in Maine is that the Park and Conservation Commissions have not had a measurable impact on the protection of agricultural land. The state Department of Agriculture, Food, and Rural Resources and other agencies with a concern for farmland protection have indicated a desire to define other possibilities through this program, but to date there has been no action in this regard.

In many ways, the adoption of the Maine Parks and Conservation Commission Act of 1965 marked the beginning of an active environmental period in Maine. In particular, public controversies over planned oil development along the coast had stimulated concern among citizens and environmental groups in the state, and the Act was the first effort to face local land use problems caused by such development. It was, however, ineffective in the cities or towns which did not actually have commissions and in the 51 percent of the state that was in the unorganized territories. Furthermore, the act had only marginal impact on the protection of agricultural land because it did not specifically confront that issue. Indeed, at that time farmland conversion was not yet considered an issue of statewide importance.

Development Control Programs (1969, 1971)

Development Control Programs instituted in 1969 and 1971 were an important step in Maine's land protection efforts. The growing awareness of environmental and land use problems in the late 1960s resulted in the passage of three major, comprehensive land use regulations known as the Site Location of Development Act, the Land Use Regulation Commission Act, and the Mandatory Shoreland Zoning and Subdivision Control Act. The Natural Resources Council again played a major role in generating citizen interest in land use laws and other environmental protection measures. These Acts suddenly placed Maine among the leading states in the "Quiet Revolution in Land Use Control."[31] Although these Acts were not expressly designed to protect agricultural land, they were vital to Maine's land conservation and protection efforts. Through these new legislative programs, land use controls were extended into areas formerly devoid of planning controls. In some cases, these programs considered farmland for the first time in their land use review procedures, although farmland protection is not their major purpose.

The Site Location of Development Act. Adopted in 1969 and since amended, this Act grants to the state of Maine the discretion "to regulate the location of developments which may substantially affect the environment." Under the Act, such developments are defined as:

> state, municipal, quasi-municipal, education, charitable, commercial or industrial developments, including subdivisions, which occupies a land or water area in excess of 20 acres, or which contemplates drilling for or excavating natural resources, on land or under water where the area affected is in excess of 60,000 square feet, or which is a mining activity, or which is a structure.[32]

All developments which meet this definition must be approved by the state Board of Environmental Protection, which is empowered to reject any development proposal which is considered to have an adverse effect on the quality of Maine's land, air, or water resources. The Board must approve a development proposal that meets the following five criteria:

1. *Financial capacity.* The developer has the financial capacity and technical ability to meet state air and water pollution control standards and has made adequate provision for solid waste disposal, the control of offensive odors, and the securing and maintenance of sufficient and healthful water supplies.
2. *Traffic movement.* The developer has made adequate provision for traffic movement of all types out of or into the development area.
3. *No adverse effect on the natural environment.* The developer has made adequate provision for fitting the development harmoniously into the existing natural environment, that the development will not adversely affect existing uses, scenic character, or natural resources in the municipality or in neighboring municipalities.
4. *Soil types.* The proposed development will be built on soil types which are suitable to the nature of the undertaking.
5. *Ground water.* The proposed development will not pose an unreasonable risk that a discharage to a significant ground water aquifer will occur.[33]

The Act has had little direct effect on preventing development on agricultural lands since the impact on farmland is not one of the enumerated criteria. Indeed, under Criterion Four (soil types), farmland often provides the most suitable soils for development since it is well-drained, ideally suited for on-site sewage disposal, and has minimal slopes. In 1979, the Maine State Planning Office recommended that the evaluation criteria be amended "to specifically enable the Board of Environmental Protection to consider development impacts on productive farmland as an important criteria in the review process for development proposals."[34] To date, however, no legislative action has taken place.

Land Use Regulation Commission Act (1969). In the same year the legislature adopted the Site Location of Development Act, it also adopted the Maine Land Use Regulation Commission Act, which extended "the principles of sound planning, zoning, and subdivision control to the unorganized and deorganized areas of the State,"[35] not including Indian reservations. The 10.5 million acres (51 percent) of the state's total area outside the jurisdiction of any city or town comes under the jurisdiction of this Act.[36] The essential purpose of the Act is to protect the formerly unregulated areas against irresponsible development which could harm water and land resources. All development in these areas requires Land Use Regulation Commission (LURC) review and approval, except those projects approved by the Board of Environmental Protection under the Site Location Act. The criteria for approval are similar under both Acts.

The LURC is empowered to classify vast tracts of land into three types of districts and establish regulations and land use guidance standards for each district. The three district types identified by the Commission were protection, management, and development. Management districts contain lands appropriate for agricultural and forestry uses.

The LURC was also directed to prepare a comprehensive land use plan by January 1, 1975, one objective of which was "to promote agricultural management as a land use, particularly in areas which are currently or potentially highly productive for crops." An important policy position of the Plan is that "major development will be permitted only where the productivity of existing agricultural and forest land is not significantly lessened."[37] Among the policies to be implemented by the LURC are the following agricultural resource policies:

1. Limit non-farm-related land uses in prime farmland areas. Support other proposals to maintain active farms.
2. Regulate agricultural practices in protecting subdistricts where these activities can cause accelerated erosion, sedimentation, pollution, siltation, or contamination.[38]

The LURC has followed these policy directives in developing its Permanent Land Use Standards and district boundaries. However, only a small part of Maine's farmland is within LURC jurisdiction since, historically, farmers tended to be the original settlers of the land who subsequently formed communities. Most of the land within the unorganized territories is not within historic agricultural communities, which are comprised almost exclusively of forest land. Thus, the LURC has had little impact on farmland presently under active agriculture.

Mandatory Shoreland Zoning and Subdivision Control Act. This program, whose title was later amended to the Shoreland Zoning Act, was the third development control program to emerge from the growing environmental concern in Maine in the late 1960s. Adopted in 1971, the law required municipalities to establish zoning controls for all areas within their jurisdiction that are within 250 feet of the shoreline of any navigable ponds, rivers, or saltwater areas within the state. In 1973 the law was amended to include land within 250 feet of any pond or lake in excess of ten acres (except when surrounded by a single landowner) as well as streams or rivers from a point where they drain twenty-five square miles or more.[39]

The purpose of the required zoning controls was to:

> Further the maintenance of safe and healthful conditions; prevent and control water pollution; protect spawning grounds, fish, aquatic life, bird and other wildlife habitat; control building sites, placement of structures and land uses; and conserve shore cover, visual as well as actual points of access to inland and coastal waters and natural beauty.[40]

If a municipality failed to adopt such controls the State Board of Environmental Protection and the Land Use Regulation Commission were empowered to adopt ordinances for them. The Attorney General can seek enforcement of such ordinances if a municipality does not.

The Shoreland Zoning Act does not require cities or towns to designate farmland as a category of their required zoning and subdivision ordinances. Municipalities subject to the Shoreland Zoning Act are required to establish a comprehensive plan and zoning and subdivision ordinances within the guidelines of the long-established enabling legislation of the Zoning Act and Municipal Subdivision Act.[41] Although state zoning enabling legislation does allow the establishment of exclusive agricultural zones within municipalities, the Municipal Subdivision Act does not allow towns to consider the impact on farmland in their subdivision review processes. Consequently, the Maine State Planning Office recommended in 1979 that the Municipal Subdivision Act be amended specifically to allow towns to consider impacts upon highly productive farmlands in their subdivision review and approval process.[42] No legislative action was taken on this recommendation, however.

Taken together, these three development control programs gave broad new powers to the state of Maine for the control of land and water uses in the early 1970s. But the impact of these laws in protecting agricultural land has not been significant, largely because these are general land use programs whose purpose was not to preserve any specific land or water use, but rather to ensure that the many impacts of land use decisions are considered in the decision-making process. In some of the programs, such as the LURC, the impact on productive agricultural lands is considered an important factor in the planning process. In the other programs, however, specific consideration of agricultural land is absent. The priorities of these general land use programs reflect the concern during the late 1960s and early 1970s to focus on the protection of the state's bountiful natural resources generally. Concern over the specific issue of farmland conversion had not yet become acute.

Nevertheless, the comprehensive nature of these land use laws provided an important background for more recent farmland protection efforts. In particular, these statewide land use control laws may prove to be a vehicle for future farmland protection efforts in Maine. Although there has been no legislative action on the 1979 farmland protection recommendations of the Office of State Planning, the Special Maine Food and Farmland Study Commission (1979) recommended that these laws be amended to allow for consideration of farmland within their approval processes. These and other recommendations of the Commission are discussed in a later section of this chapter.

Differential Tax Assessment Programs (1971, 1972)

The early conservation programs evolved from a pervasive concern in Maine over the need for land use regulations and incentives to preserve the beauty of Maine's environment. Concurrently, concerned citizens, environmental groups, and members of the agricultural community, all of whom considered agricultural land to be an important aspect of Maine's open space, began working to establish programs which would deal solely with farmland protection. In November 1970, a state referendum was passed to amend the Maine constitution, allowing legislation to tax farms, forest lands, and other desired open spaces at their current-use value. The public referendum stated that:

> Nothing shall prevent the Legislature from providing for the assessment of the following types of real estate, wherever situated, in accordance with a valuation based upon the current use thereof and in accordance with such conditions as the legislature may enact: Farms and agricultural lands, timberlands, and woodland.[43]

Subsequently, in 1971, the Maine legislature adopted the Farm and Open Space Tax Law and, in 1972, the Tree Growth Tax Law.

Although the passage of these two programs was fueled by a desire for environmental protection programs, there were more precise reasons why current-use taxation was deemed necessary to protect valued open space lands. Specifically, as in other New England states during the 1960s, land values in Maine were increasing tremendously, resulting in increased taxation and placing significant financial pressure on owners of farm and forest lands. As Vance Dearborn, Public Affairs Specialist at the University of Maine, commented:

> During the late sixties, many individuals from out of state purchased land where the price paid had no relationship to the productivity of the land. For example, a person would buy a wood lot for $2,500. He would cut off the timber, strip the land and take off, say $5,000 or $10,000 worth of wood. He would then sell the land to an out-of-state buyer for $10,000. What this activity did to local assessment ratios and the apparent value of land led to unbearable taxation levels in many towns. Maine needed a different forest land tax policy.

> Farmers also were feeling the same pressure, both from development and from local tax policies that taxed farmland equally on a front-foot basis along all roads in the town, when at no time was all the environmental land in a town in demand for development in the foreseeable future. Thus there was a tendency to tax all farmers on the basis of what one in ten farmers in the community were able to get for their land. Preferential assessment was seen to be as important for protection of the farmer from the "system" as for preservation of open space.[44]

Farm and Open Space Tax Law (1971). Like most differential tax assessment programs, the Maine Farm and Open Space Tax Law allows for the valuation of farmland or open space land based on its current use as farmland or open space, rather than at its potential fair-market value for more intensive uses. The defined purpose of the Maine program is to:

> encourage the preservation of farmland and open space land in order to maintain a readily available source of food and farm products close to metropolitan areas of the state, to conserve the State's natural resources, and to provide for the welfare and happiness of the inhabitants of the State... [and to] prevent the forced conversion of farmland and open space land to more intensive uses as the result of economic pressures caused by the assessment thereof for purposes of property taxation at values incompatible with their preservation as such farmland and open space land.[45]

The Farm and Open Space Tax Law, therefore, provides an economic incentive (or removes a disincentive) to continue farming by tying farm property values to production values.

Although the purpose of the law is to protect both farmland and open space lands, different eligibility requirements apply to the two categories of land because of the different land uses. In order to meet the eligibility requirements for use-value assessment, farmland must meet the following general conditions:

1. The tract must contain at least ten contiguous acres of land. A single application may be made for more than one tract of land, but at least one of the tracts must contain ten contiguous acres.
2. The tract must be used for farming or agricultural activities, but may include woodland and wasteland within the farm unit.
3. The tract must produce a gross income per year of $1,000 for ten acres plus $100 per acre for each acre over ten. Maximum total income is not required to exceed $2,000. In other words, for parcels of twenty acres or more, the income requirement remains at $2,000.[46]

Under the open space provisions of the law, there are no minimum acreage or income requirements to become eligible for use-value assessment.

However, eligible open space land must fall within the following categories:

1. The tract must be used to conserve scenic resources or enhance public recreation opportunities. Land designated as wildlife and management areas, sanctuaries, and preserves under Title 12 of the Maine Revised Statutes, are also eligible for open space land classification.
2. Land which is included in an area designated as open space land on a comprehensive plan or in a zoning ordinance or upon a zoning map as finally adopted shall be classified as Open Space land upon application of the owner.[47]

To place land under the Farm and Open Space Tax Law, a landowner must file an application with the local assessor by April 1 of the year in which classification first is requested. Assessors must then determine whether the land is eligible for classification as farmland based on acreage, portion actively used for agriculture, productivity, gross income, nature and value of any equipment used, contiguous nature of tracts, and other relevant criteria. The assessor and his agents may inspect the lands and investigate the information submitted by the landowner. The landowner must be notified of the decision of the local assessor by June 1 of the same year. Lack of notification constitutes denial of the application. Farmland owners that fail to meet the gross income requirement of the Act may apply for a two-year provisional classification. Participating landowners must file annually and update the initial application's information. Lack of notification to a participating landowner by June 1 constitutes approval.

Maine's Farm and Open Space Tax Law does not apply a specific formula for valuation of land after it is classified, as in Massachusetts and New Hampshire. Rather, the law merely sets forth the above-mentioned criteria and then allows the local assessors to determine the current market value for farm purposes. Leaving the valuation decision to the local assessors was considered workable in Maine because in nearly every town in the state, there is farmland whose value is not affected by development pressure. Local assessors in every town know of land that should be taxed only at its agricultural value. Therefore, when valuing agricultural land which has development potential, there is other similar land in the town for which the agricultural value has already been computed and which thus serves as a basis for valuing agricultural land in the town. Vance Dearborn commented:

> We spent a lot of time trying to develop county-wide equalized values, based on capitalized values and rents. We also did a statewide rent survey. However, when we capitalized the rents, realistic land values were not obtained. Land is considered a residual cost factor in some economic analyses. In a poor potato year in Maine, this type of analysis gives a negative value to land. We also tested the system used in New Jersey and found it did not work in Maine conditions. Thus, we recommended that local assessors determine the current market values for farm purposes. In retrospect, the assessors had already been doing

the job quite effectively. They had an agricultural background in many cases and had a pretty good idea of what the value of agricultural land was.[48]

For farmland, local assessors determine the 100 percent value of good cropland, pasture land, and orchard land. These valuations should reflect the current-use value of the parcel of land for agricultural purposes. The valuation cannot reflect the potential of the land for other uses or the value of adjacent road or shore frontage for development.

Once valuation is computed, farm and open space lands under the program are subject to the same property tax rate applicable to other property in the local jurisdiction. Any areas other than farm and open space lands, except forest land, must be assessed at fair-market value.[49] In 1975, the legislature added appeal and abatement procedures for failure to classify or for improper assessments, thus providing aggrieved landowners recourse.

There are strict penalty provisions for landowners who enroll in the Farm and Open Space tax program and fail to abide by its provisions. Many farmers have found the recapture and penalty provisions for removing previously classified land too stringent compared to the level of tax benefits they could receive.[50] They have, therefore, opted not to participate in the program at all, which partially explains why a limited amount of farmland was placed under the program in its first nine years, from 1971 to 1980.

The penalty provisions of the law require that when classified farm and open space land no longer meets eligibility requirements, the land may be withdrawn from classification by the local assessor or at the request of the landowner. A landowner must be given notice of an intention to reclassify his land. Any change in use which disqualifies land for classification will cause a penalty to be assessed, except in instances where farm or open space land is withdrawn as a result of eminent domain proceedings. The portion of the classified parcel whose use was changed is subject to the penalty.

In conformance with the state constitution, the original penalty required payment of back taxes that would have been due if the land were taxed at full value, plus 8 percent interest, up to ten prior years (fifteen prior years in cases of open space land). The 1975 recodification of the Tax Act substantially reduced this penalty to encourage greater participation. The current penalty for withdrawal from the program is an amount equal to one of the following percentages times the difference between the fair-market value of the land on the date of withdrawal and the full valuation of the land when last classified under a farm and open space classification:

- 10 percent—land taxed as farmland or open space for five years or less
- 20 percent—land taxed as farmland or open space for more than five years, but less than ten years
- 30 percent—land taxed as farmland or open space for ten years or more.[51]

The 1983 legislature further reduced the penalty for long-term participants by granting a 1 percent penalty reduction each year after the first ten years. These modifications of the penalty reflect a view that particpating landowners should be rewarded, not punished, if they have devoted their lands to open space uses.

The impact of the Farm and Open Space Tax Law in protecting Maine farmland has not been significant. In 1982, only 68,130 acres contained in 1,301 ownership parcels were classified under the law.[52] There were no lands under the program in Aroostook County, where much of Maine's farmland is located, or in Piscataquis County, a county with only 2 percent of its total land remaining in farms in 1974.[53] (Piscataquis County experienced an 81 percent decline in total farmland from 1925 to 1974.[54] Thus, it would seem to be an area ripe for participation in farmland protection programs. Unfortunately, in the case of the Farm and Open Space Tax Law, this has not occurred.) In Kennebec and Penobscot Counties, about one-half of all the acreage is enrolled. The town of Exeter has over 14,000 acres itself. In addition to the farmland classified under the law, only 13,595 acres of land were taxed under the open space provisions of the Act in 1982. Almost one-third of this acreage was in the town of Hollis alone.

As Vance Dearborn commented:

> The Farm and Open Space Tax Law is not very significant overall in the Maine agricultural sector. It is very significant, however, to an individual farmer who wishes to farm outside a community like the city of Waterville, where there has been a tremendous push for housing lots or in strip development areas along roads where property tax valuations are assessed on a front-foot basis. If the producers want to continue farming there, they need this law. Thus, one should not downplay its value, although one could downplay its total impact.[55]

As in most other New England states, therefore, farmers appear most likely to apply under the law in municipalities where revaluation has occurred and where farmland has been assessed at its development value.

Tree Growth Tax Law (1971). The results of the Farm and Open Space Tax Law sharply contrast with those of the Tree Growth Tax Law, a similar differential tax-assessment program in Maine. Adopted one year after the Farm and Open Space Tax Law, this program was designed to encourage the protection of well-maintained and managed forests on a sustained-yield basis.[56] The Tree Growth Tax Law bases its valuation on a modified-productivity formula, which is unlike the valuation system Farm and Open Space Tax Law, however. Average land values for each county are established for softwood, hardwood, and mixed-wood types, and then applied county-wide to all forest landowners who participate in the program. Also unlike the Farm and Open Space Tax Law, participation in the program became mandatory for owners of most of the forest land in the state when initially passed. All woodland parcels of 500 acres or more were automatically subject to the law. For smaller lots, application of the law was at the option of the landowner. Regardless of parcel size, however, filing an application is required of all woodland owners, because each landowner must inform the local assessor of how much softwood, hardwood, and mixed wood acreage the parcel contains. Since most of Maine's forest land is contained in large parcels which are owned by wood products and timber management companies, large, mandatory parcels accounted for 98 percent of the current land under this tax program. Only 2 percent of the applications concerned the smaller optional parcels.[57] In 1982, the legislature substantially altered the law by making the program entirely voluntary.

Predictably, the disparity of results in participation between the Farm and Open Space Tax Law and the Tree Growth Tax Law initially was marked. By 1980, approximately 11 million acres, or 75 percent of Maine's commercial forest land, was classified under the provisions of the Tree Growth Tax Law. Most analysts of the program felt that it has provided stability, uniformity, and predictability to forest land taxation by acting as a deterrent to development in isolated areas, along highways, and in other forested areas. There is general agreement in Maine among forest land use experts that positive forest taxation, combined with zoning in the municipalites and the Land Use Regulation Commission in the unorganized territories, has helped to keep forest land from being thrown onto the market and developed for unfavorable uses.[58]

Shifting tax burdens have been a major consequence of the Tree Growth Tax Law, resulting in as high as $2 million shift of revenues statewide in 1980 alone. Yet this sum represents less than 1 percent of the total property tax assessments for the state. Nevertheless, in a community where 90 percent of local property tax revenues are gained from woodland uses, the shift of burden onto home and recreational property owners has been substantial. Therefore, the state in 1980 reimbursed municipalities eleven cents per acre annually for all land taxed under the Tree Growth Tax Law. This relieved only a small fraction of the tax burden that had been shifted onto other properties.[59] With the 1981 amendments, however, less land falls under the Act so towns will have greater resources from property taxation.

In comparison, the shift of tax burden resulting from the Farm and Open Space Tax Law has not been as marked since so little of the state's farm and other open space land has been classified under the law. The potential for a significant shift of burden does exist, however, since the Farm and Open Space Tax Law makes no provision for reimbursement to municipalities for tax losses from lands classified under the law. Moreover, owners of a parcel of mixed farm and woodland that fails to qualify under the Tree Growth Tax

Law can seek to utilize the Farm and Open Space Tax Law. In fact, the Tree Growth Tax Law allows for application to the Farm and Open Space Tax Law if classification under the former law is terminated.[60]

In summary, the Farm and Open Space Tax Law and the Tree Growth Tax Law have had sharply contrasting results. Some people feel that the reason there has been such little use of the Farm and Open Space Tax Law is that farmers are reluctant to restrict their land options this way. The goals of most farmers are generational, and few farmers want to lose the option of converting their land to other uses if necessary in the future. The goals of most large forest landowners, on the other hand, are usually longer range. Generally, the paper companies are just as worried about how much wood they will have available fifty or one hundred years from now as they are today.

Another reason for the lack of participation by farmers in the Farm and Open Space Tax Law is that farmers in Maine are generally not as excessively burdened by their land property taxation as are farmers in the southern New England states. Valuation is already based on agricultural use in many local jurisdictions. Finally, many farmers have a strong interest in their local community and its financial needs, unlike many of the large forest landowners. Although a farmer may complain about his taxes, he still wants a voice in local government and is therefore willing to pay his "fair share" as long as he can afford it. Taxes may be lower in some rapidly developing areas. However, a report on farm profitability states that despite Maine's low farmland taxation in New England, tax per acre of farm real estate is much higher than the national average primarily because of the heavy use of the property tax to pay for local goods and services.[61]

Some officials feel that administrative problems also account for much of the ineffectiveness of the Farm and Open Space Tax Law. In 1978, Richard Rothe of the Maine State Planning Office prepared an evaluation report on the results of the program[62] which noted several administrative problems with the program and raised the question of how a local assessor could fairly determine current-use value versus development value. The report made several recommendations for change, among which was a proposal that guidance regarding land values be provided to local assessors. Some of the administrative recommendations, designed to make the law more usable, subsequently have been amended into the original act. Other recommendations, however, such as the proposal for state-level intervention in the assessment procedure, are less popular. Dearborn commented:

> [Some] people felt that the local assessors weren't differentiating enough between market values of land for development purposes and values of land for agricultural purposes in their current use taxation. That was when we looked at a statewide system in which the State Bureau of Taxation would set the values for local assessment. I think that type of centralized system in Maine is farther

away today than it was ten years ago. We seem to have quite a strong belief in Maine today that everything should be done locally and there's danger when we let Augusta get too deep into local affairs. I think we have a law here that will probably remain on the books and always be available to the farmer when he needs it. But it probably won't be heavily used.[63]

Purchase of Development Rights (1969, 1971)

In addition to land use regulations to protect the environment and tax incentives designed to favor farmland and open space, Maine has also enacted a purchase of development rights (PDR) program. Like Vermont's program, it is not state-funded, but rather, municipalities are authorized to accept or purchase development restrictions on farmland without state assistance.

In 1969, the legislature authorized "any governmental body" to hold a conservation restriction preventing the building of structures and other uses.[64] Two years later, as part of the Farm and Open Space Tax Law, the 1971 legislature allowed municipalities to "accept or acquire scenic easements or development rights for preserving property for the preservation of agricultural farmland or open space land" if the restriction lasts at least ten years.[65] Conservation Commissions can hold such easements in the name of a municipality if the local legislative body approves.

A few farmland parcels in Maine have been brought under this program.[66] Nevertheless, the lack of state funding and the limited resources of many communities ensure that this program has limited potential.

Recent Agricultural Land Protection Programs

Since the Farm and Open Space Tax Law was adopted in 1971, the decline in farmland acreage in Maine has been an issue of continuing concern to members of the farm community, various state agencies, and the legislature. However, there has been no significant progress in the enactment of new programs specifically designed to protect agricultural land directly. Until recently, the Farm and Open Space Tax Law stood as the only major statewide farmland protection program. However, there has been continued discussion on the subject. Several recent studies have recommended that the state government take action. By far, the most important contribution to this end was the comprehensive effort of the Special Commission established by the Maine legislature in 1977. In 1985, the Maine legislature began to act on farmland protection recommendations which originally began in 1977.

Maine Food and Farmland Study Commission (1979)

The Maine Food and Farmland Study Commission was established in 1977 by the legislature to address food and farmland issues and "recommend actions which would strengthen Maine agricultural."[67] Specifically, the legislature created the Commission "to protect the food capability of the State, to provide consumers with ready access to wholesome, locally-produced food products and to encourage greater food and agricultural self-sufficiency."[68] The twenty-four-member Commission was composied of several state legislators, the Commissioners of Agriculture, Environmental Protection, and Conservation, the Director of the Maine State Planning Office, several farmers, and citizens in the agricultural industry.

The Commission was directed to report to the legislature on policies and programs needed to:

1. Protect Maine's agricultural lands.
2. Promote agricultural use of these lands.
3. Increase self-sufficiency in food production.
4. Improve direct marketing of native foods.
5. Serve the interest of food producers and consumers in the state.[69]

As these directives indicate, the Commission's study of Maine agriculture was much broader than the specific subject of agriculture land protection. Although the primary objective of the study was to investigate the conversion of farmland acreage to other uses in the state, it was widely felt that the problems associated with agricultural land use could not be resolved without considering farm marketing, transportation, energy, research, and other similar issues. Despite this broad scope, the study was somewhat limited, however. Although the Commission was empowered to undertake one of the most comprehensive legislative studies of Maine agriculture ever conducted, limited time and resources did not allow a full investigation of important related issues such as rural development, the quality of rural life, consumer needs, and cost of land. As Chairman Frederick Hutchinson wrote in his cover letter to Governor Brennan at the close of the study:

> Early in the deliberations of the Commission, there was considerable discussion concerning the feasiblity of developing a food policy for Maine. The Commission members recognized the consumer interest in such a policy, but the majority did not feel it was feasible to undertake such a comprehensive study within the time and resources available to the project.[70]

To meet the legislature's initial mandate, the Commission held twelve public hearings throughout the state, taking testimony on agricultural issues from farmers, farm industry businesses, bankers, consumers, and other interested groups. Four subcommittees were then formed to study each of four problem areas which were identified: land protection; entrance to farming; marketing and transportation; and farm finance. In June 1979, the "Report of the Maine Food and Farmland Study Commission to the Governor and the 109th Maine Legislature" was submitted. This comprehensive report presented a synthesis of many of the major recommendations of the four subcommittee reports.

The final report of the Committee stresses that to be effective, programs must be "voluntary,...not appear to usurp individual property rights, and enhance the economic viability of farms."[71] To meet this objective, numerous proposals were made, the most important of which are the following:

Agricultural Soil and Water Conservation Districts. On the subject of agricultural land protection, the committees made several important recommendations. The most valuable of these was a proposal for new state enabling legislation which would create agricultural districts to coincide with the boundaries of the existing local Soil and Water Conservation Districts. During its research and deliberations, the subcommittee on farmland preservation had become aware of recent and serious problems of farmland loss caused by soil erosion. The Committee found soil erosion to be potentially as large a threat to Maine agriculture as urbanization and abandonment. Accordingly, the Agricultural Soil and Water Conservation District proposal was designed to combine soil conservation programs of federal and state agencies with a range of farm incentive programs similar to those found in agricultural District legislation adopted by the state of New York in 1971. The program would be administered by, and coterminous with, existing soil and water conservation districts in the state. It would also provide additional incentives for the protection of agricultural land and the use of proper soils and farm management practices to mitigate erosion and other soil depletion problems.

Although the Commission report did not specifically define the eligibility requirements to be met by landowners wishing to qualify for the program, it did outline certain eligibility criteria which new legislation should contain, including:

1. a minimum acreage in crop production
2. a residency requirement
3. evidence that the farmer is making a reasonable effort to correct any serious soil erosion problems or animal waste disposal problems[72]

Within the established districts, several incentives to encourage good farming practices were proposed by the Committee. In return for a voluntary agreement not to convert their land to nonfarm uses, farmland owners would be offered the following benefits:

1. Current-use taxation at one-half or two-thirds the current-use valuations provided in the existing Farm and Open Space Tax Law.
2. Exemption from nuisance laws and ordinances. Normal farm operations such as spreading manure and operating machinery in evening hours could not be restricted.
3. Farmland takings by state and local agencies under eminent domain proceedings would be restricted.
4. The assessment of special or ad valorem taxes on farmland for sewer, water, lighting, or nonfarm drainage would be restricted.
5. State inheritance taxes would be based on current land use, not market value of the land.
6. A five-year tax exemption would be allowed on new farm investments.
7. High priority would be given to provide funds to assist conservation practices and services.[73]

To provide an incentive for the formation of the districts, the proposal would replace the existing Farm and Open Space Tax Law. Any loss of tax revenues by a municipality as a result of the program would be partially reimbursed by the state.

Although the recommendations of the Food and Farmland Study Commission were the most comprehensive proposals in Maine in recent decades, the idea of agricultural districts to encourage farmland retention was not new to Maine. In 1972, following adoption of the first successful legislation to allow the establishment of agricultural districts in New York, the Maine State Planning Office had proposed the establishment of such districts. No legislative action was ever taken on that proposal.

In March 1979, the Maine State Planning Office again proposed the establishment of such districts in a study entitled "Agriculture in Maine: A Policy Report," prepared specifically for the Food and Farmland Study Commission as part of its study program. Many of the Commission's final recommendations were a synthesis of recommendations from the Planning Office, including the proposal for agricultural districts. In its report to the Commission, the State Planning Office discussed why the special districts were important as part of a statewide effort to protect farmland:

> Agricultural districts are proposed as a major element of a farmland preservation program for Maine because these mechanisms offer a maximum amount of flexibility to meet local and regional circumstances on a decentralized basis. It is expected that beyond their role in making preservation agreements these districts will serve as a focal point for monitoring and resolving of a great variety of land use and related agricultural economic issues in various regions of the state.[74]

The Food and Farmlands Study Commission made several additional proposals specifically tailored to the protection of agricultural lands, including:

> *Soil Erosion:* The State of Maine should encourage the USDA Agricultural Stabilization and Conservation Service to increase and allocate funds to those areas of the state with farms with the greatest agricultural erosion and water pollution problems first.
> *Federal Funding of Development:* The State should encourage the USDA Farmers Home Administration (FmHA) and other lending agencies to discontinue its policies of providing loans for development on highly productive agricultural lands.
> *Environmental Laws:* Existing federal, state, and local development-review processes should be strengthened to allow more consideration for farmland protection. Specifically, the Commission recommended that the Site Location Act and the Municipal Subdivision Review Act be amended to allow for consideration of farmland-protection objectives as part of the review process for major development permits.
> *Soil Fertility:* The University of Maine at Orono should expand research and extension activities in conservation. Priority should be given to developing viable rotation crops and a more comprehensive soil-audit program.[75]

As an integral part of a comprehensive approach to the problems of farmland protection and related farm economic issues, the Food and Farmlands Study Commission submitted recommendations in several other broad areas. As discussed in Chapter 2, much of the abandonment of active agricultural land in Maine, as with the rest of New England, has been the result of economic considerations outside the scope of local land use policies. Economic trends in recent decades have led to concentration and centralization of production inputs and marketing systems on a nationwide basis. These national economic trends often have had a negative impact on Maine agriculture, resulting in a decline of farming activity and conversion or abandonment of farmland. In 1980, the Maine Consortium for Food Self-Reliance discussed this problem as it related to Maine agriculture:

> Maine is tied into a national food system, since the vast majority of food consumed in Maine is delivered from far away farms and national and regional food businesses. This highly organized system is complex, and has replaced the earlier marketing structure which included many small local production units, several distributors, and manifold retailers.

The Consortium's report states that these and other production and marketing inputs are dominated by national agribusiness corporations. The dominance of these inputs makes local farmers dependent on market-pricing decisions of the agribusiness corporations, thereby "relinquishing Maine's ability to shape its own agriculture. Further, the money that would circulate through the local economy for the economic benefit of businesses and residents is drained by the national economy as profit for corporate stockholders."[77] This situation has influenced the growth of the average farm size and the displacement of small farms in Maine.

The comprehensive recommendations of the Food and Farmland Study Commission sought to address these economic viability problems of Maine's farm industry. The Commission suggested that the State of Maine develop a coordinated set of policies to meet the following farm economic needs:

> Marketing: market coordination, quality control and promotion, local market development, and advisory organization.
> Transportation: including railroads and intermodal transport.
> Energy: energy shortages and energy costs.
> Finance: availability especially to low-equity newcomers; farm finance/conservation considerations.
> Government: labor laws, health laws, and inflationary impacts on farming.
> Education: agricultural education programs.
> Entrance to Farming: marketing, and land availability.[78]

The comprehensive nature of the 1979 Commission report has had a significant impact on legislative and citizen awareness of the problems and solutions to the decline of Maine's agriculture industry. As a result of the recommendations of the Commission and the growing public awareness that followed, legislative focus has broadened. By 1985, new legislation and projects specifically designed to protect agricultural land had just begun, however. In fact, most recent legislative and statewide administrative activity has focused on marketing programs with a growing understanding that "few issues are as diverse, complex, and crucial to the prosperity and profitability of Maine's agricultural economy as marketing issues."[79]

Marketing Programs (1979, 1981)

Several marketing programs were adopted in Maine in an effort to preserve local agriculture. The Maine Agricultural Marketing and Bargaining Act of 1973 was designed to give individual farmers the advantage of joining together to compete in cooperative enterprises.[80] While many cooperatives formed, most observers believed that the state should become more directly involved in the marketing efforts of small farmers. In 1979 and 1981, a number of statutory changes were effected toward this end.

In response to a recommendation of the Commission that the state begin to assume more nonregulatory and promotional functions on behalf of agriculture, an act related to agricultural development was adopted in 1979.[81] Under this program, the Maine Department of Agriculture, Food, and Rural Resources was reorganized and its role was expanded beyond that of a regulatory agency. Following the lead of states such as Massachusetts, activities of the Department include an active "Buy Maine" program, the coordination of a statewide system of direct farmer-to-consumer markets, and increased promotion of rural development which includes agriculture as an important component of rural economic growth. The Department also has broadened its programs in support of communal and part-time agricultural activity. This has stimulated the growth of a community gardens movement and some 125 local food cooperatives in the state.[82]

Legislation adopted in 1979 also gave the Commissioner of the Department of Agriculture the power to enter into advertising agreements and disseminate information to increase native food consumption. A labeling law requires that only produce actually grown in Maine can be so designated. In 1981, the Maine Agricultural Commodities Marketing Act allowed the Commissioner of Agriculture to issue market orders "governing the marketing, distribution, sale or handling of 'agricultural commodities.'" Another statute allows the commissioner to appoint an agriculture Promotion Committee.[83] Additional labeling statutes were passed in 1981. Thus, Maine has focused on indirect farmland protection efforts in recent years.

Also in 1979, the legislature passed an Institutional Purchasers Act. Referred to as the "in-state producers bill," this law declares "the policy of the State to encourage food self-sufficiency"[84] and encourages state institutions to purchase more food from local suppliers. The program sets forth certain prices state institutions such as hospitals, universities, and state office cafeterias can pay for locally produced food. Moreover, the program requires that, where possible, institutions must purchase locally and if necessary at a slight premium over the market price of food produced outside the state. This premium is designed to decrease over a three-year period, and the program was scheduled to expire in 1983 unless extended by the legislature. A proposed extension did not pass, and the program was not continued.

The Food and Farmland Study Commission also recommended an increased role for the University of Maine at Orono in efforts to protect farmland and to stimulate the agricultural economy. Consequently, the Small and Part-time Advisory Committee of the College of Life Sciences and agriculture has expanded its role, focusing on new research, teaching, and outreach projects in small-farm mangement, appropriate technology, and indigenous soil amendments.

Despite all of this activity, there has been little progress on actions which the Maine State Planning Office and the Food and Farmland Study Commission identified as the most important of the farmland protection recommendations. The reasons for this lack of response are again related to the unique size and diversity of the state.

In 1985, however, the legislature adopted two significant bills. It first enacted a few of the recommendations of the Food and Farmland Study Commission: the state Department of Agriculture was restructured to expand its role, and amendments to the Finance Authority Act[85] under the Natural Resource Financing and Marketing Programs Bill gave the Finance Authority the power to secure and insure low-interest loans for active and new entrants into agriculture. The Finance Authority, advised by a seven-member Natural Resource Financing and Marketing Board, can contract with private financial institutions to secure the low-interest loans. The Authority also has the power to lend money for applicable projects if no financial institution chooses to participate. The Authority may also pay the interest on these loans over 5 percent for as long as the program participant is unable to pay. In return, the participant agrees to use specified soil conservation practices and other standards the Authority will adopt. In addition, the participant may not convey the land without the permission of the Authority and must also agree to use the land for the purpose specified in the application to the program. If these requirements are not followed, the landowner must immediately repay the balance of the loan.

In June 1985, the Maine legislature also adopted the Maine Agricultural Viability Act.[86] This Act established a seven-member committee (Maine Agricultural Viability Advisory Committee) to study and discuss agricultural viability on an ongoing basis. The Committee members are from various farm and conservation groups, and the Committee's purpose is to collect and assess information on agricultural resources and problems. The Committee will also hold public meetings to discuss the market supply, availability, and cost of production inputs; local demand for agricultural products; the land base used or suitable for agriculture, and the critical needs for agriculture's continued economic strength. The passage of the Maine Agricultural Viability Act resulted in the tabling of a bill on farmland protection which would have established a committee to actively study methods to protect farmland. Although the agricultural Land Protection Act passed both houses of the Maine legislature, it was not funded.

Because of Maine's diversity, consistent statewide programs are difficult to adopt. Unlike the five New England states to the south, the pressures to protect farmland vary dramatically from county to county and town to town. In some counties, such as Aroostook County, the physical and economic character is dominated by agriculture. The most serious problems facing Aroostook County, an area which is experiencing a population outmigration, are erosion problems. In contrast, in southern York County, the urbanization of farmland is the primary reason for farmland conversion. The erosion problems in Aroostook County do not really touch the population problems in York County. Thus, it is difficult to get a legislative consensus of the need for broad new programs such as the soil and water

conservation districts proposals. In fact, Dr. Edward Micka contends that the issue of farmland loss, particularly to urbanization and abandonment, already may be getting more attention on a statewide level than it merits. According to Micka, "the erosion problem must be improved. But from the standpoint of a lot of the rhetoric we are using regarding the loss of farmland for development purposes, I don't see it as a major statewide issue in Maine."[87]

It seems clear that in a state as large and diverse as Maine, farmland protection is most effective when undertaken locally. To that end, many of the recommendations of the Food and Farmland Study Commission and the State Planning Office were designed to encourage local and individual initiative to stimulate agriculture. Much of the recent effort in farmland protection is turning to an emphasis both on local programs and state support, with flexibility, to address local variations.

Right-to-Farm Law (1981)

To solve the problem of nuisance lawsuits state legislation was required. Newcomers to a rural area would often threaten or actually bring a nuisance lawsuit against a farmer for the effects of normal farm practices such as noise, dust, odors, and the like. Chemical spraying of blueberry patches and apple orchards often bothered newcomers. To halt the intervention of nonfarmers, the Maine legislature passed a Right-to-Farm Law in 1981.[88]

Of the New England states, Maine's version of the farm nuisance suit immunity law is the most favorable to farmers. The law states that farm operations are not a nuisance *either* if such activity existed before the interloper's use of land within a mile of the farm and the operation did not constitute a nuisance before the interloper's use, *or* if such activity currently adheres to generally accepted farm practices. Thus, unlike New Hampshire's Right-to-Farm Law, for example, the Maine statute does not require that the agricultural operation be both a prior *and* a reasonable use. This law reflects Maine's encouragement of new farm operations, expressed in the Food and Farmland Study Commission Report.

To determine which agricultural practices are generally acceptable, the statute requires the Commissioner of Agriculture, Food and Rural Resources to make such determinations on a case-by-case basis. The Commissioner issued a regulatory definition of "generally accepted agricultural practices" in December 1982, after holding two public hearings on the rule.[89] The regulation requires the farm activity to be in conformance with the "rules, regulations and published recommendations and guidelines" of various state, local, and federal agencies, land-grant universities, and farm industry organizations. Also, the farm activity must be essential to the farm operation or be in an historically conducted activity. This regulation is designed as a general framework through which the Commission will make a determination of whether a specific farm's practices fall within the immunity of the Right-to-Farm Law. However, by 1985 the Commissioners had not yet been called to make such a determination.

As in the other New England states, passage of Maine's Right-to-Farm Law sent a clear message to nonfarm rural newcomers that the state would support agriculture. Because of the liberal provisions of Maine's law, despite the somewhat technical regulations of the Commission, the vast majority of farmers could escape many court room confrontations based on nuisance.

Outside the state level, several important programs to protect agricultural land, many designed by local government and private and semiprivate organizations, show some promise of success. These programs range from local zoning for agricultural purposes and efforts in local jurisdictions to begin a transfer-of-development-rights program, to the work of private land trusts. Some of these efforts are quite recent. Others, such as the City of Auburn's exclusive agricultural zoning provisions, have been in existence for many years. The results of these programs to date have been mixed. The most notable examples are discussed below.

Zoning

Although Maine has adopted a Mandatory Shoreland Zoning Act, there has been little effort to adopt statewide zoning, as in Oregon and Hawaii, even though the Site Selection Act and the Land Use Regulation Commission vaguely resemble such programs. Becase of the strongly independent nature of local politics in Maine, it is very likely that any such statewide zoning efforts will continue to meet with stiff opposition. Therefore, as in the other New England states, Maine's efforts to use zoning to protect agriculture have been limited to enabling legislation which allows municipalities to adopt zoning regulations to protect farmland at their discretion. Moreover, while the Municipal Subdivision Act is mandatory, municipalities are not required to assess the impact on farmland in the subdivision review process.[90]

As previously mentioned, in 1979 the Maine State Planning Office and the Food and Farmland Study Commission recommended that the Municipal Subdivision Act be amended to allow towns to consider the impact of proposed new development on highly productive farmlands in their subdivision review and approval process. While no legislative action has been taken on this recommendation, several changes have been made in the comprehensive plan requirement of municipalities with zoning. These changes are significant because a locality subject to the Mandatory Shoreland Zoning Act must have a comprehensive plan.

Maine law requires that all zoning ordinances be consistent with a locally adopted comprehensive plan, a plan that accounts for all land uses in a municipality.[91] In 1979, the legislature amended the requirements of the comprehensive plan to require that all plans contain recommendations regarding the creation of land use control ordinances. The legislature also expressly provided in the new statute that zoning for open space and clustered development is permissible, although there is debate over whether these devices were allowed under the old state enabling act. In 1981, the statute was further amended to allow for local use of conditional and contract zoning, which imposes conditions or restrictions on rezoned lands. Maine state law, therefore, now encourages innovative zoning techniques which may be used to protect agricultural lands.

Previously, a 1978 Maine Supreme Court decision reaffirmed the broad powers of a municipality in Maine to zone for the encouragement of agricultural uses. In that case, the City of Auburn successfully sued for an injunction to prevent a hay storage barn from being put to a commercial use in that city's Agricultural and Resource Protection District.[92] Located on the Androscoggin River midway between Portland and Augusta, the city of Auburn is the only city in Maine which has established such an exclusive agricultural zone which prohibits development. When it was established in 1961, residences were beginning to scatter into the farmland areas. As a growth control measure, city planners therefore designated an exclusive agricultural and resource protection zone, which covers 80 percent of the city's land area and would allow only farm, forest, and open space uses.

Initially, opposition to the zone came primarily from farmers, who were not in favor of the ordinance at the time of passage. Since then, primary opposition has come from outsiders seeking to move into the area but not planning to farm. In any case, most observers, both locally and regionally, agree that the program has been modestly successful in protecting farmland. A 1979 study of the results of the program found:

> Generally, variances or exceptions are not permitted. The assessor reported that only about 300 of the approximately 21,000 acres in "farm and forest" had been converted to another use. There is no question, however, that development has been controlled. The land in the farm and forest zone has, since the passage of the ordinance been taxed by city officials at its restricted current use. The land that has been sold within the district has been sold to farmers. After passage of the ordinance, sale to someone who had merely put a few livestock on the land but did not really farm, pushed town officials in the mid 70's into (1) defining a farmer as someone who receives 50 percent or more of his income from farming or farms at least 10 acres; (2) requiring that residences cannot be occupied until farm buildings and improvements are substantially completed.[93]

Probably much of the success of Auburn's Farm and Forest Zone is because local assessments within the zone were based on current-use values long before the Farm and Open Space Tax Law was enacted. In any case, the foresight shown by local planners in Auburn several years ago has been applauded throughout the state. In 1977, the *Bangor Daily News* commented, "Auburn is one city which did not wait for a large government program to come along and preserve its farmland. The program works because both farmers and non-farmers alike contribute to the system."[94]

To date, Auburn is the best example in Maine of a specific zoning designation which has been used to preserve agriculture. As a result of recent innovations in Maine's zoning enabling legislation, a few local communities have considered transfer of development rights (TDR) as a method of farmland and open space protection. As mentioned in Chapter 3, TDR is a zoning concept which has presented an exciting possibility for the protection of farmland in local communities in Massachusetts and across the nation. To date, however, the nationwide results of TDR programs have been disappointing. Following the lead of Massachusetts and other states, in 1981 the Maine legislature revised zoning enabling legislation to allow local communities to include TDR as a provision of their local zoning bylaws.[95] Since that time, interest has grown in some Maine communities in the use of TDR for farmland protection. It would require little or no expenditure of public funds and does not limit a farmer's land use options without compensation, as exclusive agricultural zoning does. Due to the administrative and legal complexities, however, few communities actually have considered TDR programs until very recently. The town of Cape Elizabeth, just south of Portland, has pursued this possibility in depth. In a 1982 draft report entitled "A Land Use Policy for Cape Elizabeth," the Comprehensive Plan Committee recommended the development of a TDR program.[96]

Moreover, several southern Maine towns have been experimenting with density-open space proposals, linking TDR/PDR to permission to increase development densities on certain locations. The land held open in exchange for higher development elsewhere must be better agricultural land and will be protected permanently. There is some concern that these techniques, especially the negotiations with individual farmers to limit development to a less valuable (agriculturally) portion of their land, may encourage leapfrog development. This type of development would have other consequences detrimental to farming, such as right-to-farm conflicts; congested, high-volume highways; and high-speed traffic in competition with farm machinery and livestock crossings.[97]

Although there has not been a strong legislative action to protect farmland in Maine in recent years before 1985, it is clear that in certain areas of the state, strong local concern has resulted in local programs to protect farmland. Programs in cities and towns such as Auburn, Cape Elizabeth, and in southern Maine possibly have been the most effective so far in protecting farmland in Maine. In addition to these local public efforts, private actions at the local level are noteworthy.

Land Trusts

A few private land conservation and preservation trusts have become interested in farmland protection in Maine in recent years. One of thse is the Maine Community Land Trust, founded in 1972. By 1979, the Trust had acquired several parcels of farm and forest land, the first being a thirty-acre parcel of field and forest land in the rural community of Detroit, Maine. The family living on the land relies on outside income for necessities, although they have a garden, raise animals, and manage the woodland. The Maine Community Land Trust also has worked with other parcels involving several families, some relying on outside income, others setting up a number of home-based cottage industries, an orchard, and market gardens. Woodlots are managed for fuel and sale of wood products.[98]

Affiliated with the statewide Maine Community Land Trust is the Sam Ely Community Land Trust. The two groups together have worked for the establishment of local community land trusts in the state and provide information to farmers, communities, and concerned groups about innovative farming and rural programs. The latter Trust was established with several objectives, including the protection of farmland and cooperation with communities to promote rural development concerns which stimulate agriculture. The group has conducted surveys to determine the feasibility of "farmer-linkage" techniques, whereby older farmers are encouraged to sell their land to younger farmers rather than sell the land for development. Oftentimes the older farmer, seeking the best return possible for his land, is unable or unwilling to sell his farmland for a price that the prospective farmer can afford. The Sam Ely Trust attempts to bring the two together in a workable arrangement that ultimately will both preserve the farmland and assure a livelihood for active farmers. Using such techniques as purchase contracts, partnerships, or land trusts, it is possible for the seller and purchaser to develop a mutually beneficial plan. Experienced farmers may stay on the land, providing the young farmer with access to the land, or they may work together, benefiting from skills and idea-sharing.[99]

In addition to providing assistance in innovative farmer-linkage programs, the Sam Ely Trust has also been active in a range of rural and community service projects, including low-income housing design, foster-farm programs, and aid for direct acquisition or restriction of farmland for preservation purposes. However, the Trust's farmland protection activities have been limited. Only two major parcels, totalling approximately fifty to sixty acres of land, are under restriction. The Sam Ely Trust and its property holding arm, the Maine Community Land Trust, are now largely inactive, as most of their initial work stemmed from the effort of one individual who has gone "Down East" and left the Trusts.

Overall, there has been very limited activity by local land trusts in Maine. Unlike its five New England neighbors, Maine has very few private land conservation organizations. Among the eight or so land trusts that do exist, few are concerned with agricultural land, being concerned instead with coastal, historic, or scenic resources.[100]

Future of Agricultural Land Protection Programs

Maine is unique among its New England neighbors both in the concerns that have prompted legislative and private actions and in the actions that have been taken. There never has been a massive, highly visible farmland conversion problem in Maine, but there is a significant potential risk. Most of the row-crop production is in remote Aroostook County, where urbanization pressures are largely absent and the unique crops of potatoes and blueberries have kept most farmers, although often near poverty, willing to stay with the land. Erosion has been a significant cause of the decline in farmland acreage.

Various environmental concerns prompted the passage of development review legislation at the turn of the 1970s, as in Vermont. Other programs affecting farmlands also existed in the 1960s and 1970s, but only localized efforts have proved effective in farmland protection in recent years. The mid-1970s witnessed a new concern over food independence and the revitalization of the rural lifestyle. Indeed, in no state is the desire to encourage new farmers as pronounced as in Maine. This concern brought about a legislative mandate for a plethora of programs specifically designed to protect agricultural lands. This concern, as the Maine Consortium for Food Self-Reliance notes, should continue to result in the adoption of effective programs:

> In the long run, a supportive public policy for agricultural revitalization will hinge on the effectiveness of political education: convincing the public that the benefit of tax-supported programs for farmers is not just more, better, or cheaper food. It is also a more vital rural economy and community life, an economy that provides jobs and generates income and spending power, and communities with thriving schools, public services, churches, granges, newspapers, and the other institutions that make life rich without requiring material riches.[101]

In Maine, the programs which have worked and will work are those that are primarily local in nature. It seems apparent that in a state as diverse as Maine a statewide mandate can only exist for programs that can be adapted to local or regional variances. Thus, local zoning, state-assisted local marketing programs, and possibly agricultural districting will achieve the best results where localized support exists. Joyce Benson of the State Planning Office feels that protection can occur at the state level as well as the local level:

> There are state-level considerations (state policies to target its investment of public resources to prevent converting or endangering agricultural lands, for example) that need to be put in place. The state also has a role in creating the tools for local programs (enabling legislation). And the state can do much to encourage agricultural lands to be protected short of mandating (can establish the "climate" for protection by promoting agriculture, supporting local initiatives, etc.).[102]

As in other New England states, the proper mix of state and local farmland protection support has achieved profound results.

Notes

1. Peirce, *The New England States*, 363.
2. Ibid., 364.
3. Micka and Dearborn, *Where New England Counties Rank Agriculturally* (see chap. 2, note 17), 8.
4. A study by Joyce Benson and Paul Frederic, however, reports that one-third of the state's farmland is in Aroostook County, but that one-third is in the I-95 corridor and nearly all of the remaining one-third is adjacent to it. This places a large portion of Maine's best land—and land currently in production—under *urban* influence. This is higher than the national average. The study shows that 53 percent of the farmland lost in the last fifteen to twenty years is considered prime farmland and almost 20 percent was regarded as valuable for several major commodities. Joyce Benson and Paul Frederic, *A Study of Farmland Conversion in Nineteen Maine Communities* (Augusta: State Planning Office, August 1982): 5, 39-41.
5. "Land in Farms" (Grand Total), *1880 Census of Agriculture*.
6. "New England Workers Paid Less Than U.S. Average," *Boston Sunday Globe*, December 12, 1982, 53.
7. "Newcomers Fight Change," *Boston Sunday Globe*, December 27, 1980, 41, 49.
8. Joyce Benson, State Planning Office, to LILP, November 1985.
9. "Cropland Victim of Progress," *Bangor Daily News*, August 22, 1962.
10. Lapping, *The Land Base for Agriculture in New England*, 23. The study of nineteen communities, however (Benson and Frederic, *Study of Farmland Conversion*, 38) found that much open space still existed—it was just no longer in farms. Less than 10 percent of the open space had actually been lost. Benson and Frederic further suspected that a substantial share of road-front acres were scattered houselots, but this could not be substantiated by the air-photo technique. A draft follow-up study of land "parcellation" in two of the nineteen towns reveals the creation of substantial residential lots. A rough approximation would be that if the scattered houselots were added to the 10 percent loss, the actual loss would be doubled (Joyce Benson to LILP, November 1985).
11. Lapping, *The Land Base for Agriculture in New England*.

12. "Agriculture in Maine: A Policy Report" (Augusta: State Planning Office, March 1979): 112. Benson and Frederic also showed that the degree to which loss was the result of abandonment versus development over the last twenty years statewide was related to the degree of growth and development pressure; the northern and inland communities had higher rates of reforestation and the southern (I-95 corridor) towns had higher rates of conversion. The study further showed a period of idleness between cessation of farming and actual conversion. Benson and Frederic, *Study of Farmland Conversion*, 29, 45.

13. "Agriculture in Maine," 112; in development areas, however, conversion is a greater factor, but over a twenty-year period, reforestation accounts for slightly over half of the loss in growth towns and in urban areas versus 88 percent in the rural towns. Benson and Frederic, *Study of Farmland Conversion*, 29.

14. Edward S. Micka, comment, Lincoln Seminar 2.

15. Marvin W. Blumenstock, USDA Extension Forester, comment, October 27, 1981.

16. Peirce, *The New England States*, 366.

17. "Newcomers Fight Change," 41, 49.

18. Philip M. Savage, "Designing a State Land Use Program," *State Government* (Summer 1973) 157.

19. "Report on the Loss of Farmland" (Freeport: The Maine Consortium for Food Self-Reliance and the Maine Environmental Network, March 1979): 2.

20. "Report of the Maine Food and Farmland Study Commission" (Augusta: Department of Agriculture, June 1979): 17.

21. "The Maine Food System: A Time for Change" (Emmanus, Penn.: The Cornucopia Project, 1981): 2.

22. Carolyn Britt, Mike Rozzyne, Tom Roberts, and David Vaile, "The Past, the Present and the Future Competitiveness of Maine Agriculture" (Freeport: The Maine Consortium for Food Self-Reliance, December, 1980): 55.

23. Ibid., 55.

24. M.R.S.7 S.1-A.

25. M.R.S.7 S.2; H.P. 1541-L.D. 2028; recently, Maine has adopted the philosophy that a more *regional* approach to food self-sufficiency is important in addition to increasing food self-reliance. Joyce Benson to LILP, November 1985.

26. Edward S. Micka, comment, Lincoln Seminar 2.

27. Benson and Frederic, *Study of Farmland Conversion*; Joyce Benson to LILP, November 1985.

28. "Food and Farmland Study Commission," 3.

29. M.R.S.30 S.3851.

30. Duddleson, *Supplementary Report* to Scheffey, *Conservation Commissions in Massachusetts*, 191.

31. Savage, "Designing a State Land Use Program," 156.

32. M.R.S.38 S.482(2).

33. M.R.S.38 S.484.

34. "Agriculture in Maine," 117.

35. M.R.S.12 S.681.

36. Savage, "Designing a State Land Use Program," 156.

37. Ibid.

38. "Comprehensive Land Use Plan" (Land Use Regulation Commission, August 1976).

39. Richard Rothe, State Planning Office, comment, February 10, 1983.

40. M.R.S.12 S.4811.

41. M.R.S.30 S.4962; M.R.S.30 S.4956.

42. "Agriculture in Maine," 117.

43. "Comprehensive Land Use Plan."

44. Vance Dearborn, professor, University of Maine, comment, Lincoln Seminar 1.

45. M.R.S.36 S.1101.

46. M.R.S.36 S.1102(4).

47. M.R.S.36 S.1109(3).

48. Vance Dearborn, comment, Lincoln Seminar 1.

49. Property Tax Bulletin No. 18(x): 2.
50. "Report on the Loss of Farmland," 18.
51. Property Tax Bulletin No. 18, 5.
52. 1982 Municipal Valuation Return, Part IV (Augusta: Department of Taxation, March 27, 1983): 15.
53. Vance Dearborn, comment, Lincoln Seminar 1.
54. "Report on the Loss of Farmland," 3.
55. Vance Dearborn, comment, Lincoln Seminar 1.
56. M.R.S.36 S.571 et seq.
57. Vance Dearborn, comment, Lincoln Seminar 1.
58. Ibid.
59. Ibid.
60. M.R.S.36 S.579.
61. Joyce Benson, "Impact of Land Related Components on Farm Profitability for Livestock Operations" (Augusta: State Planning Office, March 1984): 3–5.
62. Richard Rothe, "The Farm and Open Space Tax Law, A Brief Analysis" (Augusta: State Planning Office, May 1978).
63. Vance Dearborn, comment, Lincoln Seminar 1.
64. M.R.S.33 S.667–68.
65. M.R.S.33 S.1111.
66. Vance Dearborn, comment, Lincoln Seminar 1.
67. "Maine Food and Farmland Study Commission," 1.
68. Chapter 65, 89, Private and Special Laws of 1977.
69. "Maine Food and Farmland Study Commission," 1.
70. Ibid.
71. Ibid., 19.
72. Ibid.
73. Ibid.
74. "Agriculture in Maine," 116.
75. "Maine Food and Farmland Study Commission," 4.
76. Britt et al., *The Past, the Present, and the Future Competitiveness of Maine Agriculture*, 35.
77. Ibid., 11.
78. "Maine Food and Farmland Study Commission," 5–12.
79. Ibid., 5.
80. "Agriculture in Maine," 117.
81. M.R.S.5 S.711; M.R.S.7 S.1 et seq.; M.R.S.7 S.956; M.R.S.7 S.958; M.R.S.7 S.2301; M.R.S.10 S.703; M.R.S.32 S.4152; M.R.S.32 S.4156; M.R.S.36 S.4693.
82. Britt et al., *The Past, the Present, and the Future Competitiveness of Maine Agriculture*, 69.
83. M.R.S.7 S.402; M.R.S.7 S.443-A; M.R.S.7 S.402-A; M.4.S.7 S.421 et seq.; M.R.S.7 S.402-A.
84. M.R.S.7 S.201.
85. M.R.S.7 S.105.956 et seq.
86. M.R.S.7 S.311 et seq.
87. Edward S. Micka, comment, Lincoln Seminar 2.
88. M.R.S.17 S.2805.
89. Maine Department of Agriculture, Food, and Rural Resources, Office of the Commissioner, Chapter 10, Definition of Generally Accepted Agricultural Practices.
90. M.R.S.30 S.1917; M.R.S.30 S.4956(3). Localities must, however, determine that a subdivision granted approval "[w]ill not have an undue adverse effect on the scenic or natural beauty of the area, aesthetics, historic sites or rare and irreplaceable natural areas."
91. M.R.S.30 S.4961–62.
92. *Giguere* v. *Inhabitants of the City of Auburn*, 398 A.2d 514 (1978). In a general decision, the Supreme Judicial Court of Maine further elaborated on the power of municipalities to create agricultural zones. The Court held that the Town of Gorham could prevent a mobile home from being located outside of a mobile home park in a Farm and Rural Residential District. *Warren* v. *Municipal Officers of the Town of Gorham*, 431 A.2d 624 (1981). The Court

established that aesthetic concerns are relevant in furthering the public welfare and that such concerns, therefore, can be used to help validate the legality of an agricultural type of zone.

93. "Report on the Loss of Farmland," 20.
94. *Bangor Daily News* (1977), reprinted in "Report on the Loss of Farmland," 23.
95. M.R.S.30 S.4962.
96. "Report on the Loss of Farmland," 20.
97. Joyce Benson to LILP, November 1985.
98. "Report on the Loss of Farmland," 22.
99. Ibid., 21.
100. *1981 National Directory of Local Land Conservation Organizations* (Land Trust Exchange, 1981): 40–41.
101. Britt et al., *The Past, the Present and the Future Competitiveness of Maine Agriculture*, 69.
102. Joyce Benson to LILP, November 1985.

9

Summary and Overview of State Experience

The foregoing discussion of the six New England states reveals wide-ranging differences in the concerns and mechanisms to protect agricultural lands. Nevertheless, several common threads can be traced. Farmland protection concerns can be categorized generally as aesthetic, environmental, food-based, and economic. Farmland protection programs exist in the form of regulatory controls, economic incentives, and direct state and private assistance. These concerns and programs comprise the fabric of the New England farmland protection issue.

Concern for Farmland Protection in New England

The motivations behind the adoption and implementation of New England's agricultural land protection programs are derived from four interrelated concerns. These concerns are based on the aesthetic value of the landscape and rustic way of life, on the quality of the environment, on the quality and quantity of food products, and on the various economic benefits of agriculture. The concerns are not distinct but subtly interwoven. Each one has proved important in all the states, although at different times and to varying degrees.

The specific concerns evolved largely in response to the more general concern over the decline of acreage devoted to active agriculture. This decline for the entire region, described in Chapter 2 and the state studies, has been marked. When citizens and government officials in each of the states became aware of the scale of conversion and abandonment of agricultural lands, specific policy goals or reasons to protect farmland were formulated. Indeed, the level of decline in active farmland acreage stands in roughly a direct relation to the level of specific concern raised and the private and public action which followed.

Aesthetic Concerns

State-by-state review of New England farmland protection efforts reveals that in every state one of the most important concerns was the desire to preserve certain aesthetic qualities which agricultural lands provide. The determination to protect open space and local community character evolved primarily from intangible motivations such as the value of farming as a lifestyle that is pleasing to the eye. The traditional Yankee farm, with its small fields surrounded by stone walls, woodlands, and rural architecture, has given the landscape a unique visual character that a majority of New Englanders, both urban and rural, want to protect.

Most of the early New England programs designed to protect various undeveloped land uses reveal this concern for aesthetics. Concerns for the quality of the landscape brought together environmentalists, farmers, and citizens of seemingly varied interests to work for legislative and private initiatives that would protect open space lands generally and farmland in particular. This concern remains most prevalent in the three southern New England states, where urbanization has been more significant in the conversion of environmental land than in the northern New England states.

Massachusetts, Connecticut, and Rhode Island all have major cities that have witnessed the virtual disappearance of farmland within and near their municipal borders. The early efforts of Massachusetts's citizen groups, which helped form the Trustees of Public Reservations, for example, were a response to the changing landsape. Today, Bay State voters regularly support farmland programs that affect lands far removed from the bulk of the population, largely because of a desire to protect rural areas from urban encroachment.

Rhode Island has few traditional farms left, although it has not become the city-state feared by some. The decline in farmland use generated concern among the citizens because they feared the possible disappearance of a traditional lifestyle and of declining open space. In Connecticut, the desire for a mixture of land uses to break up the interstate highway belt of urbanization, running from Hartford to New London to New York City, also has been based on a visual concern. This concern has played an important role in the adoption of numerous farmland protection programs.

Generally, the three northern states' concern for aesthetics has been focused more on the effect development of farmlands has had on their way of life and on tourism than on their desire for more open space. Vermont's farmland protection efforts have evolved largely in response to the perceived change in rural lifestyle which newcomers brought. Maine's legislature has moved to revitalize rural communities containing farmland.

In the late 1970s, this lifestyle concern resulted in a plethora of new state programs in both Vermont and Maine. New Hampshire's farmland policies ostensibly have been influenced less by purely aesthetic concerns given the conservative political climate of the state, although many environmental groups have worked hard for land-based programs for aesthetic reasons. In addition, Vermont and its northern neighbors have been influenced by the increased tourism revenues that a rustic, farm landscape will bring.

Environmental Concerns

Unlike the more abstract and subjective sentiments about aesthetics, farmland protection advocates in New England have sometimes marshalled together hard facts about the need for open space to curb pollution and general environmental degradation. This concern has been most prevalent in Connecticut and Rhode Island, where farmland is considered vital for ground water and air recharge areas. Vermont's Act 250, though modest in its impact on farmland protection, was adopted largely as a growth management response to sewerage and environmental problems.

Although concern over environmental quality frequently was cited as a reason to support farmland protection measures, it clearly has been a secondary issue in most of New England. Most of the environmental benefits that derive from agriculture also are provided by other open space uses, while modern farming practices often create difficult environmental problems, particularly water pollution from chemical pesticide and fertilizer use. Thus, environmental quality generally has not been a compelling reason for adopting farmland protection programs.

Food Concerns

Following initial efforts to protect New England farmlands for aesthetic and environmental quality reasons, a broad consensus spread among the states that agricultural lands should be protected to maintain or increase food production and aim at regional self-sufficiency. For a variety of reasons, the New England states are not likely to become totally self-sufficient in food; the region produces at best less than 20 percent of its food requirements. But there is now a strong demand for fresh, locally produced food. Farmland protection advocates therefore often have used this as a reason to protect the better agricultural land.

The concern over food dependence initially arose from the fear of vulnerability. The mid-1970s oil embargo made New England citizens uncomfortable with their location at the end of the national transportation and food supply line. A truckers' strike, a prolonged western drought, a blizzard, or another calamity conceivably could exhaust the rather meager stock of warehoused food in New England. Many New England citizens have come to believe that locally produced food is healthier and less expensive, that active farms are good for the economy, and that greater local production would allow for greater food exports, thereby reducing world hunger and the nation's balance of trade deficit.

The level of response to the various reasons for fostering local food production has varied from state to state. In general, however, increased local food production is a growing motivation for farmland protection programs. New Hampshire's farmland protection policies have been most influenced by this concern. The Food Policy Resolution of 1979 clearly expresses the public's desire for farmland protection programs that will increase the production of native, high-quality food products. Given the state's conservative political climate, which makes land use regulations unpopular, it is not surprising that farmland protection groups succeeded more by arguing that farmland is a food production resource than by raising aesthetic or environmental issues.

Other states have also marshalled the desire for food self-sufficiency into a farmland protection issue, perhaps to a lesser extent. State-funded purchase of development rights programs, which are established in four of the states, all consider soil productivity ratings, existing yields, and other measures of food production potential to be highly relevant in the purchase approval process. Moreover, all of the states have numerous marketing and economic incentive programs to stimulate local food production.

An issue closely related to food productivity is the problem of soil erosion and depletion. Although the problem generally is considered to be less severe in New England than in the midwestern farm belt, many tons of high-quality soils are lost each year to erosion in New England. The concern over erosion in Aroostook and Androscoggin counties in Maine has been so great that several programs have been initiated to solve the problem. Indeed, the decline of farmland use caused by erosion in Maine is a far greater concern than urban conversion and abandonment, which helps explain the adoption of relatively few farmland protection programs.

The other New England states have acted to combat erosion as well, generally in conjunction with the USDA Soil Conservation Service. The erosion problem is difficult to solve, and yet the programs proposed and adopted, such as Rhode Island's Soil Erosion and Sediment Control Act, have not encountered the roadblocks of many farmland protection measures because land use is not at issue. There is little disagreement about the need to maintain high-quality soils in farm areas, as opposed to the issue of regulating land use.

In sum, the citizens of all six states are concerned over food availability. Greater food independence is desired for reasons ranging from the fear of local food shortages to the desire to aid those people starving abroad. While not generally as important as aesthetics, food concerns have been essential in establishing a broad enough consensus to act toward protecting agricultural lands generally.

Economic Concerns

Related to the concern over aesthetics, environmental quality, and food production is a concern for the role which farming plays in the New England economy. Various costs and savings are involved in either protecting or not protecting agricultural lands, some of which are difficult to measure. Nevertheless, the bottom line has been a public perception that protecting such lands is economically worthwhile.

Several economic issues have been identified. One such issue is the rise in land values and taxes which comes from real estate speculation and development. Agricultural land protection thus keeps one aspect of the cost of living stable. Although not specific to farmland, Vermont's Act 250 and capital-gains land tax were designed to relieve such open space land conversion pressures. Another issue is tourism, which is significant to the economy of all of the states and which depends partly on the landscape. In New Hampshire, for example, tourism is the major industry. Effective farmland protection programs can help encourage more visitors.

The programs adopted in the six states also help keep the farm industry healthy. Jobs and income stay in the region, both in the farm community and in the support and supply centers that profit from farm activity. In some cases, farmland protection measures have been adopted because greater food self-sufficiency increases economic independence.

Finally, most New Englanders have strong convictions that a scenic landscape makes the region a more desirable place to live. This quality of life induces labor and industry to stay in or relocate to the region. Much of the high-technology building boom in southern New Hampshire can be attributed to the desire to live and work in a more rustic environment. Attracting industrial plants that bring jobs and income has become a persuasive economic justification for farmland protection.

Farmland protection measures also entail costs. They can drive up land values by limiting the amount of land that can be developed. In Connecticut, for example, there was substantial public debate over the protection of farmland at the urban fringe because of the impact this could have on land supply for development. Farmland protection opponents also have maintained that preventing the conversion of farmlands to different economic uses is harmful because new economic uses could generate more jobs and would produce more tax revenue. Despite these concerns, there has been no proof in any of the states that farmland protection programs have drastically increased the cost of developable land or crippled nonagricultural economic growth. The greatest cost has been in direct public outlays and in revenue bonds with PDR programs. It is not surprising, therefore, that only the more populated and wealthy states of Massachusetts and Connecticut have been able to fund their expensive PDR programs.

On balance, the states have concluded that the economic benefits to be realized from keeping agricultural lands productive far outweigh the costs. Yet because some program costs have been high, lower-cost alternatives to PDR programs and to other direct assistance programs to protect farmland have been pursued out of necessity.

Farmland Protection Programs in New England

A variety of programs have been implemented to combat the decline in farmland acreage. While these programs often have been similar, their results have varied from state to state. Taken as a whole, however, the many local and statewide farmland protection programs make New England one of the most active farmland protection regions in the country.

The range of programs in place or under consideration in New England fall into three general categories. First, there are regulatory controls which state and local governments exercise as part of their police power to determine the kinds of land uses that are allowed in specific areas. Second, economic incentives are used which encourage farmland protection by seeking to make farmland economically viable. Finally, there are direct assistance programs which enable government and private groups to become directly involved in farmland protection.

Table 9.1 summarizes the range of the major ongoing New England programs protecting farmlands. It should be noted that some of the initiatives discussed in the state studies are not listed in Table 9.1. For example, task forces directed to study farmland decline and recommend protection programs are not listed. This absence is because the programs listed only include ongoing mechanisms to protect agricultural lands. Generally policy statements, such as the New Hampshire Food Policy Resolution, and investigative reports, such as those prepared by growth management and agricultural task forces, are also not included because these are really background efforts prepared to support the new legislative programs. Although these efforts help gain support for farmland protection, they do not on their own directly protect farmland.

Table 9.1
Major New England Farmland Protection Programs

	Mass.	Vt.	Conn.	R.I.	N.H.	Maine
Regulatory Controls						
Zoning (modern)	1975	—	—	P	1981(?)	1979 1981
Agricultural districts	n/a	?	n/a	n/a	n/a	n/a
Permits	n/a	1970	(1982)	n/a	n/a	1969 1971
Right-to-farm	1979	1981	1981	1982	1981	1981
Economic Incentives						
Current-use tax	1973	1978	1963	1968	1973	1971
Contract stabilization tax and capital gains taxes	n/a	1967 1983 1974 1977	n/a	n/a	—	n/a
Institutional marketing	—	1981	n/a	n/a	—	—
Direct Assistance						
State PDR	1977	1969 1983	1978	1981	1979	n/a
Local PDR	—	1969	—	—	—	1971
Green Acres	—	?	—	1964	?	?
Private trusts	—	—	—	—	—	—
Executive order/ bond review	1981	1980	1982	P	n/a	n/a
Conservation Commissions	1957	1977	1961	1960	1963	1965

Notes:
P = under active consideration
— = exists
n/a = nonexistent and not under active consideration
? = unknown

Regulatory Controls

Regulatory controls are becoming increasingly popular nationwide as a mechanism to protect agricultural land because they generally require a limited financial commitment from the public. In New England, a number of land use regulations have been instituted or proposed to protect farmlands, including local agricultural zoning, regional agricultural districts, statewide development-permitting processes, and right-to-farm laws. Legislative task forces and other advisory bodies in each state consistently have recommended the implementation of these types of programs, and the courts generally have upheld them as constitutionally valid.

Although such regulatory devices have been attractive to the general public because of their low cost, many are opposed by farmland owners.

With the precarious economic conditions of farming in New England, farmland owners generally have been unwilling to support regulations that could reduce the market value of their land and thus reduce without compensation the value of what is usually their only real asset. Nonfarm landowners, including development interests, similarly have had the same concerns about the future value of their land. Farmers, therefore, sometimes find themselves in an uncomfortable alliance with development interests to stop new land use regulation, even though they may not actually wish to see the land developed.

Many farmland owners realize, however, that not all regulation is contradictory to their desire to continue in agriculture. Yet the fear of losing land value is the major reason farmland owners oppose exclusive agricultural zoning and many other land use controls. This opposition has made such programs difficult to adopt and relatively ineffective when implemented as a means of protecting agricultural land.

At the local level, few exclusive agricultural zones have been established in New England. Auburn, Maine is a notable exception. Many towns, such as Litchfield, New Hampshire, have sought to protect agriculture through large-lot zoning, while others have used industrial or industrial-residential zoning to promote alternative land uses.

Not all regulatory controls have been opposed by farmland owners. The most progress in implementing local controls has occurred in those states that have regulated farmland by encouraging innovative zoning and comprehensive town master planning. Cluster zoning, incentive zoning, and performance zoning, based on carrying-capacity analysis and the transfer of development rights, are regulatory devices which allow partial or controlled development of agricultural land and have been used with some success. This type of zoning allows farmland owners to realize partial development value of their land, and significant farmland acreage has been protected as a result. In these controlled development situations, farmland owners sometimes are willing to prepare a site development plan which actually results in less than maximum development value. Following the lead of Massachusetts, which adopted legislation expressly allowing use of many such innovative zoning techniques in 1975, every state in New England except Rhode Island has modernized its zoning enabling legislation. Use of these innovations are thus far limited, however. Lincoln, Massachusetts; Windsor, Connecticut; and South Kingston, Rhode Island are the best examples of towns whose innovative zoning has resulted in the protection of agricultural land.

Accompanying the change in zoning, town master planning which includes agricultural uses has been encouraged by state agencies in all of the states, especially New Hampshire and Vermont. Generally, this planning is aided by the state planning office, with the exception of Massachusetts which does not have such an agency. Maine's Shoreland Zoning Law requires such planning in some towns. A number of New England towns are better able to balance competing land uses through planning, and this trend is growing.

A regulatory program which addresses the issue of farmland protection on the regional level is the concept of agricultural districts. This scheme would designate certain areas as farm districts and give affected landowners economic and other benefits in exchange for farm use restrictions. Legislation to allow the creation of such districts has been promoted by farmland protection advocates in all of the New England states. There has not been strong support in New England for the districting concept, however. Farmland owners generally have opposed such districting legislation because it appears akin to traditional zoning. These landowners have stressed the need for farmland protection laws to be voluntary. Other analysts point out that the package of programs that participants would receive under the umbrella program are already largely in place. Also, in many areas of the New England states, and in Rhode Island and New Hampshire in particular, few contiguous parcels of farmland remain to create meaningful districts.

Finally, a variety of state regulatory programs has been enacted. Some of these programs, however, actually have stimulated the abandonment and conversion of farmland. The permit process of Vermont's Act 250 encouraged many farmers to sell their land for development before the law's effective date in 1971. More importantly, the ten-acre loophole, which was not closed until 1984 and was designed to relieve individual homeowners from the burden of obtaining permits, has encouraged conversion of farmland to ten-acre subdivisions. Act 250's problems in definition, which have prevented it from affording the degree of protection hoped for by farmland protection advocates, is symptomatic of the problem haunting Maine's statewide development permit laws—and possibly Connecticut's Upper Connecticut River Conservation Zone Commission as well. These review and permit programs were not targeted principally at the farmland decline issue when adopted. Rather, these laws were designed to assess the overall environmental impact of development on various sensitive land areas. As a result, these programs have had only a secondary impact on farmland protection. Dependence upon such programs by farmland protection advocates resulted in further declines in active farmland acreage and the relatively late adoption of more direct programs.

Although not a restrictive regulation on farmlands, right-to-farm laws affect land use by permitting farmers to conduct farming activities that may be bothersome to their nonfarm neighbors. State legislatures in each of the New England states passed one of these statutes in 1979, 1981, or 1982. Heralded as a farmer's victory against newcomers, the statutes have encountered little opposition. As a practical matter, however, their effect may be limited. Massachusetts's law, for example, only protects a farmer against nuisance suits from odors caused by livestock or spreading manure. Connecticut's and Rhode Island's statutes similarly are limited to a few specific farm activities. Moreover, the Connecticut and New Hampshire laws only protect ongoing farm activities that have been in existence for at least a year, thereby not helping a new farmer or an old farmer with a new farm practice. Vermont and Maine have the broadest right-to-farm protections, limited primarily where public health is adversely affected. These programs, therefore, may not protect against the need to change farm operations or to reactivate agricultural uses of land which has been out of production for more than a year. However, the courts have thus far broadly construed the right-to-farm laws to protect the farmers' activities further in cases in Connecticut and Rhode Island.

Despite their low public cost, regulatory controls have not been adopted widely in New England for political reasons and therefore have not protected much farmland. Regulatory controls which are part of carefully coordinated local zoning and planning efforts have had good success in retaining active agriculture in some areas, and their use is growing. Regional agricultural districts have not been implemented in any state. Regional and state programs directed at controlling changes in land uses have not had the kind of power needed to protect farmland parcels threatened by development. While regulatory devices remain available and modern planning and zoning tools have proven valuable, farmland protection advocates have found other types of programs easier to implement and more effective.

Economic Incentives

A traditional means of supporting continuing agricultural activity is the use of economic incentives. These programs provide a financial inducement to farmland owners to start or continue farming operations, thereby indirectly protecting agricultural lands. The basic premise behind these programs is to correct imperfections in land and commodity markets, as well as to reduce the cost of production. For a complete discussion of how land and commodities market imperfections affect New England agriculture, see Chapter 2.

On the federal level, dairy price supports, Farmers Home Admistration (FmHA) loans, and tax provisions are significant economic incentive programs which encourage farming in New England. Other federal programs are designed to help only larger farm operations than are found in New England, and, therefore, have had an adverse impact on the region. The six states themselves have enacted various tax measures, management and marketing assistance programs, and other devices. These programs often have been less popular to citizens than regulatory land use controls because of their cost. On the other hand, they usually are viewed as essential by farmers and farm industry supporters because they provide important assistance to a somewhat marginal industry. Although it is very difficult to measure the success of these programs in farmland protection efforts because of their indirect nature, most observers in New England conclude that they have had limited success in protecting agricultural land.

The most significant economic incentive programs fostering farmland protection in New England have attempted to reduce land cost variables. Unlike most industries in the region, land is the agricultural industry's major production cost. New England's high land costs, which usually result from competition for nonfarm uses, discourage new farmers from buying land and encourage older farmers to sell theirs. Moreover, as the experience of the six states reveals, high property taxes have been a severe problem, especially in southern New England because the limited profitability of most farms produces little cash to pay taxes.

The most widely adopted land cost reduction program is differential property tax assessment, usually referred to as current-use taxation. Such programs decrease the farmer's property tax burden by tying the assessed value of the land to its current agricultural use rather than to its speculative market value. Although generally available to owners of many types of open space land, the program was the first specific farmland protection device to be adopted in each state. All of the New England states have adopted the scheme—Connecticut was the first in 1963 and Vermont was the last in 1978. In New Hampshire, Massachusetts, and Maine, constitutional amendments were needed for the program because these states' constitutions required equal tax rates for all types of property owners.

All of the states' current-use taxation laws have a minimum acreage and gross income or sales requirement, except for Connecticut's, which allows substantial local discretion in determining eligibility. The type of criteria used in valuation procedures differ from state to state. All states but Maine provide valuation guidance to local assessors. The three southern states determine value on the actual commodity produced, while the three northern states consider the type of farmland use. There has been a general trend in the states, however, to allow for less local discretion and more statewide uniformity in administration. The penalty provisions for changing the land's use from agriculture also vary, but are all nondiscretionary—Maine's are the most restrictive and Connecticut's the least restrictive. Yet, as the experiences of each state illustrate, no penalty provision has proven an effective deterrent to conversion of farmland to other uses.

The success of these differential assessment programs has been modest for many reasons. Probably the most important reason is farmland owners' reluctance to enroll, either because they do not wish to restrict their options for the future or because they are fearful that after reassessment their tax burden will not substantially decrease. In Maine, and in other areas of New England where development pressures are weak in the major farmland areas, enrollment has been minimal. New Hampshire and Rhode Island have suffered from low enrollment because of local assessment practices which often overvalue many parcels coming under the program. Connecticut and Massachusetts have had the greatest enrollment in their programs, largely as a result of their densely urbanized character and corresponding high land values. Most applications in these two states have come from landowners in growing or suburban areas where property has been reappraised at higher market-related values. Another problem is that municipalities hesitate to support the program because of the loss of revenue from participating property and the shift of tax burden to other properties. This can be overcome by state reimbursement, but to date only Vermont has established a reimbursement fund to encourage municipal participation.

Two other important economic incentive programs which seek to protect agricultural land by focusing on land costs exist in New England. One of these is a generic type of law. Vermont's tax stabilization contract scheme and New Hampshire's discretionary easement program are examples of this type. Both allow municipalities to bargain with a farmer over his property tax bill in exchange for farmland retention. While the Vermont program has been used widely, the New Hampshire program has not. This disparity in use is because of the relatively late enactment of a current-use assessment law in Vermont and the limited applicability of New Hampshire's discretionary easement law.

A second type of program is Vermont's capital-gains transfer tax on land sales. This program also has had a modestly positive effect on farmland retention by reducing land speculation incentives.

Aside from property taxes, all of the states have designed other economic incentive programs to help farmers stay in farming and thereby protect agricultural land by reducing nonland costs and easing cash flow problems. Sales, use, estate, gift, gas, and other tax exemptions and rebates are an indirect means of helping farmers continue to work the land. Several New England states have also considered "agri-bond" type loans and direct farm financing.

Economic assistance programs aiding farmers with management and price information is the traditional activity of state departments of agriculture. In recent years, the New England states have adopted state logos and published promotional literature to encourage purchase of native produce. In addition, Massachusetts, Maine, New Hampshire, and Vermont have enacted institutional marketing laws, which encourage or require state institutions to purchase foods grown in the state. Legislation allowing formation of producer cooperatives is also common to the states. All of these programs provide a direct boost to farming which, in turn, provides an indirect boost to farmland protection.

Economic incentives exist in various forms to help New England farms stay profitable. The problem of high land costs of production has been addressed primarily through differential property taxation programs which reduce taxes. Other incentives generally are directed toward lowering other costs and maximizing marketing returns. All of these programs have helped protect farmland by keeping the industry healthy. Nevertheless, because they often have not boosted farmer and farmland owner income to a level competitive with returns from other occupations and land uses, other programs have been needed to protect active agricultural lands.

Direct Assistance

As the experience of the six states reveals, regulatory controls and economic incentives have had limited success in protecting agricultural land, although they play an important part in the range of adopted programs. In response, the executive and legislative branches of state government, together with nonfarm private groups, have taken a direct and active role in farmland retention through direct assistance programs. These efforts take the form of PDR programs, land acquisition programs, Conservation Commission activity, review of state-aided development projects, and private land trust activity. Because these programs are voluntary and give a fairly direct boost to farmland retention, they are the most popular and have achieved the most marked results. Generally, they have also been the most expensive.

All of the states have adopted some form of enabling legislation allowing PDR at the state or local level. These programs sever the bundle of ownership rights that a farmland owner has by allowing state or local goverments or a private group to purchase, or accept the donation of, the rights to develop the land. When such a sale occurs, the farmland owner retains only the right to farm or to leave his land undeveloped. Property tax benefits inure to the farmland owner because the value of his property has been limited by the development restriction. All of the states allow state and local purchases and acceptance of donations of development rights, except Maine, which only allows local purchases.

Massachusetts, Connecticut, New Hampshire, and Rhode Island have appropriated funds at the state level for their PDR programs. However, Massachusetts is the only New England state to have provided continual financial support of its program since its inception (1977). By December 1984, a total of $45 million had been appropriated for this purpose, more than the other five New England states combined.

All these states consider the critical determinations in evaluating applications to be the risk of conversion to a nonagricultural use (jeopardy), soil productivity, economic viability of the farm unit, and cost to the state of purchase. Rhode Island's and New Hampshire's programs are the only ones that give an advantage to applicants who farm the land themselves.

As the New England experience reveals, state-level PDR programs generally have entailed more administrative costs than other programs, but they have proven to be a very effective device for protecting farmland from conversion to other land uses. By mid-1985, a total of almost $75 million had been appropriated by the four states with PDR programs, with about $36 million expended so far to restrict approximately 20,000 acres. On average, the development rights have accounted for over 75 percent of a restricted parcel's value. The parcels purchased generally have been devoted to dairy operations and are well dispersed throughout New England, although Connecticut and Massachusetts officials have been concerned over the relative lack of Connecticut River Valley applicants. Also in Massachusetts and Connecticut, PDR programs have been devoted to protecting blocks of farms rather than isolated parcels, thereby solidifying the critical mass needed to sustain economic viability in less profitable farm areas.

Because of limited state funding, local and private expenditures and gifts for development rights have been a big boost to PDR programs. On Block Island in Rhode Island and in Vernon, Vermont, for example, local PDR programs have protected much in-town farmland. In addition, by mid-1983 the Ottaquechee Land Trust in Vermont had assisted in the protection of more than 5,000 acres of farmland, much of it through state and local PDR and easement programs. Some mixture of state, local, and private funding and administration, therefore, is one of the best means of utilizing PDR programs effectively. It also assures local initiatives and control.

The concept of PDR is only slightly different from purchase of the entire fee (full ownership). Although more expensive, some towns such as Farmington, Connecticut have implemented programs to purchase farmland and lease it back. These programs are similar to the Green Acres and Parks programs of the 1960s, adopted in all the New England states, whereby the state bought land for public conservation and recreational purposes. Because of concerns over cost and the propriety of state ownership of land, however, few farms have been acquired outright under direct purchase programs.

Also effective on the local level has been the work of Conservation Commissions. First established in Massachusetts in 1957, Conservation Commissions are now allowed in all the New England states, with Vermont in 1977 being the last state to adopt such enabling legislation. These commissions traditionally have served an advisory and educational role. By 1984, however, commissions in every New England state were active participants in state PDR, local PDR, and property tax programs.

On the state executive level, the governors of Massachusetts and Vermont have issued directives to state agencies to avoid the conversion of state-owned farmland whenever possible. The Vermont order is the more significant one, as it requires state agencies to follow a detailed planning process designed to ensure farmland protection. Similar to executive orders is Connecticut's bond-review program, which allows the Commissioner of Agriculture to veto state-financed development projects having a substantial adverse impact on prime farmland. These three policy directives are significant efforts that serve educational functions and work as pragmatic programs affecting numerous farmland parcels. The Vermont and Connecticut approaches are unique in their ability to affect the decision-making process of other state agencies directly whose proposed actions would affect farmland.

A discussion of farmland protection in New England would not be complete without summarizing the contribution of private groups. The direct assistance of private land trusts, for example, has been vital to farmland protection efforts in all the states. The Massachusetts Farm and Conservation Land Trust, the Ottaquechee Land Trust in Vermont, and the New Hampshire Charitable Fund are good examples of the kind of activity that is prevalent in all of the states. These groups have labored to buy farmlands under acute conversion pressure or in critical resource areas when the state could not act because of time-consuming administrative constraints or lack of funds. Private groups have also aided local governments in their planning processes. These groups have had as measurable an impact on farmland retention as any public program.

In sum, direct assistance by government and private groups has proved to be the best assurance of farmland protection in New England. By being largely voluntary and compensatory, farmland owners have provided their own needed political support for these programs. The PDR programs of the six states have been the most effective in land protection, even if limited by cost. Land purchases, land gifts, and the review of development projects involving state assistance also have had a measurable impact on farmland protection. Of vital importance has been the work of private citizens through Conservation Commissions and other local bodies and private land trusts. Indeed, without the immediate involvement of so many public servants and other interested citizens, the efforts embodied in regulatory controls and economic incentives would have proven too weak to have had a truly measurable impact on the retention of active agriculture.

Despite their effectiveness, direct assistance programs do not guarantee that the land will remain in agriculture production. It is too early to know whether PDR and easement programs will foster agricultural use or, because of other economic pressures on farming, merely preserve open space.

Summary

The four specific concerns leading to the adoption of farmland protection programs in New England evolved in response to the general decline in farmland acreage. Undoubtedly, aesthetic concerns of various types have been the most critical. Environmental quality and economic concerns are secondary to aesthetics. In recent years, food production has become a significant issue prompting legislative action.

A range of current programs have begun to have a substantial impact on farmland retention in numerous locations. Regulatory controls sometimes have been effective in protecting agriculture, but their unattractiveness to farmers and other farmland owners has limited their application. Economic incentives have eased the costs associated with New England farming, but farms continue to operate with limited profitability. Direct assistance programs have proven the most successful devices because both private and public organizations involved have provided substantial, noncoercive assistance. Local planning and implementation, when combined with state-supported incentives and technical assistance, seems to work best in all the states, particularly when there is a plurality of programs to encounter all the contingencies in farmland protection efforts.

The following chapter discusses issues and options for the future of farmland protection in New England. The success of future programs will depend largely on how they evolve from existing programs and the mandate that such programs receive from first farmers and nonfarmers.

10

Issues and Options for the Future

Despite many social, political, and economic differences among and within the New England states, the basic concerns which have prompted public and private activity to protect farmland are similar. While each state has developed its programs in response to the unique circumstances within its borders, there are common approaches to farmland protection. And certain general approaches have proven more successful across the board than others, largely because of the common problems that New England farmers and farmland owners face.

Based upon the summary of farmland protection concerns and programs presented in Chapter 9, this chapter will analyze the issues and options New England will face in the future and critically evaluate the issues which prompted action in the past. This discussion should have relevance beyond the New England region, especially in places where natural and man-made constraints have made modern farming practices difficult. In this respect, the New England experience may be viewed as a crystal ball to the future for areas of the country seeking to protect a changing farm landscape.

Issues

The experience of the six states indicates that public concern to protect agriculture land has grown in response to declining farmland acreage and generally has focused on aesthetics, environmental quality, food production, and economic value issues. Based on past experience, a broad coalition of support for future action can be formed only in response to these types of concerns. State, regional, and local policy, therefore, should be based on a clear understanding of the comparative importance of these concerns and on which concerns can be addressed effectively.

Aesthetics

Aesthetic concerns probably will continue to generate the widest coalition for farmland protection. Although aesthetic value can be considered highly subjective, most citizens can agree on the aesthetic value of actively worked farmlands. As nonmetropolitan areas continue to feel urbanization pressures in coming years, concern will grow over community character, scenic values, and rural ways of life.

Farmland conversion has become a lifestyle issue largely because of the way and degree to which urbanization spreads. In addressing aesthetic concerns, therefore, state and local farmland policies should be part of a coordinated package of programs dealing with rural issues, not isolated programs which may be inconsistent with other aesthetically motivated efforts.

Environmental Quality

In some New England states, notably Connecticut and Rhode Island, the protection of basic environmental processes has been cited as an important basis for farmland protection efforts. Farmland protection will undoubtedly continue to be attractive to those citizens concerned about open land areas for aquifer recharge, air recharge, and wildlife habitat.

The protection of farmland as opposed to other types of open space, however, may not result in any greater degree of environmental protection. In addition, as in many industries, agricultural practices can be a major source of environmental pollution—including soil erosion and stream sedimentation, water pollution from manure, pesticides, and other chemicals. Thus, farmland protection efforts sometime can be at odds with efforts to protect the quality of natural resources.

A delicate balance must be maintained between positive and negative environmental effects to ensure that a healthy landscape is protected and that the regulatory burdens on New England farmers do not result in farmland conversion. State and local farmland protection policies which ensure this balance should be supported.

Food Production

Concern for local food production has been an important reason that many farmland protection programs have been adopted in New England. Increasing consumption of locally grown food can increase food production and improve the local farm economics picture, thereby maintaining a supply of active farmland in the region.

Food production issues probably will continue to be an important farmland protection concern for several reasons. First of all, there is a growing concern over the quality of food in New England, as opposed to the quantity. For example, the potential for food contamination, such as from the use of pesticides on large western farms, may stimulate the desire to protect local farmland. In addition, there is the prudent desire to protect the agricultural land resource base in case it should be needed for future food production. Negative short-term farm economic conditions of the type experienced in the early 1980s should not result in the permanent destruction of these nonrenewable resources if there are reasonable alternatives for the location of urban uses. Finally, there is a long-standing Yankee pride in regional and local self-sufficiency: a desire not to be dependent and thereby under control of other regions of the country, particularly for basic necessities. This spiritually based concern is akin to the aesthetic pleasure of purchasing food directly from its source.

Food production for self-sufficiency may not be a strong argument in the future, however, because there are so many natural limitations on the amount of food that can be produced in New England. The region is densely populated. There are too many people for the amount of good farmland. Furthermore, modern food consumption habits are so diverse that New England farmers cannot satisfy all food needs. Climate, soil, and other natural constraints do not allow the production of commodities such as avocados and oranges which New Englanders have come to expect all year.

In addition, there are limits to the impact which local food production can have on the competition that exists in the national food supply system. This problem is augmented by the nationwide surplus of agricultural produce and agricultural land in production.

Finally, increased food production may not lead necessarily to greater farmland areas. Production efficiencies will continue to reduce the acreage required for certain commodities. Advances in technology and genetic engineering portend even greater yields per acre, resulting in the need for less land to produce a given quantity of an agricultural comodity. These advances are even more likely for the large farms of the West and may further reduce the economic competitiveness of the New England family farm.

As an issue of state and local concern, therefore, the protection of farmland resources for local food production may well continue to be a popular theme. Yet there is a clear limit to the extent to which regional policies can address these concerns. Efforts which seek to protect agricultural land based primarily on the need for a local food supply base must be evaluated in light of national and international food policies, which may be inconsistent with local goals.

Economic Value

Active farmland areas are important to the New England states because of the employment opportunities and tourism value they provide. Red and white farmhouses, varied livestock, and tilled fields—and their activities—will continue to generate income.

The farm-related economy can be enhanced by addressing certain cost factors of New England farming. The production cost input of land (including its acquisition, taxation, fertility, location, and size) and marketing inputs (such as advertising and transportation) can cost the farmer and farmland owner less if prudent programs are adopted. Other boosts to the farm industry may be needed with changing economic trends.

Unfortunately, public and private resources likely will be too limited to address all major cost variables effectively. However, the view taken toward agricultural land must be consistent with the best interests of the state. Treating farmland as a marketable commodity is an idea that welcomes wholesale development. Treating farmland as a valuable resource can allow for balanced growth. Nonetheless, it must be recognized that competing land uses sometimes produce a greater economic benefit to a state because the employment needs and taxable gross income of alternative industries can be far greater than the economic contribution of farms. Therefore, there may not be public support for expensive programs. A good spectrum of economic industries, however, would provide both alternative industries and a healthy agricultural economy.

The states should recognize that the underlying issue of farmland protection is the link between the economics of agriculture and active agricultural areas. Public concern for farm economic issues will result in more successful farmland protection strategies. This approach will enlist the support of farmers and farmland owners who often have been reluctant to support farmland protection measures which appear to have been based on the aesthetic and environmental concerns of the nonfarm majority. A more limited approach, based on aesthetics and environmental concerns, could result in farmland without farmers—inactive open space.

In short, there probably will continue to be broad public suport for farmland protection programs which protect the aesthetic values New England farms provide. Environmental quality and food production are public concerns which are important goals but of questionable merit for farmland protection. The economic value of selected agricultural areas will become more important as successful programs are adopted because of the

link between the economic prosperity of the farm industry and the effective protection of farmlands. Other concerns may arise in the future, and the efforts to protect agricultural land will be as strong as the bases for concern.

Options

The New England experience illustrates that farmland protection is a multifaceted problem which requires a multifaceted solution. The broad range of concerns requires that many perspectives and points of view be considered. A broad range of initiatives by state and local government, private groups, and individuals must be undertaken to ensure success.

The types of programs that most likely will be available fall into the same three groups discussed in Chapter 9. Regulatory controls implement land management policies by establishing statewide or regional planning objectives, granting localities the power to implement tools for local planning, and safeguarding farmers from nuisance suits. Economic incentive programs aid the farm industry financially. Direct economic assistance or direct government actions protect farmland in a simple and forthright fashion.

State Government

Statewide regulatory controls can address the farmland conversion issue by planning and providing for a balance of all land uses with a minimum of conflict. Such programs can meet the open space, community character, and quality-of-life needs of all citizens in a state, while also addressing environmental, food production, and economic concerns.

Regulatory controls which provide for land management and planning can regulate all publicly owned land. This type of program can be accomplished without much effort and is apt to provide an example of leadership in farmland protection at the state level for local government and private landowners to follow.

To have a significant impact, however, a state land management program must regulate all land in the state, including privately owned lands. However, political problems associated with the regulation of private lands makes the adoption of such programs unlikely. Private landowners are likely to be skeptical about programs that could force them to alter their future desired land uses. The broad scope of such plans usually renders them subject to the criticism that they are unable to respond to the specific circumstances of local communities and landowners. Nevertheless, statewide land use planning is very important to the success of farmland protection efforts to the extent that it determines which state resources will be allocated for protecting agricultural land and for providing incentives for development in appropriate areas (areas which have the appropriate physical and social resources, i.e., critical mass for agriculture). Such planning should be coupled with local regulatory programs.

The state also should seek to regulate private disputes between a farmland owner and others. Some New England states could amend their antinuisance laws to protect all farming activities unless public health becomes endangered. The state can also assist local regulatory efforts by revising eminent domain laws and sewer bond assessment procedures.

In addition to regulatory programs which encourage sound planning for agriculture, the state may choose to initiate economic incentive and development programs to aid individuals and landowners within the industry. Improving the economic viability of agriculture generally will, in turn, encourage the use of agricultural lands.

The most common and effective economic assistance programs for agriculture have been property tax incentives. Future efforts to reduce the impact of property taxes on farm production costs must deal with the problem that the financial benefits from these taxes often are not sufficient to make farming profitable. Moreover, the purpose of the penalty provisions of these taxes must be clarified and conflicts resolved. Most programs are perceived by farmland owners as restricting future land options during both good and bad economic times. Thus, some farmland owners do not enroll in the program because the tax benefits do not outweigh the potential liabilities. In some cases, development opportunities of enrolled farmland can be so lucrative that farmland owners or developers can afford to pay the penalty. Public policy should assess whether a weak penalty provision which encourages greater short-term use of the program offsets the problems associated with a delayed conversion.

Other statewide tax incentive programs include changes in inheritance and ad valorem taxes. Other statewide farm economic development programs which reduce a farmer's production costs are marketing programs and institutional purchase laws which expand demand by encouraging greater consumption of locally produced food.

As the New England experience indicates, however, there is a limit to the economic benefits states can provide because of limited public resources and the need for tax revenues. In addition, many of the economic problems facing New England farmers are national economic trends outside the control of the individual states. At the same time as a state's policy may increase demand and thus production, federal farm economic programs may reduce production and the acreage of cultivated land. Thus, it is doubtful that economic incentive programs alone can protect farmland, but they have proven quite successful in conjunction with other programs.

The third category of state programs is direct assistance or action measures. The most promising of these programs involve the purchase of development rights (PDR) and the executive review of state or local decisions affecting farmland.

Given the high land values generated by the land development market, PDR programs require a major funding source. If funded over many years, the purchase of farmland development rights can be highly successful in protecting agricultural land. It may be in fact the only cost-effective way, short of outright purchase, to protect farmland that is in the direct path of urban growth and in immediate jeopardy of development. Because PDR programs do not guarantee agricultural operations, however, they could be improved by targeting specific farmland zones or districts, especially if coordinated with other protection programs and with a landscape inventory and other critical food areas. Statewide planning can be useful in determining where state funds will be spent to maximum benefit. Purchase should be focused in areas where a critical mass for farming can be created and into ares where other environmental benefits will accrue. In addition, greater involvement by local organizations willing to participate, such as land trusts, Conservation Commissions, and private individuals, should be encouraged, perhaps by granting them broader powers.

Executive review programs can take the form of executive orders relating to state-owned land and state agency policies and actions which indirectly convert agricultural land. To be effective, however, such programs must be broad enough to touch upon most of a state's farmland. Bond review programs, for example, can limit or prevent state-financed development on farmland, whether the development is public or private.

Direct involvement by the state often is constrained by high costs, political disfavor, and the problem that future farming cannot be guaranteed. The state, however, can give technical assistance to local communities in their planning efforts to encourage cities and towns to identify their agricultural lands and plan for agriculture.

The results of state actions to protect farmland in New England suggest that options are limited at the state level. The Massachusetts experience, however, indicates that a "web" or wide range of farmland protection programs is needed to assure success when confronted with the many and varied problems of farmland protection. Regardless of the approach selected, most of the successful programs have been backed by a well-developed Food Policy Statement or Land Use Policy for the state. These well-researched studies or declarations served to help gain important legislative support. State programs should be part of an integrated land use program which coordinates open space inventories, targets developable land, and is part of a much larger program which plans for growth rather than only the protection of agricultural land. A good urban policy which plans for urban as well as rural uses is a key to the solution of the farmland conversion problem. Such coordinated planning is best accomplished at the state level. Of great importance, however, is local involvement in the planning and implementation of specific farmland protection programs.

Local Community

Farmland protection is primarily a local issue. The New England experience illustrates that there is generally greater consensus to protect agricultural land locally than statewide. Much of this local consensus is the result of a concern among both farmers and nonfarmers over open space and specific community character. Thus, the programs that have the most potential for success are those implemented at the local level with strong local participation. Farmland protection efforts should have a local emphasis.

Local governments can protect farmland by first setting community goals and then working, within the authority granted them by the state, to achieve those goals. The eventual outcome of local efforts will vary from town to town depending upon the concerns of the citizens, the quality of in-town farmland, the resources available to the community, the need for other land uses, and local expertise. Every New England municipality can make some kind of contribution.

As a first step, local governments seeking to protect agricultural land should inventory existing regulatory farmland protection tools. In general, the legislatures of the New England states already have made available the necessary tools for sound local planning. Where they have not, the affected locality should actively petition for new legislation.

For example, the passage of the Agricultural Incentive Areas Act in Massachusetts represents the culmination of important efforts to develop a comprehensive agricultural land protection program in Massachusetts. The main purpose of the Act—to encourage cities and towns to identify their agricultural lands and to give more local control over their protection—is consistent with statewide goals. Regardless of whether the purpose is to protect farmland for open space, for food production potential, or to preserve community character, the Massachusetts experience indicates greater possibility of success in farmland protection at the local level than through expensive statewide programs. Effective coordination at the local level to protect farmland can be achieved at a minimum cost, without placing unfair economic burdens on farmland owners, and with a small increment of political controversy. Programs such as the Incentive Areas Program may be a useful vehicle for such local control. Also, in Vermont, the Agricultural Land Task Force (1983) recognized the primary importance of local planning for farmland protection and recommended that local towns be strongly encouraged to develop their own plans.

Throughout New England, tools which are likely to prove most effective include general land planning and management powers which can be used to plan for all land uses in the town. Local policy should identify important farmland areas and recognize that farmland is more than "vacant" land on town land use plans. Agriculture must be considered a "use," not a holding zone for future development. If farmland protection is a community objective, growth should be provided for inadequate areas and should be encouraged and directed away from farmland areas. The most important local regulatory powers include zoning, especially cluster, incentive, bonus and TDR zoning, and easements or title restrictions such as PDR.

Municipalities can be assisted by state government in local regulatory efforts by revising eminent domain laws and sewer bond assessment procedures and granting broader powers to Conservation Commissions and other local bodies. Sound local planning can be aided by state technical assistance and the contribution of private organizations and individuals.

Regarding economic incentives, local communities can do little to address the macroeconomic issues which often hurt the small family-run farm industry of New England, not to mention the basic costs of production and commodities market aspects of local farm industries. Nevertheless, localities can adjust their tax policies, particularly with respect to the property tax. Towns should encourage local assessors to respect farming and not to overvalue farmlands near residential areas.

Protecting farmland at the local level, especially with direct assistance programs, also should entail assessing the role of local facilitators for farmland protection. Public facilitators include planning boards, Conservation Commissions, local public land trusts, and perhaps an agricultural development board. Private facilitators include private land trusts, foundations, farm industry groups, such as product cooperatives, and private individuals, including farmland owners. Finally, there can be partnership facilitators which use seed money from a public source for private farm enterprises.

More than any other area, greater effort by state and local government should be expended to implement farmland protection policies at the local level. When community concern is strong, and coordinated with state conservation and farmland protection programs and policies, sound local land use planning, zoning, and other initiatives will be the most effective way to protect farmland. To be successful, however, the techniques implemented must carefully consider the interests of the farmland owner. Policies such as exclusive agricultural zones, which reduce the potential market value of a farmer's land, for example, will be much less successful than policies which allow a farmland owner to realize some of the potential value of the land and still retain farmland, such as cluster, bonus, and TDR zoning innovations.

Private Sector

The New England experience indicates that the protection of farmland is not a problem which government alone can solve. The private sector has an important role to play in partnership with public objectives, a role which has been neglected to some degree in farmland protection strategies to date. Indeed, the results of private efforts can be as good as or better than public efforts; private landowners and organizations not only can fill in the gaps of state and local farmland protection activities, but also can take the lead with new programs and efforts.

The private sector is composed primarily of land trusts, conservation and environmental groups, farm industry organizations, farmers, and nonfarm private groups and individuals. Each one of these farmland protection facilitators can contribute to the retention of agricultural land in unique ways.

Land trusts can be established by philanthropic individuals, groups of concerned citizens, or private corporations. Their objective may be to protect a variety of open space areas or a designated territory for a specific use. They can work effectively with public agencies in implementing direct assistance programs. Because these trusts can be involved with the land itself, they can assist government in PDR programs. Greater consideration should be given to the idea of making these trusts and government entities joint owners of development rights. Such an approach would enable greater involvement with limited revenues by both groups and may help to ensure that the retention of the development restrictions is more permanent since both owners would have to assent to a release of the development rights.

Conservation and environmental groups likewise may have either a broad or a limited purpose. These groups often serve several agendas, such as wildlife, open space, and recreational goals. They are able to generate citizen awareness of the value of agricultural land because they usually serve an educational function. These groups often can work with state and local government in the implementation of all types of farmland protection programs. Government agencies should welcome their assistance.

Other organizations can assume a variety of functions. Large corporations can donate land for protection. More likely candidates for involvement are the farm industry groups. Producer cooperatives, for example, can provide a substantial economic benefit to their participants. Development or other investors in farmland, however, can pose more of a threat to farmland retention than a comfort.

Certain farm tax shelters provide tax benefits by which farm operations produce a loss. These operations can depress commodity values, thereby hurting farmland owners who rely on a positive annual cash flow to continue farming. Investors also may artificially increase the value of farmland in certain areas, thereby preventing potential long-term farmers from starting or continuing operations. Speculation should be discouraged by public actions at the state and local levels.

Public actions, therefore, should encourage broad private sector participation in farmland protection efforts. At the same time, individuals and institutions with short-term or narrowly focused aims should be regulated so that their actions alone do not thwart well-planned public projects designed to protect agricultural land.

Farmland Owners

The farmland protection programs which have had the best results in New England are those that have had the active support of farmland owners. These landowners must be active participants in the process or success in future farmland-protection efforts will be very limited.

Farmland owner participation is important because of the high percentage of farmland owner/operators in New England. Most owner/operators are not receptive to farmland-protection programs, particularly regulatory controls, that restrict the use and thereby reduce the value of their land. Farmland owners are not supportive of regulatory programs and plans which suddenly reduce the value of their only asset and guaranteed hope for the future—their land—particularly when farm economic conditions are most difficult. Farmer/lessees, on the other hand, generally wish to protect the land to assure their future livelihood. Absentee landlords are not always able to prevent the adoption of regulatory controls in distant communities. Because of the high percentage of owner/operators, therefore, few regulatory controls are likely to be adopted. Without the support of farmers, it is difficult to develop a meaningful farmland protection program.

Noncoercive and equitable regulatory controls which farmland owners are liable to support include local planning and innovative zoning such as cluster zoning with density bonuses and transfer of development rights (TDR), which allows owners to realize all or part of the market value of their land while still protecting viable agricultural tracts. Such landowners also will support economic incentive programs, such as loans and marketing, provided few strings are attached. Tax incentives also are likely to be favored by farmland owners, particularly if the benefits received are commensurate with the liabilities of enrollment. Such benefits usually are attainable only in urban or suburban areas where property taxes are high and the owner wishes to keep the land in agriculture. Finally, farmland owners often welcome involvement in direct actions by state and local government, such as the PDR programs, because they compensate the landowner while allowing the land to continue to be farmed.

Farmland owners should recognize that farmland protection advocates who are not farmland owners are not adversaries. Achieving public and private goals is possible only through cooperation. A farmland owner must compete within the farm industry, and yet his agricultural survival is dependent upon enough cooperation for a critical mass of farmland to exist. Similarly, a farmland owner can benefit in the long run if he helps formulate farmland protection policies and programs in a give-and-take atmosphere of common purpose. Outsiders to the process will either lose out themselves or cause an unnecessary public loss.

Summary

Several conclusions can be drawn based on the experience of the six New England states. First, a range of programs is necessary to protect agricultural land in order to confront all the opposing contingencies. All facilitators of farmland protection should evaluate the issues underlying protection efforts. To select the appropriate options for the future, these issues should be prioritized.

Once a program is implemented, it should be tailored to the program's objectives on an ongoing basis. Each state, muncipality, and private sector participant should mold its programs according to its current needs. Participants would achieve the best results if they could learn from and improve upon each others' experiences, and not merely copy them.

It seems evident that aesthetics will continue to be the overriding issue prompting farmland protection action in New England. It is also clear that the various participants should share their resources on an ongoing basis. In addition, the most effective program is likely to be sound local planning with equitable controls implemented by the municipality. The state should clearly focus on improving awareness and use of these local programs. In addition, state governments should provide financial and technical assistance. Moreover, private sector participants should devote their resources to the farmland experiencing the greatest threat of conversion to other uses. Finally, central to farmland protection is the active participation of farmland owners and a broad coalition of support from the entire community.

Index